Pollination Ecology

Pollination Ecology

Editor: Toby Hernandez

Callisto Reference
www.callistoreference.com

Callisto Reference,
118-35 Queens Blvd., Suite 400,
Forest Hills, NY 11375, USA

Visit us on the World Wide Web at:
www.callistoreference.com

ISBN: 978-1-64116-264-7 (Hardback)

The publisher's policy is to use permanent paper from mills that operate a sustainable forestry policy. Furthermore, the publisher ensures that the text paper and cover boards used have met acceptable environmental accreditation standards.

Printed in China.

Cataloging-in-Publication Data

Pollination ecology / edited by Toby Hernandez.
p. cm.
Includes bibliographical references and index.
ISBN 978-1-64116-264-7
1. Pollination. 2. Plant ecology. 3. Plants--Reproduction. I. Hernandez, Toby.
QK926 .P65 2020
571.864 2--dc23

Table of Contents

Preface

The purpose of the book is to provide a glimpse into the dynamics and to present opinions and studies of some of the scientists engaged in the development of new ideas in the field from very different standpoints. This book will prove useful to students and researchers owing to its high content quality.

Pollination is the process by which pollen is transferred from the anther to the stigma. This process can be affected by abiotic vectors like water and wind, and animals such as insects, birds and mammals. The major flower-frequenting insect taxa are flies, bees, beetles, ants, butterflies and moths. The field of pollination ecology is concerned with the study of reciprocal interactions between pollinators and flowering plants, along with their abiotic modes of pollination. Pollination-related interactions occurring between plants and pollinators are mutualistic. Insects initiate pollination when they visit flowers either to prey on other species, or to obtain nectar. Certain flowers like orchids also attract insects by mimicking female insects. Due to the rapid pollinator decline across the world, the study of pollination biology and ecology has emerged as an important field. The topics included in this book on pollination ecology are of utmost significance and bound to provide incredible insights to readers. It presents researches and studies performed by experts across the globe on pollinators. Through this book, we attempt to further enlighten the readers about the emerging concepts in pollination ecology.

At the end, I would like to appreciate all the efforts made by the authors in completing their chapters professionally. I express my deepest gratitude to all of them for contributing to this book by sharing their valuable works. A special thanks to my family and friends for their constant support in this journey.

Editor

1

FORAGING DYNAMICS AND POLLINATION EFFICIENCY OF *APIS MELLIFERA* AND *XYLOCOPA OLIVACEA* ON *LUFFA AEGYPTIACA* MILL (CUCURBITACEAE) IN SOUTHERN GHANA

Ben A. Mensah* and Andreas A. Kudom

Department of Entomology & Wildlife, University of Cape Coast, Cape Coast, Ghana

Abstract—As a result of different levels of pollination efficiency of pollinators, knowledge on appropriate pollinators of a plant has become important, especially in the management and conservation of both the pollinators and the plants. In this study, the pollination efficiency of *Apis mellifera* and *Xylocopa olivacea*, important pollinators of *Luffa aegyptiaca*, were assessed in the southern coastal part of Ghana from June 2009 to September 2010. Pollination efficiency of *A. mellifera* and *X. olivacea* was estimated in terms of fruit set and fruit size. Further, data on daily and seasonal nectar dynamics of *Luffa aegyptiaca* were collected. In the early mornings (0600-0700), *X. olivacea* was the most frequent visitor (0.47 min^{-1}) on the female flowers compared to *A. mellifera* (0.13 min^{-1}). The mean nectar (sugar) concentration in the dry season was 36.58 ± 0.55 %, which was higher than the 34.03 ± 0.38 % obtained for the rainy season (F = 14.986; df = 2; $P < 0.0001$). Total amount of sugar in the early mornings was 1.88 ± 0.37 mg which was higher than 0.28 ± 0.04 mg in the mid mornings ($\chi 2 = 14.33$, df = 1, $P < 0.0001$). Fruits that developed from flowers that had received a single visit from *X. olivacea* had a mean weight of 428.7g and were 1.5 times heavier than fruits from flowers visited by *A. mellifera* (286.76 g). *X. olivacea* was more efficient than *A. mellifera* in terms of number of fruit set per single visit. This study has provided some knowledge on pollination ecology of *L. aegyptiaca*, which can be exploited to improve fruit production in commercially grown vine crops.

Keywords: *nectar concentration, nectar energetics, foraging dynamics, cucurbit plants*

INTRODUCTION

Luffa aegyptiaca Mill (Cucurbitaceae) is a vigorous, annual climbing herb found in tropical and subtropical regions and in the same family with most of the commercially grown vine crops. *L. aegyptiaca* is monoecious, having separate male and female flowers on the same plant, a characteristic of many species in the family, that makes them even more dependent on pollinators, especially on bee species. Cucurbit plants do not just rely on pollinators but also require repeated visits from pollinators (Collinson 1976; Stanghellini et al. 1997; Meléndez, 2002). Furthermore, on the same plant, some bee species have been shown to be more efficient pollinators than others (Stanghellini et al. 2002). Therefore knowledge on appropriate pollinators of a plant has become important especially in the management and conservation of both the pollinators and the plants.

One of the factors determining pollinator effectiveness is the rate of pollinator visitation (Fishbein & Venable 1996). Measurements of visitation rates can identify the linkages amongst pollinators, plants and subsequent fruit set. Additionally, temporal variation in production of nectar may influence the rate of visitation (Thomson et al. 1989). Nectar volume and concentration are the basis upon which nectar energetics are calculated, and the abundances of the dominant species of flower visitors within some ecosystems are linked to the amount of energy provided by nectar (Roubik 1989). Daily changes in available nectar clearly affect the identity and abundance of flower feeders (Potts et al. 2001; 2004), as do seasonal changes (Petanidou & Ellis 1996; Bosch et al. 1997).

In an earlier study, the importance of *Apis mellifera adansonii* Linnaeus and *Xylocopa* species to pollination of *L. aegyptiaca* was demonstrated. No fruits were formed when flowers were bagged to exclude visits by bees. A single visit from *Xylocopa* species was enough to cause pollination while more than one visit of *A. mellifera* was needed (Mensah & Kudom 2010). That study, however, failed to demonstrate whether the single visit that resulted in pollination was enough to set fruit. In this study, pollinating efficiency of both *A. mellifera* and *X. olivacea* (Fabricius) were assessed. Furthermore, daily dynamics of nectar energetics were explored to explain temporal foraging behaviour of the bees on flowers of *L. aegyptiaca* and the effect of bee visits on fruit development. It is expected that the results will contribute to the knowledge of pollination ecology of *L. aegyptiaca* and to the improvement of fruit production among commercially grown vine crops.

METHODOLOGY

Study sites

The project was carried out in two contrasting patches of wild *L. aegyptiaca* in Biriwa (N05° 09.956', W01° 08.900) and Cape Coast (N05° 06.473', W01° 17.431') in the

*Corresponding author; email: b2mensah@yahoo.com

southern coastal part of Ghana. At Biriwa, the study site was a small plot of *L. aegyptiaca* isolated from other vegetation by roads and buildings, with < 100 flowers produced daily. At Cape Coast, the study site was a larger plot with > 500 flowers of *L. aegyptiaca* produced per day. This site was swampy with other vegetation surrounding it. From June 2009 to September 2010, data were collected on the frequency of bee pollinators on female flowers and daily and seasonal nectar dynamics. In addition, pollination efficiency of *A. mellifera* and *X. olivacea* were assessed in terms of fruit set and size.

Nectar survey

Nectar was collected in the two major seasons in Ghana, from February to March 2010 (dry season) and July to September 2010 (rainy season). In each season, nectar was collected from 60 open flowers exposed to floral visitors at different times of the day between 0600hrs and 1400hrs. The time for nectar collection was chosen after a preliminary survey on the period of activity of floral visitors and the temporal pattern of opening and closing of flowers of *L. aegyptiaca,* as described in Mensah and Kudom (2010). Each observation day was divided into three sessions: early morning (0600-0700), mid morning (1000-1100) and afternoon (1300-1400) and nectar was collected from four flowers in each session for five weeks. Nectar was taken from the flowers with 1 μL micro capillary tube (Drummond, U.S.A) and nectar concentration was measured with a sugar refractometer modified for small volumes (Bellingham and Stanley, UK) as described in Kearns & Inouye (1993). Nectar volume was determined by measuring the length of the nectar column in the micro capillary tube with a veneer calliper. After each use, the refractometer was thoroughly washed with distilled water and dried with tissue paper. Temperature and relative humidity at the time of nectar collection were measured as close as possible to the flowers with a hand-held thermohygrometer (Kestrel, U.S.A). Any day with particularly aberrant weather (e.g. heavy rainfall) was skipped to ensure that all data were collected under 'typical' microclimatic conditions. Total amount of sugar in nectar and nectar energetics were calculated as described in Galetto & Bernardello (2005) and Kudom & Kwapong (2010). ANOVA was used to compare the difference in nectar concentration between the two seasons while Kruskal Wallis non parametric test was used to analyze the daily changes (SPSS, Version 16).

Temporal pattern of bee visitation

Number of bee visits per minute to a *L. aegyptiaca* flower was recorded between 0600 and 1400. A visit was counted when *X. olivacea* or *A. mellifera* made contact with the stigma of a flower. Data were collected at the two study sites for three days each week, during the peak flowering season between July and September 2010. As for nectar survey, each observation day was divided into three sessions; early morning (0600-0700), mid morning (1000-1100) and afternoon (1300-1400). Observations were made every 15 minutes in each session and visitation rate was calculated by dividing the number of bee visits by observation time. Again, any day with particularly aberrant weather (e.g. heavy rainfall) was skipped to ensure all data were collected under 'typical' microclimatic conditions.

Pollination efficiency per visit

At the Cape Coast study site, 30 bagged female flowers each were open to *A. mellifera* and *X. olivacea* for a single visit and the flowers were re-bagged and the numbers that set fruit were recorded. Female flower buds were bagged as described in Mensah and Kudom (2010). Fruits were harvested after two weeks and their length and weight measured.

RESULTS

Throughout the study period, female flowers at the Biriwa study site did not receive visits from any *Xylocopa* species. The only bee visitors to flowers at that study site were *A. mellifera.* However, *Xylocopa* species (*X. varipes* Smith, *X. olivacea* and *X. imitator* Smith), *Amegilla calens* (Lepeletier) and *A. mellifera* were the bee visitors to the Cape Coast study site. Among the *Xylocopa* species at this site, *X. olivacea* was the most abundant. Due to the absence of *Xylocopa* from the Biriwa site, data are presented and comparisons between *A. mellifera* and *X. olivacea* for the Cape Coast site only.

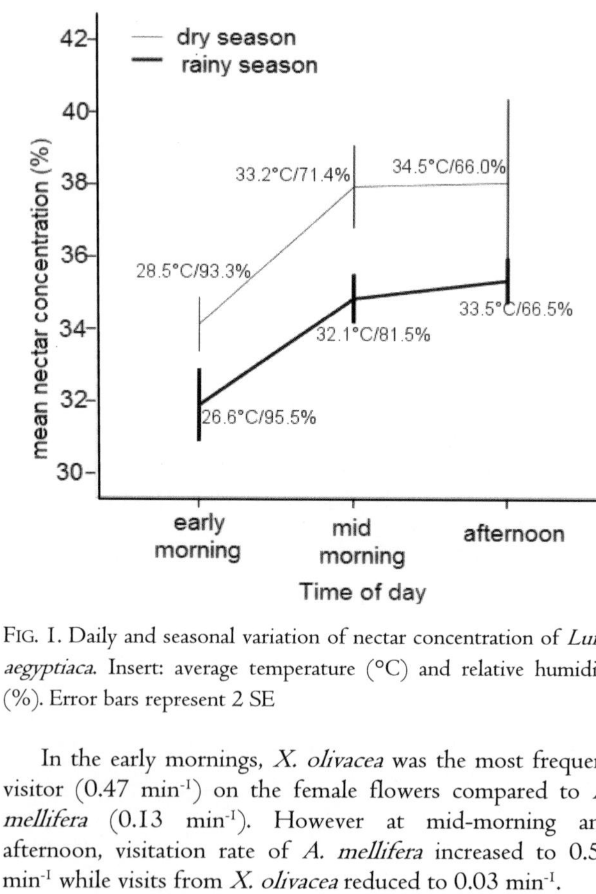

FIG. 1. Daily and seasonal variation of nectar concentration of *Luffa aegyptiaca.* Insert: average temperature (°C) and relative humidity (%). Error bars represent 2 SE

In the early mornings, *X. olivacea* was the most frequent visitor (0.47 min^{-1}) on the female flowers compared to *A. mellifera* (0.13 min^{-1}). However at mid-morning and afternoon, visitation rate of *A. mellifera* increased to 0.53 min^{-1} while visits from *X. olivacea* reduced to 0.03 min^{-1}.

Nectar survey

The mean nectar concentration (± SE) in the dry season was 36.58 ± 0.55 %, which was higher than the 34.03 ± 0.38 % in the rainy season (F = 14.986, df = 2, P < 0.0001). In both seasons, nectar concentration increased as temperature increased and relative humidity decreased through the day. In the dry season, nectar concentration in early morning was lower than in the mid morning ($\chi2$ = 13.94, df = 1, $P < 0.0001$). However, nectar concentration at mid-morning was similar to nectar concentration in the afternoon, ($\chi2$ = 0.38, df = 1, $P = 0.54$; Fig. 1). There was a similar trend during the rainy season, nectar concentration at mid morning was higher than the concentration in early mornings ($\chi2$ = 12.05, df = 1, $P = 0.001$) but similar in the afternoon ($\chi2$ = 1.93, df = 1. $P = 0.17$; Fig. 1).

As the concentration of nectar increased through the day, nectar volume decreased and it affected the amount of sugar in the nectar. Total amount of sugar in the early morning was 1.88 ± 0.37 mg which was higher than 0.28 ± 0.04 mg in the mid morning ($\chi2$ = 14.33, df = 2, $P < 0.0001$). However the amount of sugar (mg) in nectar measured in the afternoon (0.26 ± 0.04 mg) was similar to amounts measured in the mid mornings ($\chi2$ = 0.669, df = 2, $P = 0.41$). Nectar energetics were highest in the early mornings (Fig. 2).

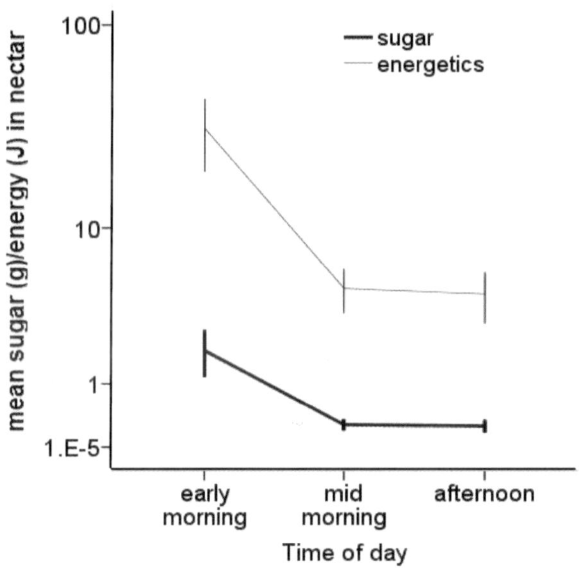

FIG. 2. Daily variation of sugar content (mg) in nectar and nectar energetic (J) of *Luffa aegyptiaca*. Error bars represent 2 SE

Pollination efficiency

All (100%) flowers visited by *X. olivacea* set fruit while only 55% of flowers visited by *Apis mellifera* set fruit. Fruits that developed from flowers that received a single visit from *X. olivacea* had a mean weight of 428.7 ± 47.76 g and were 1.5 times heavier than fruits from flowers visited by *A. mellifera* (286.76 ± 10.53 g). Furthermore, fruits developed from flowers pollinated by *X. olivacea* were longer (29.5 ± 2.5 cm) than fruits from flowers pollinated by *Apis mellifera* (22.9 ± 0.4 cm). The non overlapping standard error bars (Fig. 3) indicate that the differences in weight and length were significant. A Mann-Whitney U test, however, shows otherwise ($P = 0.121$)

DISCUSSION

This study showed that the daily foraging dynamics of bee pollinators of *L. aegyptiaca* were as a result of daily changes in nectar volume and energy. This change in nectar resources was also as a result of daily and seasonal changes in temperature and humidity. This is consistent with previous studies that have linked dynamics of floral visitors to daily changes in nectar volume and energy (Roubik 1989; Abrol 2005).

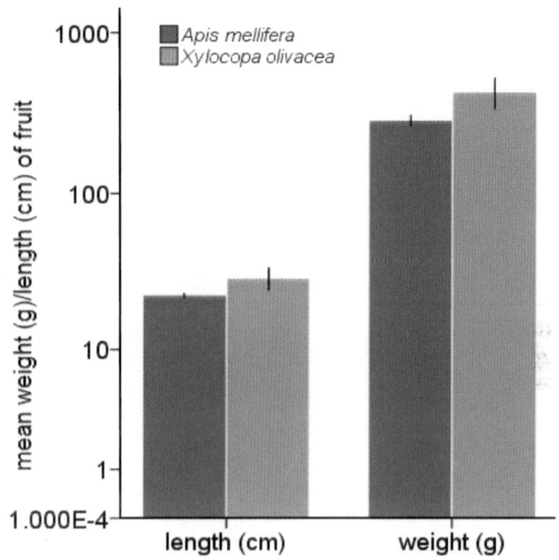

FIG. 3. The weight and length of fruits developed from a single visit of *Xylocopa olivacea* and *Apis mellifera*. Error bars represent 2 SE.

Weather is known to have marked effect on nectar concentration (Corbet et al. 1979), which may be a result of evaporation. Thus, daily and seasonal increase in temperature and decrease in relative humidity may have been the cause of the daily and seasonal variation in nectar concentration. The structure of a flower may also contribute to the degree of the effect of temperature and humidity on the nectar. Experimental studies have shown the direct effect of corolla depth on nectar concentration (Corbet et al. 1979; Galetto & Bernardello 2005). For instance, the long corolla tubes of *Ananas comosus* flowers reduced the effect of temperature and relative humidity on the nectar. Thus nectar concentration remained fairly constant even when temperature increased during the day (Kudom & Kwapong 2010). However, flowers of *L. aegyptiaca* have short corolla tubes, and are subject to greater effect of temperature and humidity.

The frequency of visits to female flowers by *X. olivacea* was high early in the morning which coincides with the time the plant produced high volumes of nectar and amounts of sugar, whereas visitation rate decreased as nectar volume and amount of sugar decreased. Large bees such as bumble bees have been shown to be dependent upon flowers providing relatively high volumes of nectar to support their foraging activities (Heinrich 1976) whereas smaller bees specialize on flowers with lower nectar volumes (Proctor et al. 1996). Hence, high volumes of nectar coupled with high nectar energy in early morning make the flowers more attractive to *X. olivacea* during that period. Though both pollinators were found at the same time at one of the study sites, the visit of *A. mellifera* to the flowers appeared to be impeded or the smaller bees were 'bullied' by the larger bees in early mornings. Furthermore, due to the high volumes of nectar in the early mornings, *A. mellifera* probably obtained the amount of nectar needed upon visiting few flowers. However, to increase the chance of a pollinator encountering a female flower, it may have to visit a lot of flowers, due to the high ratio of male to female flowers (Mensah & Kudom 2010). The increase in frequency of *A. mellifera* visits to female flowers in the late mornings and in the afternoons may be a result of low nectar volumes and low encounter rates with *X. olivacea.* During the period when nectar quantity was small, *A. mellifera* might have had to increase the number of flowers visited to get the amount of nectar needed.

A high visitation rate is essential for fruit set and optimum fruit size of *L. aegyptiaca,* especially for flowers visited only by *A. mellifera.* This became evident since only half of the virgin flowers were able to fruit after a single visit from *A. mellifera.* Moreover, fruits that developed from flowers pollinated by *A. mellifera* were comparatively smaller in size and weighed less than fruits developed from flowers pollinated by *X. olivacea.* Further investigation of this apparent difference in size is however needed, as it was not supported by the statistical test performed. This difference might be accounted for by a combination of behaviour and body size. *Xylocopa* bees are larger and by virtue of their size, they are exposed to a larger surface area of either the anther or the stigma. Therefore *Xylocopa* can either take more pollen from the anther or deposit more pollen on the stigma than *A. mellifera.* The behaviour of *A. mellifera* during nectar collection may also affect the number of pollen taken from anthers or deposited on stigmas of a flower. When honey bees stand on the petals to collect nectar, they do not make much contact with anthers or the stigma, which reduces efficiency in transferring pollen (Mensah & Kudom 2010).

It can be inferred from the results that *X. olivacea* was more efficient than *A. mellifera* as a pollinator of *L. aegyptiaca* in terms of fruit set resulting from a single visit. However, large bees such as *X. olivacea* may be attracted to forage in plots with large number of flowers so as to support their need for relatively high volumes of nectar (Heinrich 1976). There was an indication of this when no record was made of any *Xylocopa* species visiting female flowers at the Biriwa study site, where daily *L. aegyptiaca* flowers were less than 100. This observation supports an earlier submission for the need to maximize floral diversity in agro-ecosystems to maintain the abundance and diversity of pollinators for adequate pollination of crops (Steffan-Dewenter et al. 2006; Mensah and Kudom 2010). The result also suggests that in pollination management, especially in an agro-ecosystem, temporal foraging behaviour of pollinators must be considered so as to know the appropriate time to engage in some farm practices (e.g. insecticide application) that can affect activities of pollinators.

Conclusion

The daily foraging dynamics of the two important bee pollinators of *L. aegyptiaca* were a result of daily variation in nectar energetics. Seasonal and daily dynamics of nectar concentration were also due to effects of temperature and humidity. *X. olivacea* foraged early in the morning when nectar energetic was highest while *A. mellifera* still foraged effectively even at low nectar energetics. Although both *X. olivacea* and *A. mellifera* are pollinators of *L. aegyptiaca, X. olivacea* is more efficient than *A. mellifera* in terms of number of fruit set per single visit and probably the resulting fruit size. However, *X. olivacea* was attracted to plots with only large number of flowers. This study has provided some knowledge on pollination ecology of *L. aegyptiaca,* which can be exploited to improve fruit production in commercially grown vine crops. Also, the results showed that *L. aegyptiaca* provides an important nectar resource for its foragers throughout the year, and this finding can be useful in agro- and natural ecosystem management.

References

Abrol DP (2005) Pollination Energetics. Journal of Asia-Pacific Entomology 8: 3-14.

Bosch J, Retana J, Cerda X (1997) Flowering phenology, floral traits and pollinator composition in a herbaceous Mediterranean plant community. Oecologia 109: 583–591.

Collinson CH (1976) The interrelationships of honey bee activity, foraging behaviour, climatic conditions and foraging in the pollination of pickling cucumbers *Cucumis sativus* L. PhD thesis: Michigan State University, USA

Corbet SA, Willmer PG, Beament JWL, Unwin DM, Prys-Jones OE (1979) Post-secretory determinants of sugar concentration in nectar. Plant, Cell and Environment 2: 293–300.

Fishbein M, Venable DL (1996) Diversity and temporal change in the effective pollinators of *Asclepias tuberosa.* Ecology 77:1061–1073.

Galetto L, Bernardello G (2005) Nectar. In: Dafni A, Kevan PG, Husband BC (eds) Practical Pollination Biology. Enviroquest Ltd, Ontario, Canada, pp 261-313.

Heinrich B (1976) Resource partitioning among some eusocial insects: bumblebees. Science 57: 874–889.

Kearns CA, Inouye DW (1993) Techniques for pollination biologists. University Press of Colorado Press, Niwot, Colorado, USA

Kudom AA, Kwapong PK (2010) Floral visitors of *Ananas comosus* in Ghana: a preliminary assessment. Journal of Pollination Ecology 2:27-32

Meléndez RV, Magaña RS, Parra TV, Ayala BR, y Navarro GJ (2002) Diversity of native bee visitor of cucurbit crops (Cucurbitaceae) in Yucatán, México. Journal of Insect Conservation 6 (3): 135-147.

Mensah BA, Kudom AA (2010) Relating bee activity to pollination of *Luffa aegyptiaca* Mill. in Ghana. Journal of Apicultural Research and Bee World 42:192-196.

Petanidou T, Ellis WN (1996) Interdependence of native bee faunas and floras in changing Mediterranean communities. The Conservation of Bees. Linnean Society Symposium Series 18 (ed. by A Matheson, SL Buchmann, C O'Toole, P Westrich & IH Williams), pp. 210–226, Academic Press, London, UK.

Potts SG, Dafni A, Ne'eman G (2001) Pollination of a core flowering shrub species in Mediterranean phrygana: variation in pollinator diversity, abundance and effectiveness in response to fire. Oikos 92: 71–80.

Potts SG, Vulliamy B, Roberts S, O' Toole C, Dafni A, Ne'eman G, Willmer P G (2004) Nectar resource diversity organises flower-visitor community structure. Entomologia Experimentalis et Applicata 113: 103–107

Proctor M, Yeo P, Lack A (1996) The Natural History of Pollination. Harper Collins Publishers, London, UK.

Roubik DW (1989) Ecology and Natural History of Tropical Bees. Cambridge University Press, Cambridge, UK.

Stanghellini MS, Ambrose J, Schultheis JR (1997) The effects of honey bee and bumble bee pollination on the fruit set and abortion of cucumber and water melon. American Bee Journal 137: 386-391.

Stanghellini MS, Ambrose J, Schultheis JR (2002) Diurnal activity, floral visitation and pollen deposition by honey bees and bumble bees on field-grown cucumber and watermelon. Journal of Apicultural Research 41(1-2): 27–34.

Steffan-Dewenter I, Klein, AM, Gaebele V, Alfert T, Tscharntke T (2006) Bee diversity and plant-pollinator interactions in fragmented landscapes. In Waser, NM and Ollerton, J (Eds). Plant-pollinator interactions: From Specialization to Generalization in. Chicago Press; Chicago, USA. pp 387-410.

Thomson JD, Mckenna MA, Cruzan MB (1989) Temporal patterns of nectar and pollen production in *Aralia hispida*: implications for reproductive success. Ecology 70:1061–1068.

2

FLORAL SYMMETRY AFFECTS BUMBLEBEE APPROACH CONSISTENCY IN ARTIFICIAL FLOWERS

Brett M. Culbert and Jessica R.K. Forrest*

Department of Biology, University of Ottawa, 30 Marie Curie, Ottawa, ON, K1N 6N5, Canada

Abstract—Bilateral symmetry has evolved from radial symmetry in several floral lineages, and multiple hypotheses have been proposed to account for the success of this floral plan. One of these hypotheses posits that bilateral symmetry (or, more generally, a reduced number of planes of floral symmetry) allows for more precise pollen placement on pollinators. Greater precision would maximize the efficacy of pollen transfer to conspecifics, while minimizing reproductive interference amongst plant species. Despite the intuitiveness of this hypothesis, it has little experimental support. Here, we tested whether a reduction in the number of floral planes of symmetry (as in the transition from radial to bilateral symmetry) increases the potential precision of pollen placement. We analyzed video recordings of bumblebees (*Bombus impatiens*) visiting artificial flowers to determine whether consistency in flower entry angle differed between radial (round) and disymmetric (rectangular) "flowers". We observed more consistent entry angles for disymmetric flowers than for radial flowers, with entry angles to radial flowers 43% more variable on average (standard deviations of 30° vs. 21°). Bees trained on flowers with an intermediate (square) morphology exhibited a slight, non-significant preference for radial symmetry over disymmetry. Our results show that disymmetry—an evolutionarily intermediate form of floral symmetry—has the potential to increase pollen transfer to conspecific stigmas, relative to radial symmetry. Thus, evolutionary reduction in the number of planes of floral symmetry likely provides benefits in terms of pollen delivery, as suggested by the pollen-placement-accuracy hypothesis. These findings offer insight into the evolution of floral symmetry.

Keywords: artificial flowers, Bombus, floral symmetry, pollen placement, preference, zygomorphy

INTRODUCTION

Bilateral floral symmetry (zygomorphy) has evolved from the ancestral state of radial symmetry (actinomorphy) multiple times across the angiosperms (Donoghue et al. 1998; Damerval & Nadot 2007; Hileman 2014; reviewed by Citerne et al. 2010). Various hypotheses have been put forth to explain this evolutionary trend (reviewed by Neal et al. 1998). For example, greater distinctiveness of zygomorphic flowers may promote pollinator fidelity (Ostler & Harper 1978), or pollinators may have an innate preference for bilateral symmetry (Leppik 1953). While greater pollinator fidelity towards bilateral flowers has been observed (Fenster et al. 2004), evidence for floral-symmetry preference is mixed (e.g., West & Laverty 1998), with previous studies finding pollinator preferences for both radial (Free 1970; Wignall et al. 2006) and bilateral (Leppik 1953; Rodríguez et al. 2004; Lázaro et al. 2008) floral symmetry.

Another hypothesis, the pollen-placement-accuracy (or pollen position) hypothesis, states that flowers with fewer planes of symmetry promote more consistent entry by pollinators into the flower, such that the plant's reproductive parts—both male and female—make more consistent contact with certain parts of pollinators' bodies (Leppik 1972; Macior 1974). This consistency would maximize the efficiency of pollen transfer while minimizing reproductive interference amongst pollinator-sharing floral species (Muchhala & Thomson 2010). Precise placement of reproductive parts reduces pollen waste, increases the accuracy of stigma contact with pollinators (Armbruster et al. 2009), and reduces the chances of stigma-clogging by heterospecific pollen, thereby potentially increasing reproductive output of both the pollen parent and the seed parent (Morales & Traveset 2008; Armbruster & Muchhala 2009). This should benefit plant fitness by maximizing the amount of potential reproduction per unit of floral reward produced (Muchhala et al. 2010; Muchhala & Thomson 2012; Armbruster 2014).

Despite the logic underlying the pollen-placement-accuracy hypothesis, there is little experimental evidence to support it. We therefore set out to test this hypothesis experimentally, by training laboratory-reared bumblebees to visit artificial flowers, and measuring the angle at which these bees entered both radially symmetric and disymmetric "flowers". We used disymmetry, in which flowers possess only two planes of symmetry, to represent an evolutionary step from radial towards bilateral symmetry, in which flowers possess a single plane of symmetry. Indeed, disymmetry appears to be an evolutionarily intermediate stage between radial and bilateral floral symmetries in some plant lineages (Endress 1999; Friis et al. 2011). We expected pollinators to enter disymmetric flowers at a more consistent angle than radially symmetric flowers. We also tested for a symmetry preference in our bees, as a preference for radial symmetry

*Corresponding author: jforrest@uottawa.ca

across several pollinator groups could help explain the persistence of radial symmetry in spite of less efficient pollen transfer.

MATERIALS AND METHODS

Experimental design

Colonies of captive-reared bumblebees (*B. impatiens* Cresson) were connected to a 60 × 60 × 180 cm flight cage by a gated tunnel that allowed for controlled entry and exit of individual bees. During initial training, bees were allowed to forage freely on artificial flowers until consistent foraging began and reliable foragers could be individually marked using dots of coloured paint applied to the dorsum of the thorax. Flowers consisted of 1.5 mL Eppendorf tubes with the lids removed, ringed by blue plastic 11.4 cm^2 "corollas" (details below). Training flowers were filled with 30% (*w/w*) sucrose solution ("nectar") and were presented in the horizontal plane (i.e., corollas parallel, and Eppendorf tubes perpendicular, to the floor of the cage) to allow us to fill Eppendorf tubes to the brim during initial training. Training flowers were replaced every second day to avoid build-up of sucrose residue or bee scent marks on corollas.

Two colonies were used in this study. The first colony was used to compare entry-angle variability between radial and disymmetric flowers. The second colony was used both to measure entry-angle variability and to assess floral symmetry preference. For the first colony, training flowers were identical to our radially symmetric flowers (described below). For the second colony, we used 3.4 × 3.4 cm square corollas for training in an effort to minimize bee bias towards a particular symmetry before completion of the preference test. Pollen was provided to the colonies daily.

For testing entry-angle consistency, 5.5 × 2.1 cm blue plastic rectangles were used to represent corollas of disymmetric flowers, and circles with radii of 1.9 cm represented corollas of radially symmetric flowers (Fig. 1). The surface area of all corollas (both experimental and training) was the same (11.4 cm^2). Flowers (corollas and Eppendorf tubes) were washed with warm water and detergent, rinsed, and dried prior to each foraging bout. The first colony, of which 6 bees were individually observed (15 Dec. 2014–30 Jan. 2015), foraged on a setup consisting of 9 flowers that were either entirely radial or entirely disymmetric. The centre of each flower was 8–10 cm from the centre of the closest neighbouring flower. The second colony, of which 10 bees were individually observed (13 Mar.–2 Apr. 2015), foraged on a setup consisting of a combination of radial and disymmetric flowers (10 of each symmetry) in a 5 × 4 matrix. Flowers were set up in a pattern that allowed for an equal ratio of neighbouring radial and disymmetric flowers. For both colonies, flowers were presented in a single vertical plane (i.e., corollas perpendicular, and Eppendorf tubes parallel, to the floor of the cage) (Fig. 1). Each flower contained 3.5 μL of nectar at the start of a foraging bout and was refilled as soon as it was emptied by a bee. For the first colony, each bee performed 2–3 foraging bouts per flower symmetry (approximately 60 visits per symmetry per bee; Appendix 1), with alternation of symmetries between foraging bouts and alternation of starting symmetries among individual bees. For the second colony, three foraging bouts per bee were performed overall (approximately 44 visits per symmetry per bee; Appendix 1). Each individual bee's foraging bouts were recorded by video camera and still photographs were extracted as the bee entered a flower. Entry was identified as insertion of the entire thorax of the bee into the Eppendorf tube. Any entry attempts or movements that did not meet this criterion were ignored and not measured. ImageJ (National Institutes of Health, Version 1.49k, 2014) was used to measure the angle at which bees entered flowers by drawing two lines (one for the vertical midline of the flower and one for the midline of the bee) and measuring the angle between these lines (Fig. 2). The midline from the centre of the flower to the bottom of the flower was considered to be 0°, and arbitrary designations of positive for left of 0° and negative for right of 0° were used to measure directionality. This meant that entries left of the midline theoretically ranged from 0°–180°, and entries right of the midline could range from −180°–0°. For testing flower symmetry preference, we counted the number of flower entries per symmetry by the 10 bees (from the second colony) that foraged on the mixed (10 radial: 10 disymmetric) array.

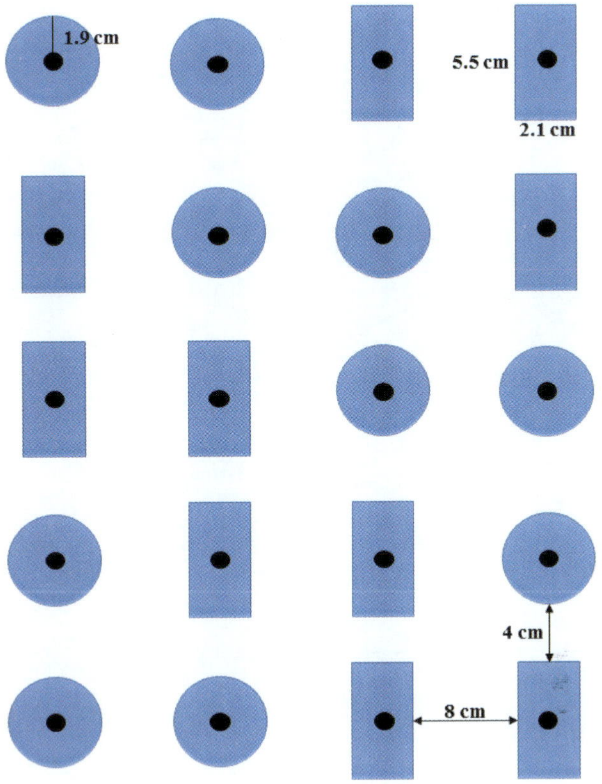

FIGURE 1. Experimental setup of artificial flowers for the second *Bombus impatiens* colony. Note the presence of both radial and disymmetric (rectangular) "flowers". The array was set up in a vertical configuration (i.e., the "flowers" in the top row of the array were closest to the roof of the flight cage; those in the bottom row were closest to the floor; and rectangular flowers were oriented so as to be longest top-to-bottom).

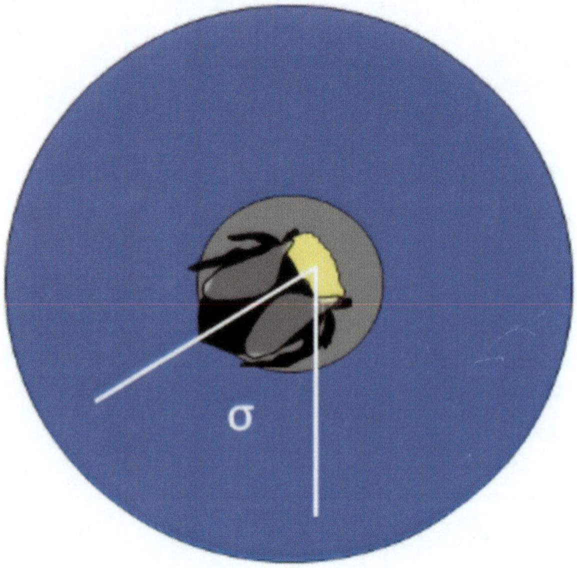

FIGURE 2. Schematic representation of an entry-angle measurement, as performed for each flower visit. Entry angle was measured as the bee's thorax completely entered the flower.

Statistical analyses

For each bee on each symmetry, the standard deviation of entry angle was calculated separately for each foraging bout. We ran linear mixed models using the R (R Core Team, Version 3.1.1, 2014) package lme4 (Bates et al. 2015) to test for an effect of floral symmetry on these standard deviation values, which were normally distributed. We included colony and bee nested within colony as random factors in the models. To test for symmetry preference, we modelled the number of flower visits received (per foraging bout) from bees in the second colony, as a function of floral symmetry and bout number (a categorical factor). Because these visit counts were normally distributed, we again used linear mixed-effects models, with bee as the random factor. For all models, we tested the significance of predictor variables using likelihood-ratio tests of nested models.

RESULTS

Observed flower-entry angles ranged from −140° to 159° for radially symmetric flowers and from −60° to 169° for disymmetric flowers. However, bees typically entered flowers from below: only 12 of 1585 entry angles (< 1%) had absolute values ≥ 90°. Entry angles were more consistent for disymmetric flowers than radial flowers for 15 of the 16 bees tested (Fig. 3). Standard deviations of entry angles were approximately 9° lower (95% C.I. = 6.59–11.32) for bees entering the disymmetric flowers (mean SD = 21°) than for bees entering radial flowers (mean SD = 30°; LMM, X_1^2 = 24.1, N = 16 bees, P < 0.0001). Excluding visits with extreme entry angles (those with absolute values ≥ 90°) and recalculating the standard deviations only strengthened this result (LMM, X_1^2 = 47.4, N = 16 bees, P < 0.0001). There was no significant difference in entry-angle variability between the two colonies, despite their different training regimens (X_1^2 = 0.10, N= 16 bees, P = 0.75), nor did mean entry angles differ significantly between floral symmetries (1.8° for disymmetric; 3.0° for radial; LMM, X_1^2 = 0.06, N = 16 bees, P= 0.80). Bees showed a non-significant preference for radial flowers (average proportion of visits to radial flowers = 0.53; X_1^2 = 0.9, N= 10 bees, P= 0.34; Fig 4).

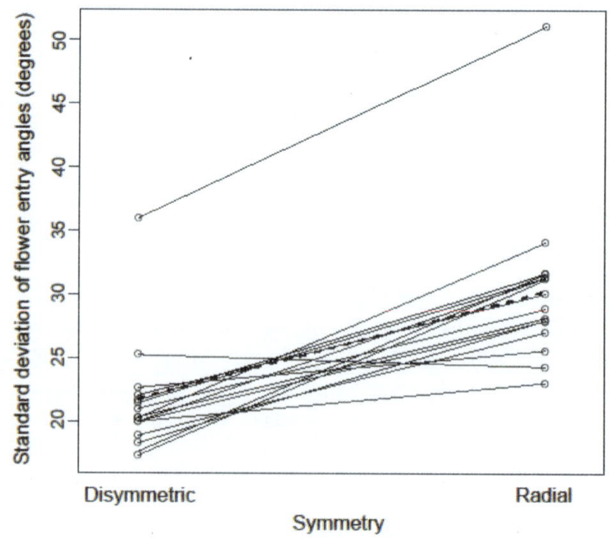

FIGURE 3. Standard deviations of flower entry angles for all bees on disymmetric and radial symmetries. Lines connect radial and disymmetric values for a given individual. The average trend across all bees is depicted by the thick, dashed line. N = 16 bees; note overlap between several points and lines.

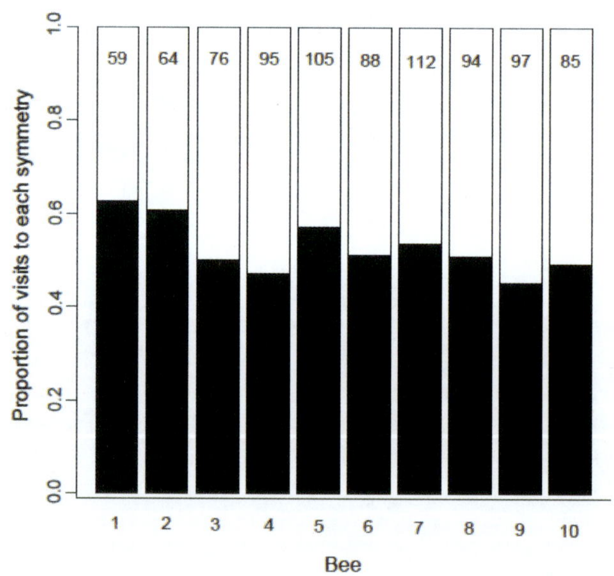

FIGURE 4. Proportion of visits made by bees to flowers of each symmetry. The lower portion of each column (black) represents visits to radial flowers; the upper portion of each column (white) represents visits to disymmetric flowers. Total number of flower visits made by each bee is given at the top of the appropriate column.

DISCUSSION

Floral symmetry affects the consistency with which bees enter flowers. We found more consistent floral entry angles by bumblebees on disymmetric artificial flowers than on radially symmetric flowers. Assuming that more consistent floral entry allows plants to more precisely place pollen onto (and receive pollen from) their pollinators (Armbruster et al. 1994; Wang et al. 2014), disymmetric flowers should have a fitness advantage over radial flowers—perhaps especially in species-rich plant communities where plants share pollinators. If, more generally, flowers with fewer planes of symmetry exhibit more precise pollen placement and therefore more efficient deployment of resources, then the evolution of bilateral symmetry should be favoured. Thus, our results may help explain the evolution, in many plant lineages, of bilateral floral symmetry from ancestrally radial symmetry.

Our artificial flowers did not have reproductive structures, and we did not measure actual pollen transfer. Artificial flowers with male and female "reproductive organs" and dye as a pollen analogue (see Fenster et al. 1996; Adler & Irwin 2006; Thomson et al. 2012) could be developed in the future to determine whether transfer rates differ between radial and bilateral flowers. In this study, we have imagined our flowers as having few or bilaterally arranged reproductive structures (e.g., the conjoined stamens or staminal columns of orchids, lobelioids, and many legumes), such that consistent pollinator entry angles would allow consistent dorsal or ventral pollen placement. (Disymmetric corollas combined with numerous, radially arranged reproductive organs would not be expected to increase pollen-placement precision, regardless of their effect on pollinator behaviour.) In fact, Rudall & Bateman (2004) have noted that the evolution of bilateral symmetry is associated with stamen suppression (the development of fewer stamens per flower), and suggested that the latter was a preliminary step towards the development of bilateral symmetry. Stamen suppression frequently results in flowers with a single functional stamen, most commonly positioned at the top of the flower. Alternatively, the evolution towards zygomorphy may have been initiated by a shift in placement of stamens to one side of the flower (e.g., as seems to have occurred in *Passiflora* and in *Schlumbergera* cacti). This evolutionary scenario (reduction or rearrangement of reproductive organs first; selection on corolla symmetry second) was the premise for our current study and would be the logical starting point for future studies measuring actual pollen transfer.

Although several angiosperm lineages have evolved bilateral symmetry, many others have not, and some may have even reverted to radial from bilateral symmetry (Donoghue et al. 1998). This suggests there are benefits to radial symmetry. One possible explanation for the persistence of radial symmetry is that pollinators have a preference for radial flowers. Our bees did show a slight, though non-significant, preference for radial flowers (Fig. 4); but our bees were not completely naïve (having been trained on flowers with square corollas, which they may have perceived as more similar to our disymmetric flowers). In general, although many types of bees have a strong preference for symmetric over asymmetric flowers (Møller 1995; Møller & Eriksson 1995; but see West & Laverty 1998), preferences for a particular symmetry type seem to be idiosyncratic. Previous studies have found no innate preference for radial over bilateral symmetry in *Bombus impatiens* (West & Laverty 1998); a preference for bilateral symmetry in *B. terrestris* (Rodríguez et al. 2004) and other bumblebee species (Leppik 1953; Lázaro et al. 2008); and a preference for radial symmetry in honeybees (*Apis mellifera*) (Free 1970; Wignall et al. 2006).

Even if bumblebees do not consistently prefer radial symmetry, radial flowers may attract more functional groups of pollinators (Fenster et al. 2004). Radially symmetrical flowers may therefore benefit from high and relatively stable visitation rates, at the expense of low pollination efficiency. For this reason, it has been suggested that plants with radially symmetric flowers might be less pollen limited (McLernon et al. 1996; Wolowski et al. 2014); however, syntheses of the available data show no general trend for greater pollen limitation in plants with zygomorphic flowers (Knight et al. 2005; Vamosi et al. 2013). Regardless, radial flowers may have advantages over bilateral flowers that would be apparent only in a more diverse field setting.

In nature, radial flowers are generally oriented in the horizontal plane (parallel to the ground), whereas bilateral flowers are oriented in the vertical plane (Neal et al. 1998). In our study, both symmetries were presented in the vertical plane. It is therefore possible that radial flowers have a pollen-transmission advantage when oriented horizontally, such that pollinators may approach from any direction (Ushimaru & Hyodo 2005; Fenster et al. 2009; Ushimaru et al. 2009; Wang 2010; Wang et al. 2014a; Wang et al. 2014b). It seems likely that bilateral symmetry would have evolved on a background of vertical floral orientation—the background we simulated here. However, floral orientation is an additional factor that could be manipulated in future studies to investigate its functional interaction with floral symmetry. It should also be noted that since we deployed our disymmetric flowers in a vertical orientation, they presented an enhanced "landing platform", or lower lip, compared to the radial flowers. The presence of this landing platform may have contributed to the increased entry-angle consistency that we observed for disymmetric flowers. Bilateral symmetry is often associated with presence of a landing platform in nature—an association we mimicked in this study—but any landing-platform effect could be experimentally decoupled in the future from the effect of floral symmetry by rotating the corollas 90°, such that they are wider than long.

In conclusion, bumblebees (*Bombus impatiens*) have a more consistent angle of entry into disymmetric flowers than radially symmetric artificial flowers. They do not, however, have a significant preference for one symmetry over the other. Our findings provide the first experimental support for the pollen-placement-accuracy hypothesis, and shed light on why bilateral symmetry has evolved from radial symmetry so frequently.

Acknowledgements

We thank the University of Ottawa's Undergraduate Research Opportunity Program (UROP) for facilitating this collaboration. We also thank C. Darveau for allowing access to his bee facilities. Conversation with J. Ogilvie and thoughtful comments from N. Muchhala on an earlier draft improved the manuscript. Bees were provided by Biobest Canada Ltd.

Appendices

Additional supporting information may be found in the online version of this article:

Appendix I. Summary of entry angle standard deviation (SD) and preference data for *B. impatiens* on artificial flowers.

References

Adler LS, Irwin RE (2006) Comparison of pollen transfer dynamics by multiple floral visitors: Experiments with pollen and fluorescent dye. Annals of Botany 97:141-150.

Armbruster WS, Edwards ME, Debevec EM (1994) Floral character displacement generates assemblage structure of Western Australian triggerplants (*Stylidium*). Ecology 75:315-329.

Armbruster WS, Hansen TF, Pélabon C, Pérez-Barrales R, Maad J (2009) The adaptive accuracy of flowers: measurement and microevolutionary patterns. Annals of Botany 103:1529-1545.

Armbruster WS, Muchhala N (2009) Associations between floral specialization and species diversity: cause, effect, or correlation? Evolutionary Ecology 23:159-179.

Armbruster WS (2014) Floral specialization and angiosperm diversity: phenotypic divergence, fitness trade-offs and realized pollination accuracy. Annals of Botany Plants 6:1-24.

Bates D, Maechler M, Bolker B, Walker S (2015) Fitting linear mixed-effects models using lme4. Journal of Statistical Software 67:1-48.

Citerne H, Jabbour F, Nadot S, Damerval C (2010). The evolution of floral symmetry. Advances in Botanical Research 54:85-137.

Damerval C, Nadot S (2007) Evolution of perianth and stamen characteristics with respect to floral symmetry in Ranunculales. Annals of Botany 100:631-640.

Damerval C, Citerne H, Le Guilloux M, Domenichini S, Dutheil J, Ronse De Craene L, Nadot S (2013) Asymmetric morphogenetic cues along the transverse plane: shift from disymmetry to zygomorphy in the flower of Fumarioideae. American Journal of Botany 100:391-402.

Donoghue MJ, Ree RH, Baum DA (1998) Phylogeny and the evolution of flower symmetry in the Asteridae. Trends in Plant Science 3:311-317.

Endress, PK (1999) Symmetry in flowers: diversity and evolution. International Journal of Plant Sciences 160:S3-S23.

Fenster CB, Hassler CL, Dudash MR (1996) Flourescent dye particles are good pollen analogs for hummingbird-pollinated *Silene virginica* (Caryophyllaceae). Canadian Journal of Botany 75:189-193.

Fenster CB, Armbruster WS, Wilson P, Dudash MR, Thomson JD (2004) Pollination syndromes and floral specialization. Annual Review of Ecology, Evolution and Systematics 35:375-403.

Fenster CB, Armbruster WS, Dudash MR (2009) Specialization of flowers: is floral orientation an overlooked first step? New Phytologist 183:502-506.

Free JB (1970) Effect of flower shapes and nectar guides on the behaviour of foraging honeybees. Behaviour 37:269-285.

Friis EM, Crane PR, Pedersen KR (2011) Patterns of structural diversification in angiosperm reproductive organs. In: Early Flowers and Angiosperm Evolution. Cambridge University Press, Cambridge, pp 387-414.

Hileman LC (2014) Trends in flower symmetry evolution revealed through phylogenetic and developmental genetic advances. Philosophical Transactions of the Royal Society B 369:20130348.

Jabbour F, Nadot S, Damerval C (2009) Evolution of floral symmetry: a state of the art. Comptes Rendus Biologies 332:219-231.

Knight TM, Sheets JA, Vamosi JC, Mazer SJ, Burd M, Campbell DR, Dudash MR, Johnston MO, Mitchell RJ, Ashman TL (2005) Pollen limitation of plant reproduction: pattern and process. Annual Review of Ecology, Evolution, and Systematics 36:467-497.

Lázaro A, Hegland SJ, Totland Ø (2008) The relationships between floral traits and specificity of pollination systems in three Scandinavian plant communities. Oecologia 157:249-257.

Leppik, EE (1953) The ability of insects to distinguish number. American Naturalist 87:229-236.

Leppik EE (1972) Origin and evolution of bilateral symmetry in flowers. Evolutionary Biology 5:49-85.

Macior LW (1974) Behavioral aspects of coadaptations between flowers and insect pollinators. Annals of the Missouri Botanical Garden 61:760-769.

McLernon SM, Murphy SD, Aarssen LW (1996) Heterospecific pollen transfer between sympatric species in a midsuccessional old-field community. American Journal of Botany 83:1168-1174.

Møller AP (1995) Bumblebee preference for symmetrical flowers. Proceedings of the National Academy of Sciences of the United States of America 92:2288-2292.

Møller AP, Eriksson M (1995) Pollinator preference for symmetrical flowers and sexual selection in plants. Oikos 73:15-22.

Morales CL, Traveset A (2008) Interspecific pollen transfer: Magnitude, prevalence and consequences for plant fitness. Critical Reviews in Plant Sciences 27:221-238.

Muchhala N, Thomson JD (2010) Fur versus feathers: pollen delivery by bats and hummingbirds and consequences for pollen production. American Naturalist 175:717-726.

Muchhala N, Brown Z, Armbruster WS, Potts MD (2010) Competition drives specialization in pollination systems through costs to male fitness. American Naturalist 176:732-743.

Muchhala N, Thomson JD (2012) Interspecific competition in pollination systems: costs to male fitness via pollen misplacement. Functional Ecology 26:476-482.

Neal PR, Dafni A, Giurfa M (1998) Floral symmetry and its role in plant-pollinator systems: Terminology, distribution, and hypotheses. Annual Review of Ecology and Systematics 29:345-373.

Ostler WK, Harper KT (1978) Floral ecology in relation to plant species diversity in the Wasatch Mountains of Utah and Idaho. Ecology 59:848-861.

R Core Team (2014) R: A language and environment for statistical computing. R Foundation for Statistical Computing, Vienna, Austria. URL: http://www.R-project.org/

Rodríguez I, Gumbert A, Hempel de Ibarra N, Kunze J, Giurfa M (2004) Symmetry is in the eye of the 'beeholder': innate preference for bilateral symmetry in flower-naïve bumblebees. Naturwissenschaften 91:374-377.

Rudall PJ, Bateman RM (2004) Evolution of zygomorphy in monocot flowers: Iterative patterns and developmental constraints. New Phytologist 162:25-44.

Thomson JD, Ogilvie JE, Makino TT, Arisz A, Raju S, Rojas-Luengas V, Guo Rui Tan M (2012) Estimating pollination success with novel artificial flowers: Effects on nectar concentration. Journal of Pollination Ecology 9:108-114.

Ushimaru A, Hyodo F (2005) Why do bilaterally symmetric flowers orient vertically? Flower orientation influences pollinator landing behaviour. Evolutionary Ecology Research 7:151-160.

Ushimaru A, Dohzono I, Takami Y, Hyodo F (2009) Flower orientation enhances pollen transfer in bilaterally symmetrical flowers. Oecologia 160:667-674.

Vamosi JC, Sheets JA, Ashman TL (2013) Drivers of pollen limitation: macroecological interactions between breeding systems, rarity, and diversity. Plant Ecology & Diversity 6:171-180.

Wang Y, Meng LL, Yang YP, Duan YW (2010) Change in floral orientation in *Anisodus luridus* (Solanaceae) protects pollen grains and facilitates development of fertilized ovules. American Journal of Botany 97:1618-1624.

Wang H, Tie S, Yu D, Guo YH, Yang CF (2014a) Change of floral orientation within an inflorescence affects pollinator behavior and pollination efficiency in a bee-pollinated plant, *Corydalis sheareri*. PLoS ONE 9:e95381.

Wang H, Xiao CL, Gitruru RW, Xiong Z, Yu D, Guo YH, Yang CF (2014b) Change of floral orientation affects pollinator diversity and their relative importance in an alpine plant with generalized pollination system, *Geranium refractum* (Geraniaceae). Plant Ecology 215:1211-1219.

West EL, Laverty TM (1998) Effect of floral symmetry on flower choice and foraging behaviour of bumble bees. Canadian Journal of Zoology 76:730-739.

Wignall AE, Heiling AM, Cheng K, Herberstein ME (2006) Flower symmetry preferences in honeybees and their crab spider predators. Ethology 112:510-518.

Wolowski M, Ashman TL, Freitas L (2014) Meta-analysis of pollen limitation reveals the relevance of pollination generalization in the Atlantic forest of Brazil. PLoS ONE 9:e89498.

3

The large carpenter bees (Hymenoptera: Apidae: Xylocopa spp.) of Nuevo León, México

Liliana Ramírez-Freire[1*], Glafiro José Alanís Flores[1], Ricardo Ayala Barajas[2], Humberto Quiroz Martínez[1] & Carlos Gerardo Velazco Macías[3]

[1]*Facultad de Ciencias Biológicas, Universidad Autónoma de Nuevo León, 66451 San Nicolás de los Garza, Nuevo León, México.*
[2]*Estación de Biología Chamela (Sede Colima) Instituto de Biología, UNAM*
[3]*Parques y Vida Silvestre. Av. Eugenio Garza Sada 6604, Col. Lagos del Bosque, 64890 Monterrey, Nuevo León, México.*

Abstract—There is a severe information gap regarding wild bees in northern Mexico. The present study is a contribution to knowledge of the distribution, habitat preferences and floral usage patterns of bees of the genus *Xylocopa* in the state of Nuevo León. Field sampling was done using aerial net and pan traps (yellow, blue, white and pink) at 35 sites throughout the state. *Xylocopa* species were found at only seven of these sites. Four of the five species, previously reported for the state were collected plus two new state records (*Xylocopa micans* and *X. strandi*), bringing the total number of species in the state to seven. Individuals were collected visiting only flowers of the Fabaceae and Bignoniaceae families, and they occurred primarily in shrub lands and disturbed areas.

Keywords: *pollination, native bees, biodiversity, conservation, inventory*

Introduction

Xylocopa is a cosmopolitan genus of approximately 470 species of medium to very large bees (Michener 2007). All *Xylocopa* are believed to be polylectic and are commonly associated with large and showy flowers with abundant pollen and nectar (Hurd 1956). *Xylocopa* is currently considered to be the sole member of the tribe Xylocopini in the subfamily Xylocopinae of the family (Michener 2007).

With 36 species reported from Mexico, *Xylocopa* is one of the more diverse Mexican genera of bees (Ayala et al. 1996). Five species of *Xylocopa* have been reported from the state of Nuevo León: *X. fimbriata, X. mexicanorum* and *X. tabaniformis, X. lateralis* and *Xylocopa californica* (Ayala et al. 1996; Ascher & Pickering 2011). Unfortunately, most of these records lack precise locality data.

It is alarming for a country such as Mexico with high level of complexity of topography, climate and biodiversity to find large zones where little is known about the diversity and distribution of its native bees (Ayala et al. 1996). For example, we have very few bee specimens in entomological collections from the state of Nuevo Leon (Ramírez-Freire 2008). The objective of the present work is to contribute to our knowledge of *Xylocopa* species for Nuevo Leon, providing data about vegetation on in which they occur and the plant species they visit, with the ultimate goal of assembling a formal collection of native bees for the state.

*Corresponding author, email: biolily@gmail.com

Material and methods

Area of study

The state of Nuevo León is located in the northeast of the Mexican republic (Figure 1), total extension of the state is 64 555.94 km², with elevations ranging from 200 to 3 700 meters. It is divided into three physiographic regions: Planicie Costera del Golfo (Gulf coastal plains), Sierra Madre Oriental (Eastern main mountains) and Altiplano Mexicano (Mexican high land plateau) (Alanís-Flores et al. 1996). The main climatic types for the state are dry and semi-dry with highly variable precipitation ranging from 300 mm to 600 mm. The Sierra Madre Oriental in the central part of the state is relatively temperate and cool while the regions in the north and south are more arid. The main vegetation type for the state is desert shrub, with high variability on its composition while forest can only be found in the Sierra Madre Oriental area. There are also several sites with saline or gypsophilous vegetation within the state with high levels of plant endemism (Treviño-Garza et al. 2001). For a more detailed vegetation description, see Velazco-Macías et al. (2011).

The study was carried out between February 2009 and May 2010. Taxonomic literature about the genus *Xylocopa* was gathered and the following data base were consulted: Colección Hymenoptera del Museo de Zoología "Alfonso L. Herrera" de la Facultad de Ciencias, Universidad Nacional Autónoma de México, México, Colección Nacional de Insectos (CNIN), Bee Species Database, American Museum of Natural History, from Snow Entomology Collection,

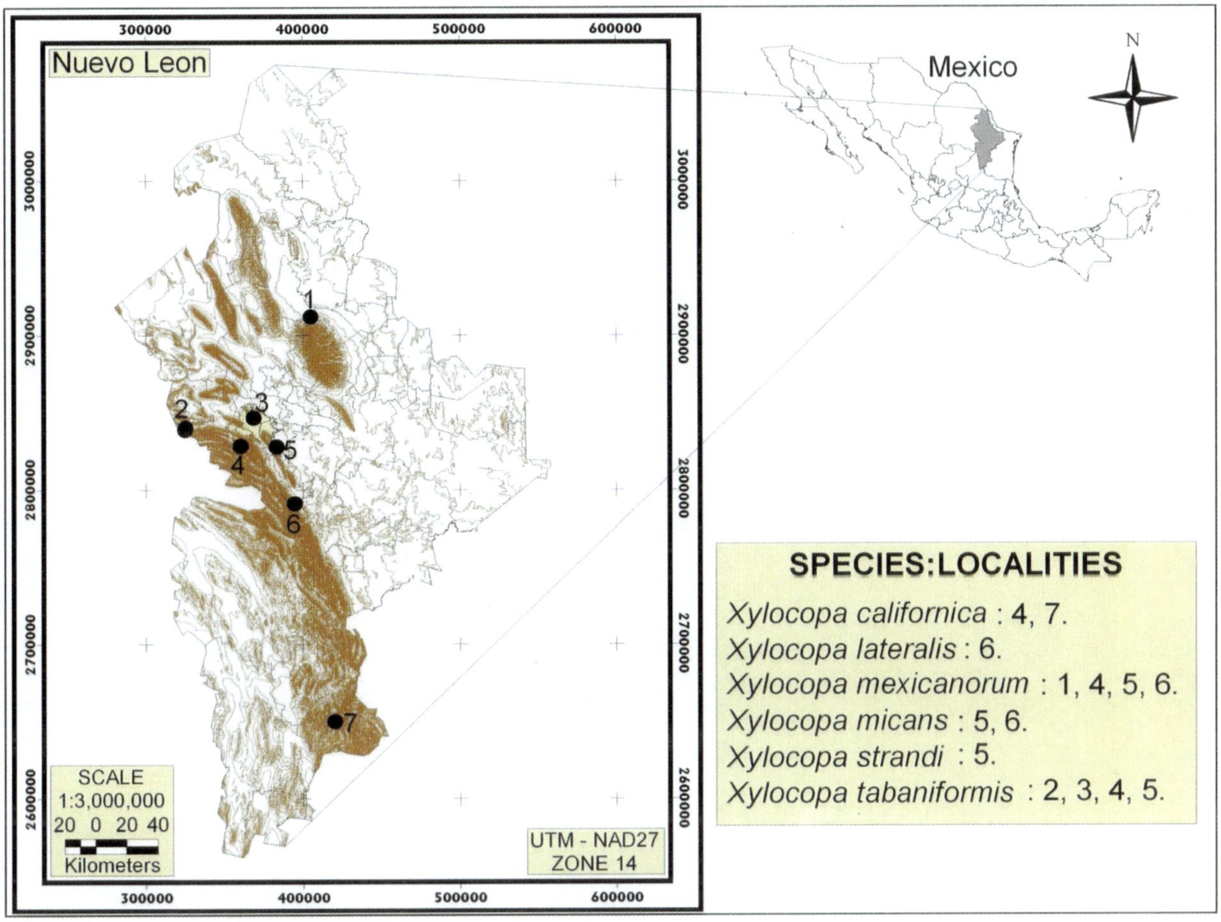

Figure 1. Sampling sites and distribution of *Xylocopa* species at Nuevo Leon, México.

University of Kansas Natural History Museum (Ascher & Pickering 2011). Field work was done in 35 localities at 20 municipalities within the state of Nuevo León. Sites were distributed across the state from north to south and were picked, in part, on the basis of accessibility. None of these sites had been previously sampled for bee species. Sampling was done using aerial entomological nets for 30 minutes periods on each flowering plant species at the sampling site. Pan traps were also used at each site, with 120 traps of four different colours (yellow, blue, white and pink / 30 traps each) with soapy water set in an 1 000 m^2 area and with a trap sampling period from 8:00 am to 6:00 pm (Kaerns & Inouye 1993; McIntyre & Hostetler 2001; Campbell & Hanula 2007; Wilson et al. 2008; Droege 2008; Droege et al. 2009). Bee specimens collected by net were killed with acetone and mounted on entomological pins on the same day. Pan trap specimens were processed according to Droege (2008). Bee identification was made using keys from Hurd (1955), Michener et al. (1994) and Ascher & Pickering (2011); two species (*X. lateralis* and *X. strandi*) were identified by Dr. John L. Neff at Central Texas Melittological Institute. Along with bee specimens, plant material was collected and processed to get proper identification of species visited by bees, vegetation type at the sampling sites were defined according to INEGI (1986).

Results

Xylocopa was collected at only seven of the 35 sites. A total of 38 specimens were collected: 25 females and 13 males. Six species were identified from field collected specimens: *Xylocopa californica* (1♀, 2♂), *X. mexicanorum* (16♀, 4♂), *X. tabaniformis* (4♀, 3♂), *X. lateralis* (1♂) which had previously been reported from the state, and two additional species, *X. micans* (3♀, 3♂) and *X. strandi* (1 ♀), as new state records. With adding *X. fimbriata* which has been reported by literature for Nuevo Leon, but not collected during this survey, this brings the total number of species of *Xylocopa* in the state of Nuevo León to seven.

All specimens but one (collected in a yellow pan trap), were collected at flowers: *Caesalpinia mexicana, Dalea formosa, Sophora secundiflora* (Fabaceae) and *Tecoma stans* (Bignoniaceae) (Table 1). In addition *X. mexicanorum* was observed twice to rob nectar from *Cordia boissieri* and *Nicotiana glauca.* Distribution of the genus appears restricted to the Sierra Madre Oriental (brown parts Figure 1), and preferred vegetation types are shrub and disturbed sites with secondary vegetation (Table 2).

TABLE 1. Plant species visited by *Xylocopa* on sampled sites.

Bee species	Plant species
X. californica	*Caesalpinia mexicana, Tecoma stans*
X. lateralis	*Caesalpinia mexicana*
X. mexicanorum	*Caesalpinia mexicana*
X. micans	*Caesalpinia mexicana*
X. tabaniformis	*Dalea sp, Sophora secundiflora, Caesalpinia mexicana*
X. strandi	*Caesalpinia mexicana*

DISCUSSION

There is little and scattered information regarding *Xylocopa* in northern Mexico, for example, even if most species were previously reported for the state, all specimen records from database lacked information regarding specific locality and municipality, only a *X. tabaniformis* record from "Colección Nacional de Insectos" collected ten years ago in Cola de Caballo, Santiago municipality. No specimen record from American Museum of Natural History had any information on locality. The previously reported species *Xylocopa fimbriata* (Ayala et al. 1996), which we were not able to observe, has no data about the collection. The two new state records are reported by the same authors for the neighboring state of Tamaulipas. Other species like *X. muscaria* and *X. nautlana* may occur in Nuevo Leon, since they are common in the neighboring states of Tamaulipas and San Luis Potosí (Ascher & Pickering, 2011).

While *Xylocopa* is known as polylectic (Porter, 1981), in this study the bees were mainly collected from species of the Fabaceae family (Table 1). Fabaceae being the third largest family within the Nuevo Leon state (Villarreal & Estrada, 2008; Velazco-Macías 2009), this could place *Xylocopa* as one of the most important pollinators for plant species in shrub vegetation in northeast Mexico. We can also conclude the best way to collect *Xylocopa* is areal nets instead of bow traps.

ACKNOWLEDGMENTS

Dr. John L. Neff for helping with identification of some specimens and manuscript review, also to Dra. Virginia Meléndez Ramírez and Dr. Carlos Vergara Briceño by their comments and reviewing.

REFERENCES

Alanís-Flores GJ, Cano G, Rovalo MM (1996) Vegetación y flora de Nuevo León, una guía botánico ecológica. Impresora Monterrey, S.A. de C.V., México.

Ascher J, Pickering J (2011) Discover Life's Bee species guide. [online] URL: www.discoverlife.org/mp/20q?search=Xylocopa (accessed December 2011).

Ayala R, Griswold TL, Yanega D (1996) Apoidea (Hymenoptera). In: Llorente BJ, García AN, Soriano E (eds) Biodiversidad, Taxonomía y Biogeografía de Artrópodos de México. Hacia una síntesis de su conocimiento. UNAM-CONABIO, México. pp 423-464.

Campbell JW, Hanula JL (2007) Efficiency of Malaise traps and colored pan traps for collecting flower visiting insects from three forested ecosystems. Journal Insect Conservation, 11(4): 399-408.

Droege S (2008) The Very Handy Manual: How to Catch and Identify Bees and Manage a Collection. [online] URL: www.nbii.gov/images/uploaded/152986_1215796993080_Handy_Bee_Manual_Jun_2008.pdf; (accessed April 2009).

Droege S, Tepedino VJ, Lebuhn G, Link W, Minckley RL, Chen Q, Conrad C (2009) Spatial patterns of bee captures in North American bowl trapping surveys. Insect Conservation and Diversity, The Royal Entomological Society. 3:15-23.

Hurd PD (1955) The Carpenter Bees of California (Hymenoptera: Apoidea). Bulletin of the California Insect survey. University of California Press. 4(2): 31-72.

Hurd PD (1956) Xylocopa (Notoxylocopa). Notes on the subgenera of the New World carpenter bees of the genus *Xylocopa* (Hymenoptera, Apoidea). *American. Museum Novitates.* 1776: 1-7 [2]

INEGI (1986) Síntesis geográfica del estado de Nuevo León. Secretaría de Programación y Presupuesto. México, D.F.

TABLE 2. *Xylocopa* species according to collection site, vegetation types and number of individuals at each site.

Locality #	Locality name/municipality	Vegetation type	Species present	Number of individuals
1	Ejido Rancho Nuevo, Agualeguas	Tamaulipan thorn shrub	*X. mexicanorum*	1
2	Corral de los Bandidos, García	Microphyllous desert shrub	*X. tabaniformis*	2
3	Desert Botanic Garden UANL, San Nicolás de los Garza	Disturbed area (Secondary vegetation)	*X. tabaniformis*	1
4	La Huasteca Canyon, Sta. Catarina	Pied mont shrub	*X. californica,*	2
			X. mexicanorum	14
			X. tabaniformis	3
5	Santa Ana Canyon, Juárez	Disturbed area (Secondary vegetation)	*X. mexicanorum*	4
			X. micans	3
			X. tabaniformis	1
			X. *strandi*	1
6	Road Raíces-La Trinidad, Allende	Disturbed area (Secondary vegetation)	*X. mexicanorum*	2
			X. micans	3
			X. lateralis	1
7	Road to La Encantada, Zaragoza	Conifer forest	*X. californica*	1

Kearns CA, Inouye DW (1993) Techniques for Pollination Biologists. University Press of Colorado. U.S.A.

Mclntyre NE, Hostetler ME (2001) Effects of urban land use on pollinator (Hymenoptera: Apoidea) communities in a desert metropolis. Basic and Applied Ecology. 2:209-218.

Michener CD, McGlinley RJ, Danforth B (1994) The Bee Genera of North and Central America (Hymenoptera:Apoidea). Smithsonian Institution Press. Washington y London.

Michener CD (2007) The Bees of the World, Second Edition. The Johns Hopkins University Press. USA.

NBII. 2011. Carpenter Bees - Both Pollinators and Nectar Robbers. [online] URL: http://www.nbii.gov/portal/server.pt/community/bees_and_wasps/857/carpenter_bees/5745 (accessed August 2011).

Porter CHC (1981) Ecological notes on Lower Rio Grande Valley Xylocopa (Hymenoptera: Anthophoridae). Florida Entomologist. 64 (1): 175-182.

Ramírez-Freire L (2008) La floración de cinco especies de cactáceas y sus insectos asociados en el area natural protegida "Sierra Corral de los Bandidos" municipio de García, Nuevo León. Tesis de Maestría. Facultad de Ciencias Biológicas, U. A. N. L.

Treviño-Garza EJ, Cavazos-Camacho C, Aguirre-Calderón OA (2001) Distribución y estructura de los bosques de galería en dos ríos del centro sur de Nuevo León. Madera y Bosques, 7(1): 13-25.

Velazco-Macías CG (2009) Flora del estado de Nuevo León: diversidad y análisis espacio-temporal. Tesis doctoral. Facultad de Ciencias Biológicas, Universidad Autónoma de Nuevo León.

Velazco-Macías CG, Alanis FGJ, Alvarado VMA, Ramírez FL, Foroughbakchk PR (2011) Flora endémica de Nuevo León, México y estados colindantes. Jornal of the Botanical Research Institute of Texas 5(1): 275 – 298.

Villarreal Quintanilla JA, Estrada C E (2008) Flora de Nuevo León. Listados florísticos de México, XXIV. Instituto de Biología, Universidad Nacional Autónoma de México.

Wilson JS, Griswold T, Messinger O (2008) Sampling Bee Communities (Hymenoptera: Apiformes) in a Desert Landscape: Are Pan Traps Sufficient? Journal of Kansas Entomological Society. 81(3): 288-300.

4

SPRING FORAGING RESOURCES AND THE BEHAVIOUR OF POLLINATING INSECTS IN FIXED DUNE ECOSYSTEMS

Aoife T. O'Rourke[1,2]*, Úna Fitzpatrick[3], Jane C. Stout[1,2]

[1]*School of Natural Sciences, Botany Department, Trinity College Dublin, College Green, Dublin 2, Ireland*
[2]*Trinity Centre for Biodiversity Research (TCBR), Trinity College Dublin, College Green, Dublin 2, Ireland*
[3]*National Biodiversity Data Centre, Beechfield house, Carriganore WIT West Campus, County Waterford, Ireland*

Abstract—In temperate climates, foraging resources for pollinating insects are especially important in early spring when animals emerge from hibernation and initiate annual life cycles. One habitat, protected under EU law, which provides resources for a range of pollinating insects, but has received little research attention, is fixed (grey) dunes. Fixed dunes often contain creeping willow (*Salix repens*, Salicaceae), which may be an important early season resource for obligate flower visitors. We examined the springtime activity of flower visitors in fixed dune ecosystems in relation to sugar concentration and composition in nectar, composition of essential amino acids in pollen, and floral abundance. We also investigated whether the presence or absence of *S. repens* influenced the abundance and species richness of three obligate flower visiting guilds (solitary bees, bumblebees and hoverflies) in eight sites along the eastern and southern coasts of Ireland.

Higher insect visitation rates were observed to species whose nectar contained greater concentrations of glucose and fructose. Solitary bee visitation rates were related to % Essential Amino Acid (EAA) in pollen and floral species richness. *Ulex europeaus*, and *S. repens* were the most abundant flowering species, but visitation rates were not related to floral abundance. Higher abundances of bumblebees and hoverflies were discovered at sites where *S. repens* was present. This study raises further questions about the nutritional requirements and preferences of obligate flower visitors in fixed dune ecosystems in spring time.

Keywords: pollinators, amino acids, sugars, visitation rates, fixed dunes, spring

INTRODUCTION

The Global declines in both wild and domesticated pollinators have been documented (Biesmeijer et al. 2006; Cameron et al. 2011; Kearns et al. 1998; Kremen & Ricketts 2000; Potts et al. 2010), which has led to concern for the pollination of crops and wild plants (Garibaldi et al. 2013). Multiple drivers have been implicated, but one of the primary factors causing pollinator decline is habitat loss and fragmentation, often driven by agricultural intensification and a lack of appropriate land-use management (Giannini et al. 2012; Potts et al. 2011). This has resulted in a scarcity of high quality habitats for invertebrates (Diacon-Bolli et al. 2012; Klein et al. 2012; Kohler et al. 2008; Tscharntke et al. 2002b). Although there is a good knowledge on the biology of a handful of commercially utilised pollinating insect species, in particular the honey bee (*Apis mellifera*) and some bumblebees (e.g. *Bombus terrestris*), we know remarkably little about the foraging ecology and habitat use of the majority of the world's insect pollinators, including relatively common species (Mayer et al. 2011). Many of the best pollinating species are obligate flower visitors, reliant on the floral resources available in a habitat, or in several habitats within a landscape (Junker & Blüthgen 2010). Therefore, habitat type, the quantity and quality of the resources available (e.g. in terms of the flowering plant species available, and their relative abundance and composition of nectar and pollen rewards), and heterogeneity of available habitats (both within a given site and in the landscape surrounding that site) can have a significant impact on pollinating insect communities (Krauss et al. 2010; Krauss et al. 2003; Murray et al. 2012; Tscharntke & Brandl 2004; Tscharntke et al. 2002a; Tscharntke et al. 2002b). Much of the research on habitat use by pollinators has focused on fragments of semi-natural habitats within or bordering agricultural landscapes (Kennedy et al. 2013; Munyuli et al. 2013; Öckinger & Smith 2007), or on large areas of natural or semi-natural habitat such as national parks (Forster 1989; Herrera 1988; Inouye & Pyke 1988; Pascarella et al. 1999; Primack 1983). There has been little attention paid to coastal ecosystems (but see Howe et al. 2010; Redpath 2010; Redpath et al. 2010).

One particularly high quality habitat for pollinators in temperate coastal areas is fixed (grey) dune. Fixed dunes are the mature sections of dune systems, which are relatively stable, and are characterised by an almost complete carpet of vegetation (Curtis 1991). They are home to a range of invertebrate species (Archer 1994; Everard et al. 2010; Howe et al. 2010). Fixed dunes are an Annex I priority habitat under the EU Habitat Directive (Commission 1992) and thus legally protected. In Ireland, they are the most

*Corresponding author; email: orourkat@tcd.ie

florally diverse section of the dune system (Curtis 1991), which, accompanied by their areas of open ground and favourable microclimate (Howe et al. 2010), make them important for meeting the demands of pollinating insects in terms of both forage and nesting/larval habitat. Some of the plant species available in these habitats are important during different life stages of pollinating insects (i.e. early-flowering species are important in the diet of early-emerging pollinators). Female insects often require pollen in order to complete the maturation of their ovaries (Branquart & Hemptinne 2000a) and for nest provisioning (Cane et al. 2011; Tepedino & Parker 1982). They also need nectar to maintain their metabolism (Crane 1990), especially in sub-optimal spring temperatures and unpredictable weather patterns.

Although advances have been made into understanding the links between the nutritional properties of floral resources available to insects and the floral preferences of obligate flower visitors, previous research has largely focused on the nutritional resources of nectar (Cane et al. 2011; Chalcoff et al. 2006; Herrera et al. 2006; Krömer et al. 2008; Nocentini et al. 2012; Perret et al. 2001; Roubik et al. 1995; Sokal & Rohlf 2005; Southwick et al. 1981; Wacht et al. 2000). Sucrose-rich nectar has been associated with visits by long-tongued insects, e.g. butterflies, whereas nectars rich in fructose or glucose have been associated with short-tongued bees and flies (Baker & Baker 1883; 1990; Baker et al. 2006; Perret et al. 2001). However, pollen is arguably the most important nutritional resource required by pollinators in spring (Branquart & Hemptinne 2000b). Studies suggest that the nutritional value of pollen for bees may be defined more accurately by the amino acid composition than by protein content (Cook et al. 2003; Loper & Cohen 1987; McCaughey 1980) and floral selection by honeybees seems to reflect pollen quality (Cook et al. 2003). de Groot (1953) determined ten essential amino acids for honeybees, and it is assumed that pollen containing more of these essential amino acids (EAAs) are of greater nutritional value for bees than those containing a lower proportion of EAAs (Cook et al. 2003). Roulston and Cane (2000) state that most plant species contain the full range of EAAs, with tryptophan and phenylalanine being the only two EAAs sometimes absent from pollen. EAAs are important for the early development of bees (de Groot 1953; Hanley et al. 2008; Roulston & Cane 2000) and the absence of certain EAAs can be detrimental to larval development (Herbert et al. 1970; Roulston & Cane 2002).

Willow (*Salix* spp.) has long been recognised as an important forage resource for spring pollinators, particularly bees (Kearns et al. 1998; Sladen 1912; Sommerville 1992; Stelfox 1927). Recently, several studies have highlighted an abundance of *Salix* pollen grains in honey from different regions around the globe (Salonen et al. 2009; Wróblewska & Stawiarz 2004). *Salix* is also known to host numerous oligolectic solitary bee species (Dötterl et al. 2005). European fixed dunes often contain creeping willow (*Salix repens* L., Salicaceae), which, like other *Salix* species, can provide important early season resources for bees and other obligate flower visitors (de Jong et al. 2005; Goulson 2010; Stelfox 1927), particularly female bees emerging from hibernation, when few other resources exist in the landscape (de Jong 2011). *S. repens* is a low growing shrubby plant, which is highly variable in growth form (Stace 2010), and can become dominant in these habitats. It is dioecious, with catkins producing a large quantity of pollen (Fox 1992), and appearing before the leaves in April and May (Meikle 1984). Both male and female flowers produce nectar and the species is pollinated by both wind and insects (Totland & Sottocornola 2001), but few studies have examined the relationship between *S. repens* and early-emerging pollinating insects (but see de Jong et al. 2005; Füssel 2007; Hornoy et al. 2011).

In this study, we examined the springtime activity of obligate flower visitors in fixed dune ecosystems in relation to nectar sugar concentration and composition, relative abundance (%) of pollen essential amino acids and floral abundance of flowering plant species. We also investigated the effect of *S. repens* on the abundance and diversity of three obligate flower visiting guilds: solitary bees, bumblebees and hoverflies.

MATERIALS AND METHODS

Study sites and experimental design

The study was conducted within the fixed dune habitat, EU Habitats Directive Feature 2130 (Commission 1992), along the South-eastern coast of Ireland. Eight sites were selected, four of which contained populations of *S. repens*, and four paired sites without *S. repens* (Fig. 1). The majority of the sites are candidate Special Areas of Conservation (cSACs), with the remaining being proposed Natural Heritage Areas (pNHAs). The sites vary substantially in area of fixed dune habitat, and management regime (Tab. 1).

Within each site, floral abundance was recorded in a 50 × 100 m focal study area. The focal area was selected to encompass the most florally rich area of the fixed dune, and contained *S. repens* when present at a site. Floral abundance was recorded as the total number of floral units (FU) of each plant species in the focal area (Tab. 1), measured in 4m2 quadrats ($N = 13$ per site: a row of four equally spaced quadrats along both long edges of the focal area, and five in the middle of the focal area). Floral units were defined for each plant species (Dicks et al. 2002), and consisted of either individual flowers or groups of flowers in the case of inflorescences (Appendix 1, supplementary material). As *S. repens* is dioecious, and it is known that the two sexes of this species produce chemically distinct nectar (Füssel 2007), the sex of each individual was recorded.

Rewards

Nectar was extracted from 20-100 FU of each plant species that had >10 individuals in the focal area (10 species in total at all sites: *Glechoma hederacea, Hyacinthoides non-scripta, Lotus corniculatus, Pilosella officinarum, Ranunculus bulbosus, Salix repens* (Female), *Salix repens* (Male), *Taraxacum* agg., *Vicia sativa, Viola lutea, Viola riviniana*) using 0.5 µl and 1 µl Drummond glass microcapillary tubes. For logistical reasons most samples were taken from two sites, Brittas, and Raven Point. Nectar

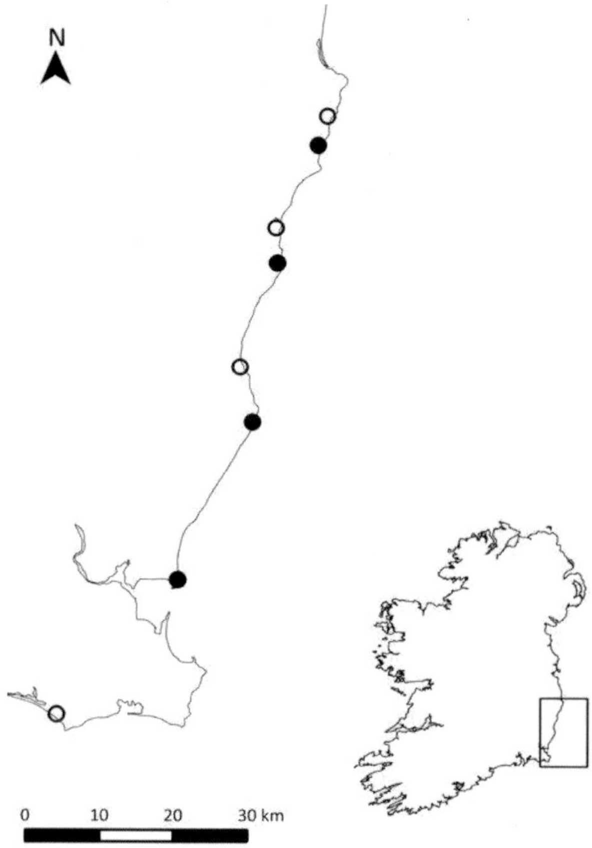

FIGURE 1. Distribution of study sites in SE Ireland. Closed circles indicate sites where *S. repens* was present; open circles indicate sites where it was not. Site names (North-South) = Magherabeg, Brittas, Arklow, Kilpatrick, Courtown, Cahore, Raven Point, and Ballyteige Burrows.

from 30 FU was pooled to create six composite nectar samples of ≥5 µl per species for sugar analysis (see Appendix II). Pure nectar was diluted 1:1999 with milli-Q H_2O before sugar analysis.

Glucose, fructose and sucrose in the diluted extracts were directly analysed by liquid chromatography-tandem mass spectrometry (LC-MS/MS) (see additional methods and supplementary information in Appendix III).

In order to investigate the relative abundance of EAA in pollen of the same 10 plant species, plus *B. perennis* and *U. europeaus*, pollen was hand-collected from 20-100 individuals in the same two sites where nectar was collected. Pollen was extracted from anthers using forceps or a paint brush. A sterilised forceps and separate paintbrush was used for each pollen sample. Pollen was pooled per species in order to produce a large enough quantity for analysis. Thus it was not possible to ascertain the concentrations of EAAs in pollen from each individual, and so relative abundance of EAA (%) (see Appendix IV) was determined for each species from the pooled sample. The relative abundance of EAA in plant species tend to be very constant (see table 1 in Nocentini et al. 2012). It was subsequently diluted with 300µl methanol to break down the exine and outer protein binding materials of the pollen grains. The resulting product was then filtered through a 0.2 µl nano syringe filter to ensure purity of amino acid extract.

Amino acids in the diluted extracts were directly analysed by LC-MS/MS. The analysis method was modified from a protocol described by (Jander et al. 2004) (see additional methods and supplementary information in Appendix V).

TABLE 1. Location and properties of the eight study sites in Ireland

Name	Status	Area (ha)	Location	Site details	*S. repens* (Y/N)	Site pairs	Ownership	Plant species present
Magherabeg	cSAC	7.591	52°55'24"N 06°01'52"W	Undergrazed	N	1	Private	4,8,9,10
Brittas Bay	cSAC	44.94	52° 88'33"N 06° 05'00"W	Highly impacted recreationally	Y	1	Wicklow CoCo	1,7,8,9,10,12
Arklow North	pNHA	1.606	52° 48'57"N 06°07'49"W	Eroded	N	2	Private	1,2,3,8,910,12
Kilpatrick	cSAC	12.992	52°44'00"N 06°09'48"W	Highly impacted recreationally	Y	2	Private	1,7,8,9,12
Courtown	pNHA	0.543	52° 64'50"N 06°22'90"W	Invaded heavily	N	3	Wexford CoCo	1,8
Cahore North	cSAC	78.303	52°32'45"N 06°12'57"W	Grazed	Y	3	Private	4,7,8,10,11
Raven Point	cSAC	26.937	52°21'29"N 06°02'24"W	Managed effectively for recreation	Y	4	NPWS	4,5,7,8,10,12
Ballyteige Burrows	cSAC	238.638	52°11'19"N 06°36'39"W	Grazed	N	4	NPWS	1,4,6,8,10,12

Legend: cSAC = candidate Special Area of Conservation, pNHA = proposed Natural Heritage Area, CoCo = County Council, NPWS = National Parks and Wildlife Service. Plant species present at each site: 1 = *Bellis perennis*, 2 = *Glechoma hederacea*, 3 = *Hyacinthoides non-scripta*, 4 = *Lotus corniculatus*, 5 = *Pilosella officinarum*, 6 = *Ranunculus bulbosus*, 7 = *Salix repens* (Female), *Salix repens* (Male), 8 = *Taraxacum* agg., 9 = *Ulex europeaus*, 10 = *Vicia sativa*, 11 = *Viola lutea*, 12 = *Viola riviniana*

Insect observations

Observations of insect visitation to the flowers of 12 plant species from which rewards were sampled, were made at all sites from early April to early May 2012, encompassing the full flowering period of *S. repens*. Not all sites contained all flower species (Tab. 1), so not all species were observed at all sites. Both sites in a pair were visited on consecutive days, and pairs of sites were visited in temporal succession. All plant species observed were in flower throughout April and May. Data were only collected when temperatures ≥13 °C (mean 14.8 °C) and wind speeds were less than Beaufort force 3 (mean 2.16). Each plant species was observed for a total of 90 min over a two day period (30 min morning Day 1, 30 min afternoon Day 1, 30 min afternoon Day 2), and we selected a new 1 m^2 patch for each observation. The number of FU in the observation patch were recorded prior to recording insect visitation, the number of visitors, visitor identity and number of FU visited were also recorded. Only insects in three obligate flower visiting guilds, bumblebees (BB), solitary bees (SB) and hoverflies (HF) (Appendix VI), that touched the flowers' reproductive organs whilst foraging, were recorded. These three guilds were chosen as they were the most common flower visitors and were sufficiently abundant for statistical analysis.

Total visits per FU per hour were first calculated on a site by site basis for each plant species as: the sum of visits by obligate flower at the site, divided by the sum of FU per plant species observed over the total 90min period (three 30 min observation period per plant species pooled per site) divided by 1.5. Visits per FU per hour to each plant species were calculated for each insect guild separately. Site visitation rates were then summed for each plant species per guild, to give a total visitation rate per plant species.

In addition, pan traps were also employed as they allow simultaneous sampling of the abundance and richness (i.e. number of species) of pollinators at multiple locations, laboratory identification of specimens to finer taxonomic resolution, and are the most efficient method for sampling bees (Westphal et al. 2008). All eight sites were sampled twice; once in each of two sampling periods (beginning of April, and beginning of May), in 2012. To maximise comparability among sites, all sites were sampled during a 48 hour time-frame in each sampling period. Traps were polypropylene plastic bowls painted with white, yellow and blue UV paint (LeBuhn et al. 2003). One bowl of each colour was attached to a post using a metal clamp and the rim of the bowls adjusted to vegetation height. Three posts were placed at each site on each sampling day in the most flower rich areas within each site. At sites containing *S. repens*, these were within areas that the shrub was growing.

Statistical analyses

Differences in concentrations of glucose, fructose and sucrose between each plant species were tested using three separate one-way ANOVA, carried out using R software (version 3.0.1, 2013). Where transformation failed to reduce heterogeneity of variance, analyses were carried out on untransformed data (Underwood 1997). Tukey's HSD tests (Hsu 1996) were used to determine significant differences between group means when an effect was detected.

Pearson's product moment correlation coefficients were calculated to quantify the relationships between concentrations of glucose, fructose and sucrose in nectar in plant species. Data were transformed (log_{10}) to meet assumptions of normality.

Correlation analyses were also performed to test relationships between insect guild visitation rates (visits per FU per hour) and floral rewards (i.e. sugar concentration of nectar and relative abundance of EAAs (%) in pollen). Transformed data failed to meet assumptions of normality and so Spearman's rank correlation was used. Correlation analyses were also carried out in order to test relationships between total insect visits (per FU per hour) and total site-level floral richness of co-flowering plant species, and, separately, total insect visits and total site-level FU abundance (Spearman's rank) (R, version 3.0.1, 2013).

Insect abundance and richness from pan-traps, and plant abundance and richness, were compared among sites using ANOVA with a randomised complete block design (Southwick et al. 1981), where each block represented a pair of sites, one with and one without *S. repens* (henceforth "treatment") (N = 4 per treatment). Prior to analysis, all data were tested for heterogeneity of variance, and Log_{10} transformations where used.

RESULTS

Forage resources

Mean glucose and fructose concentrations varied significantly among plant species; with mean sucrose concentrations displaying the highest variability between species. *S. repens* had a higher mean glucose concentration than all other species except *V. riviniana* (Fig. 2A-C). Concentrations of fructose in *S. repens* nectar were similar to five of the other plant species, and sucrose lower than four species, and *S. repens* male flowers were similar to *G. hederacea* and female flowers to *V. sativa* (Fig. 2A-C). The concentration of fructose and glucose were highly correlated in all plant species ($r = 0.72$, $N = 78$, $P < 0.05$).

Eighteen out of the total range of 35 (Weiner et al. 2010) amino acids were found to be present within the pollen of the plant species analysed, ten of which are classified as essential amino acids (EAA) (de Groot, 1952). *Bellis perennis* and *Hyacinthoides non-scripta* contained the highest proportion of EAA in their pollen (>25%), whereas *R. bulbosus* contained the lowest (<10%) (Fig. 3, Appendix IV) and *S. repens* trended in the middle.

U. europeaus, *S. repens* (F), and *S. repens* (M) were the most abundant flowering species in the focal areas (Fig. 3). When FU of both sexes of *S. repens* were added together, this species was more abundant than *U. europeaus* (summed mean floral abundance of *S. repens* = 1829.75 FU per focal area). *V. sativa* had the lowest floral abundance, with the remaining plant species being found to have similar abundances to each other (Fig. 3).

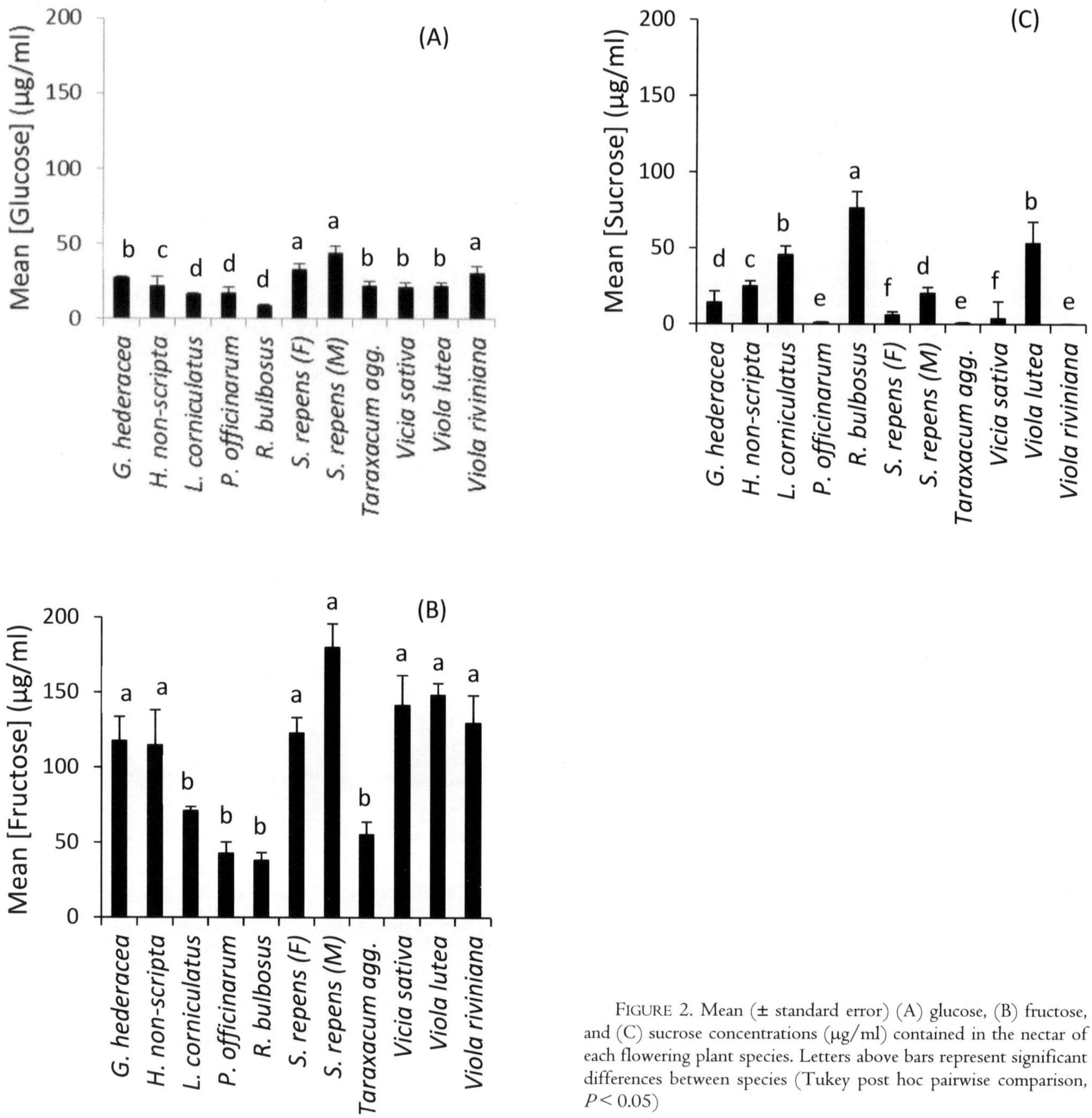

FIGURE 2. Mean (± standard error) (A) glucose, (B) fructose, and (C) sucrose concentrations (µg/ml) contained in the nectar of each flowering plant species. Letters above bars represent significant differences between species (Tukey post hoc pairwise comparison, $P < 0.05$)

Insect visitation rates

Visits by bumblebees, solitary bees and hoverflies (per FU per hour) to each plant species were relatively low to most species, with the exception of visits to *Taraxacum* agg., *S. repens* (F) and *U. europeaus* (Fig. 4). *Taraxacum* agg. received the highest number of visits (per FU per hour) from bumblebees, solitary bees and hoverflies, followed by female *S. repens* (Fig. 4). Most visits to *S. repens* were from bumblebees; with female flowers receiving more visits than the males (Fig. 5).

There was a significant positive correlation between total solitary bee visitation rate and total site-level floral richness ($r = 0.72$, $df = 7$, $P = 0.045$) (Fig. 6A). However, there was no significant correlation between total site-level floral richness and visits by bumblebees ($r = 0.49$, $df = 7$, $P = 0.217$) or hoverflies ($r = 0.58$, $df = 7$, $P = 0.130$). There were also no significant correlations between visits per insect guild or total site-level floral abundance (bumblebees: $r = 0.479$, $df = 7$, $P = 0.229$; solitary bees: $r = 0.599$, $df = 7$, $P = 0.117$; hoverflies: $r = 0.590$, $df = 7$, $P = 0.123$).

There were several significant correlations between insect visitation rates and EAAs and sugar content of the different plant species (Tab. 2). However, relationships involving bumblebee visitation were driven by high visitation rates to *Salix* and *Taraxacum* only. For solitary bees, there were positive relationships between visitation rates and EAAs (Fig. 6B) and glucose content of nectar, but negative relationships between visitation and sucrose content of nectar (Fig. 6C). The relationship between solitary bee visitation rates and the glucose content of nectar also appears to be driven by high visitation rates to *Salix* and *Taraxacum*. There was a notable significant negative correlation between sucrose concentration and EAA abundance (r = 0.558, df = 7, $P <$ 0.0001).

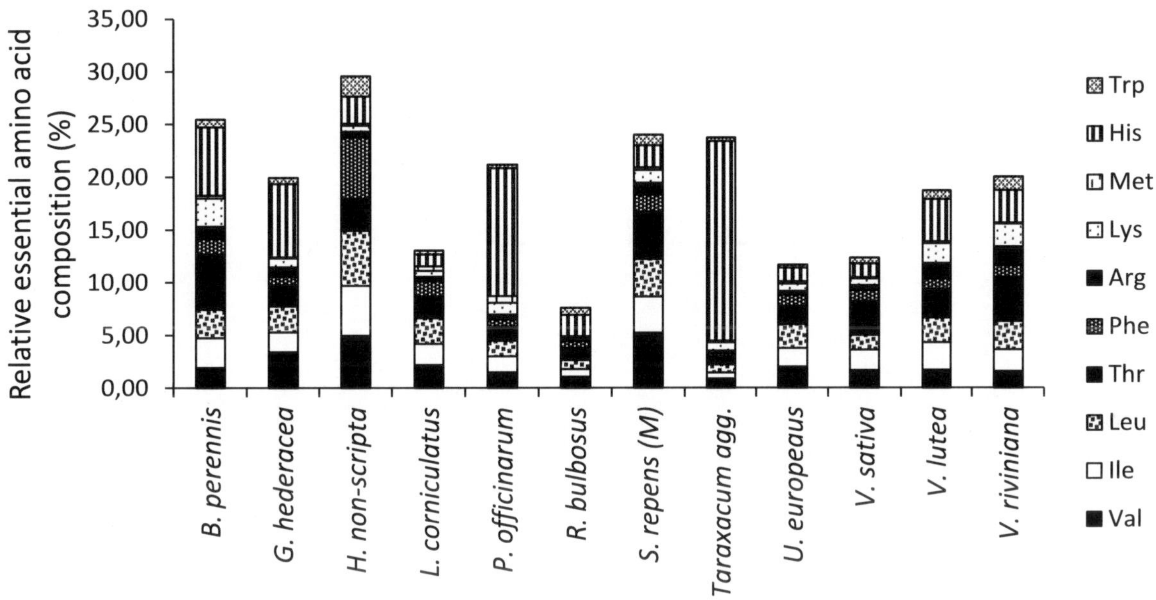

FIGURE 3. Relative essential amino acid composition (%) in pollen for the following species: *Bellis perennis, Glechoma hederacea, Hyacinthoides non-scripta, Lotus corniculatus, Pilosella officinarum, Ranunculus bulbosus, Salix repens* (Female), *Salix repens* (Male), *Taraxacum* agg., *Vicia sativa, Viola lutea, Viola riviniana.* Legend abbreviations: Val = Valine, Ile = Isoleucine, Leu = Leucine, Thr = Threonine, Phe = Phenylalanine, Arg = Arginine, Lys = Lysine, Met = Metionine, His = Histidine, Trp = Tryptophan

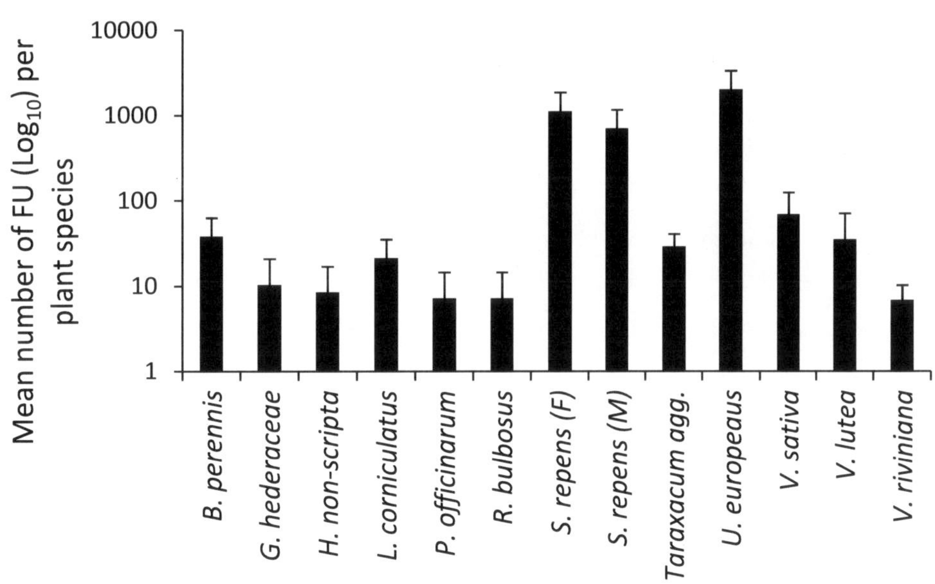

FIGURE 4. Mean floral abundance ($\log_{10}$) $\pm$ standard error, measured by number of floral units present on average per 52 m^2 focal area in each site

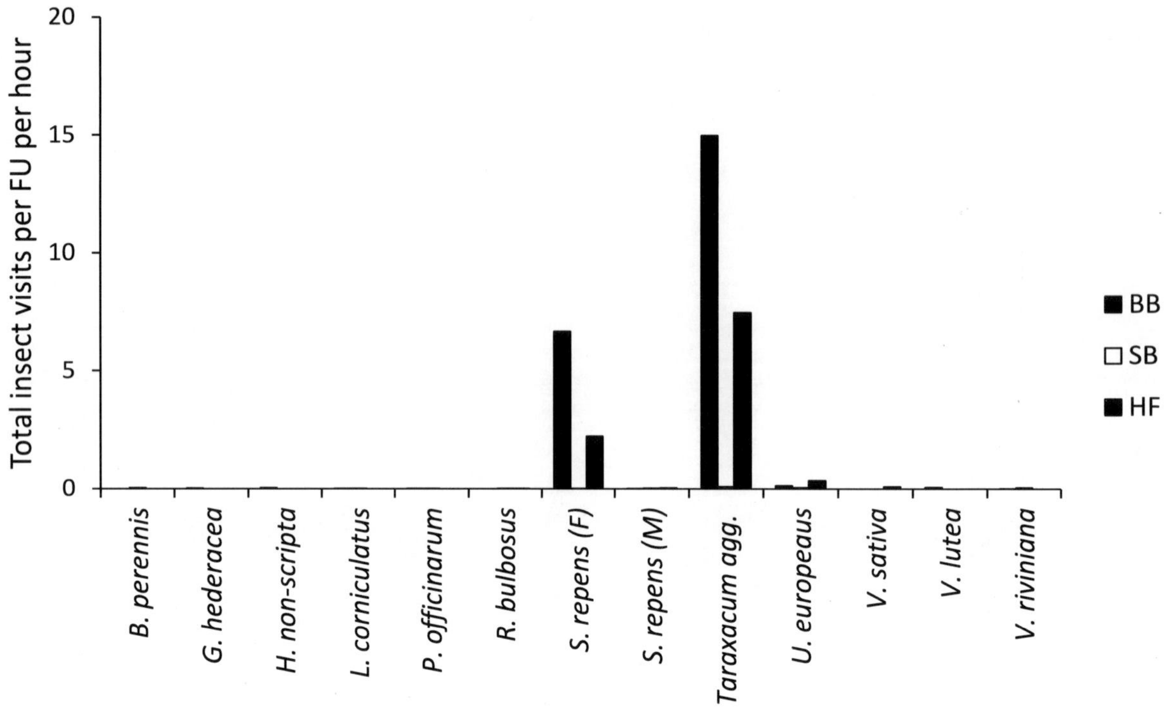

FIGURE 5. Total insect visits per FU per hour for each plant species. BB=bumblebees, SB=solitary bees, HF= hoverflies

Table 2. Relationships between plant reward (sugar concentrations/relative abundance of EAA (%) and visits (per flower per hour) for each insect guild

Visits	[Glucose] (µg/ml)		[Fructose] (µg/ml)		[Sucrose] (µg/ml)		EAA (%)	
	coeff	*P*-value	coeff	*P*-value	coeff	*P*-value	coeff	*P*-value
Bumblebee	0.328	0.007	0.133	0.287	-0.275	0.026	0.571	≤0.0001
Solitary bee	0.285	0.020	-0.146	0.242	-0.380	0.002	0.300	0.020
Hoverfly	0.211	0.089	0.073	0.562	-0.099	0.428	-0.041	0.756

Richness and abundance of insects according to presence of S. repens

In sites where *S. repens* was present, significantly higher abundances of both bumblebees and hoverflies were captured in the pan-traps (Tab. 3, Fig. 7A,E). A similar trend was observed for solitary bees (Tab. 3, Fig. 7C), but their numbers (compounded by our low number of replicate sites) were too few to detect differences. Similarly, a non-significant trend for greater species richness of bumblebees in sites where *S. repens* was present was also detected (Tab. 3). Bumblebee abundance was also significantly different among pairs of sites (blocks).

DISCUSSION

In all plant species sampled, nectar sugar composition was dominated by fructose. This could explain why we observed short-tongued insects (e.g. short-tongued bumblebees, solitary bees which have relatively short tongues, and flies, i.e. muscids) as the dominant flower visitors during our sampling period, as these insects have been associated with fructose-rich nectar in the past (Baker & Baker 1983; Baker & Baker 1990; Baker et al. 2006; Perret et al. 2001).

Although the traditional view is that bumblebees exhibit a strong preference for sucrose, this is not well supported in the literature. In our study, *Taraxacum* and *Salix* (F) nectar contained very low levels of sucrose, and yet the highest visitation rates were to these two species. A recent study by Mommaerts et al. (2013) compared the gustatory responses of bumblebees to glucose, fructose and sucrose, and concluded bees have a plastic response to sugar, preferring glucose and fructose when harnessed, and sucrose when free-flying, and that responses are modified by experience. As bees and other pollinators become familiarised with the nectar composition in flowers that are of greatest abundance in the environment when they first emerge, this may affect their later preferences. In our study, this may explain the preference for *S. repens*: early-emerging bees might use these resources and respond plastically by foraging predominantly on this plant species because of its massive floral abundance (especially in their naïve stages), regardless of its nectar sugar composition.

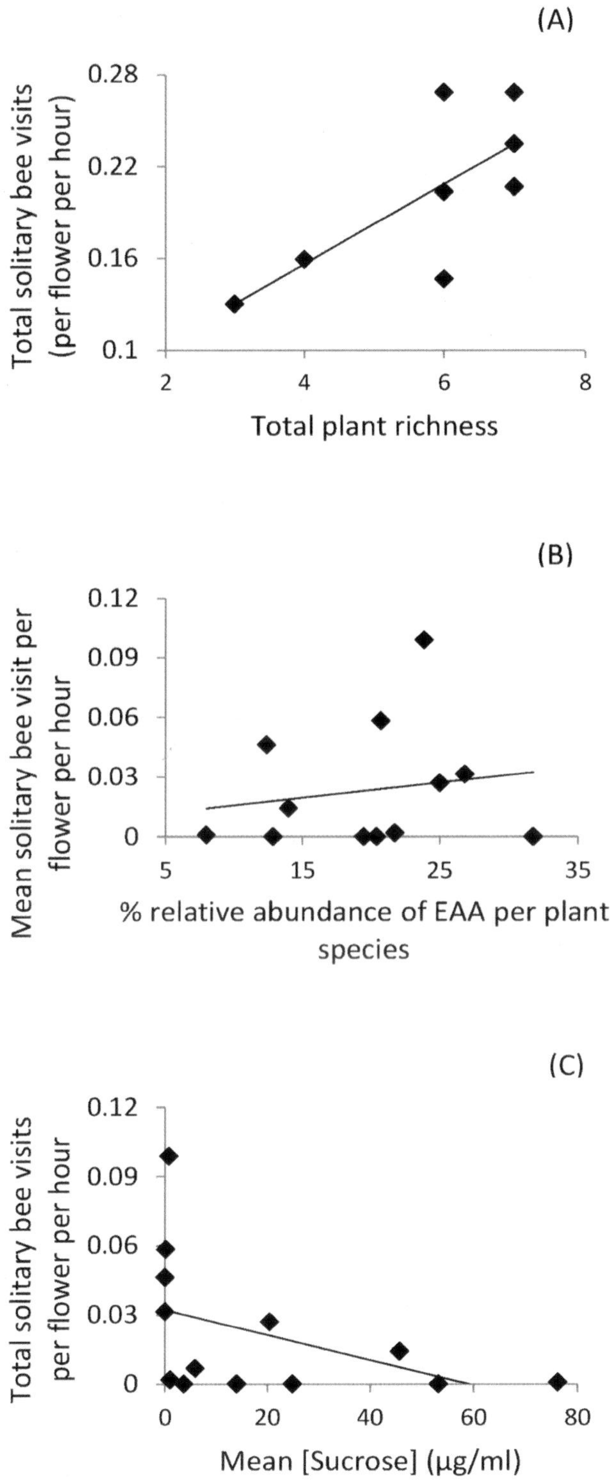

FIGURE 6. The relationship between total solitary bee visits per flower per hour and (A) total floral richness ($r = 0.72$, $P = 0.045$), (B) total relative abundance of EAA (%) ($r = 0.30$, $P = 0.02$), and (C) mean sucrose concentration (μg/ml) ($r = 0.38$, $P = 0.002$).

Previous studies have shown that nectar composition may vary depending on region, locality and between times of flowering period (Herrera et al. 2006), and so it is possible that further investigations into spatial and temporal patterns of nectar production and visitation may reveal patterns not detected here. Coastal sand dunes are extreme environments, exposed and relatively dry, and this may have affected floral rewards.

It is notable that EAA composition and abundance appears to display phylogenetic constancy, consistent with findings by Roulston & Cane (2000). The Asteraceae have higher abundances of EAAs than the Fabaceae and Ranunculaceae. It is difficult to compare the relative abundance of EAAs sampled from the plants within this study compared with other similar studies, as the extraction methods differ substantially. Few studies appear to use the same protocol, making it difficult to draw meaningful parallels.

Although honeybees have shown a distinct preference for pollen that is richer in EAAs (Cook et al. 2003), it is unclear from the literature whether or not high total abundance of EAAs influences obligate flower visitor foraging rates (Thomson et al. 2000). We found a significant positive relationship between solitary bee visitation rates to plant species and total relative abundance of EAAs (%) in the pollen of those species. Pollen stimulates egg maturation of the ovaries (Branquart & Hemptinne 2000b; Yuan et al. 2007) and is believed to be more important than nectar as a food source for solitary bees compared to social bees, as the energetic requirements of solitary bees are lower, because they do not have to expend energy heating a nest (Heinrich 1975). Thus solitary bees gather proportionally more pollen than nectar compared with bumblebees (Cane et al. 2011). In addition, pollen sources gathered by oligolectic solitary bee species have been found to contain significantly lower concentrations of EAAs than those gathered by polylectic species (Weiner et al. 2010). The solitary bee species surveyed over the course of this study where predominantly polylectic, which may explain the positive relationship we found between visitation and % EAAs. However, there is a paucity of information on the EAA requirements and preferences of solitary bees (Cane et al. 2011; Weiner et al. 2010) and the majority of other pollinating insects. Further research is required in order to gain a greater understanding of the nutritional requirements of most pollinating insects.

Salix repens and *U. europeaus* both have peak flowering times in spring (Hornoy et al. 2011) and were the most abundant flowers in our sites. However, few studies have examined the floral visitors associated with *S. repens* (de Jong et al. 2005); most have focused on visitors to other species of the *Salix* genus, and show that bumblebees are the most common visitors, particularly in spring time (Füssel 2007). Our observations of insect visitors to *S. repens* are consistent with de Jong et al. (2005), however, we also commonly observed syrphids visiting *S. repens*, particularly of the genus *Eristalis*. Although there were no significant differences in floral abundance or richness between sites with and without *S. repens*, there was a significantly higher abundance of both bumblebees and hoverflies at sites where *S. repens* was present. This concurs with the findings of previous studies that showed that *S. repens* is a valuable

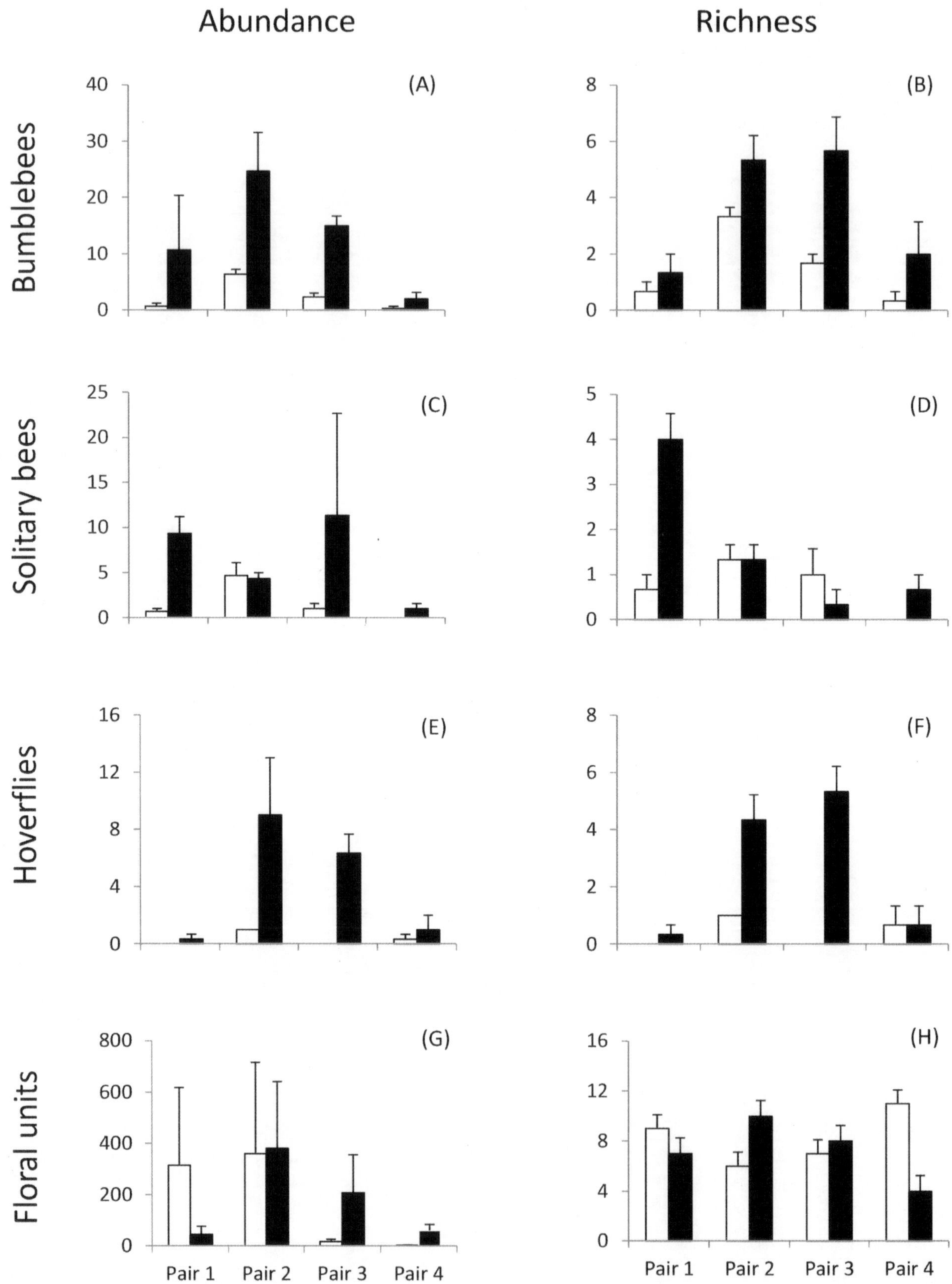

FIGURE 7. (A-H) Mean abundance and richness (± standard error) of each taxonomic group based on treatment (*S. repens* vs. no *S. repens*). Pair 1 = Arklow; Kilpatrick, pair 2 = Magherabeg; Brittas pair 3 = Ballyteige; Raven Point, pair 4 = Courtown; Cahore. Empty bars = *S.repens* absent, filled bars = *S. repens* present; richness here means number of species occurring.

TABLE 3. Results from ANOVA randomised complete block to compare abundance and species richness of insects captured in pan traps between treatments (*S. repens* present vs. *S. repens* absent), in four replicate blocks (pairs of sites). BB = bumblebees, SB = solitary bees, HF = hoverflies, PL = plants, * = $P < 0.05$.

	Abundance				Species richness			
	BB	SB	HF	PL	BB	SB	HF	PL
Treatment	0.01*	0.09	0.04*	0.29	0.06	0.38	0.13	0.47
Block	0.01*	0.19	0.08	0.99	0.07	0.56	0.41	0.38

forage resource for bumblebees in spring (de Jong et al. 2005; Füssel 2007). Although we cannot rule out that there are third causal factors influencing the abundance of insects and presence or absence of *S. repens* at a site (e.g. site size, presence or absence of grazing, degree of disturbance due to recreational use), these findings do suggest that the presence of *S. repens* can be beneficial for not only bumblebees, but also other pollinating insects, including hoverflies. This may have a beneficial knock-on effect to the pollination of flowering plants later in the season, after *Salix* has finished flowering.

Taraxacum agg. was also an abundant flowering plant at our study sites and has been suggested as an important food source in spring when the flowering period of *Salix* has come to an end (Alanen 2009; Fussell & Corbet 1992; Teräs 1985). Our findings concur with these previous studies: *Taraxacum* agg. provides forage for a range of obligate flower visitors.

We also found a significant relationship between floral richness at a site and solitary bee visitation rates, which is consistent with other studies (Ebeling et al. 2008; Holzschuh et al. 2007; Klein et al. 2003; Tscharntke et al. 1998). In addition, the total number of solitary bee visits increased linearly with increasing floral richness. This is in agreement with findings by Ebeling et al. (2008), but contrasts Ebeling et al. (2012), where increasing floral cover was the only factor which increased the visitation rate of both solitary bees and hoverflies.

In conclusion, flower-visiting insects, foraging in fixed dunes early in the season, are influenced by the nectar and pollen composition of plant species, as well as their relative abundance. Our research has highlighted some of the complexities associated with reward choice by pollinating insects. Future research should address the chemistry of nectar (including nectar sugar concentrations, nectar protein content, trace elements in nectar, nectar temperature and viscosity, microbial content of nectar, secondary plant compounds in nectar, nectar quantity and secretion rate) and pollen (including pollen protein concentrations, pollen protein sugars and trace elements, pollen fatty acids, digestibility of various pollen grains, and quantity of pollen), as well as flower structure (shape, colour including UV-reward guides, and micro-topography), in relation to insect preferences through the season. Laboratory-based research is important and often necessary, yet must be combined with field experiments, as ecological context (relative abundance, co-flowering species, local and landscape factors) can influence choice.

ACKNOWLEDGEMENTS

We wish to thank Dr Michael Reichelt (Max-Planck Institute of Chemical Ecology) for kindly analysing our plant extracts, Dr Katharina Schramm (Max-Planck Institute of Chemical Ecology) and Paul Egan (TCD, Ireland) for amino acid extraction advice, and Dr Eileen Power (University of Newcastle) for advice on sampling nectar and pollen, sugar analysis and amino acid extraction. Thanks also to Dr Dave Bourke (TCD, Ireland), Dr Serena Dool (NBDC, Ireland), and Miles Newman (TCD, Ireland) (pan-trapping assistance), Dr Tom Gittings, and Dr Martin Speight (hoverfly verification). We are very grateful to private landowners and the NPWS for giving us permission to carry out fieldwork on their land. This study was funded by a postgraduate research scholarship awarded to Aoife O'Rourke by Trinity College Dublin (TCD), and the Botany Department (TCD).

APPENDICES

Additional supporting information may be found in the online version of this article:

APPENDIX I. Floral unit description

APPENDIX II. Sugar concentration of plants

APPENDIX III. Protocol for nectar sugar analysis

APPENDIX IV. Amino acids in pollen

APPENDIX V. Protocol for pollen amino acid analysis

APPENDIX VI. Insect species list

REFERENCES

Alanen E-L (2009) Bumblebee density in agroecosystems during the starting stage of the colonies and its implications for pollination services. MSc thesis, University of Helsinki, Finland

Archer ME (1994) Survey of Aculeate wasps and bees (Hymenoptera) on the sand dune systems of South Wales during 1994. Countryside Council for Wales, UK

Baker H, Baker I (1983) Floral nectar sugar constituents in relation to pollinator type. Handbook of experimental pollination biology. S & AE, New York, pp 117-141

Baker H, Baker I (1990) The predictive value of nectar chemistry to the recognition of pollinator types. Israel Journal of Botany 39:157-166

Baker HG, Baker I, Hodges SA (2006) Sugar composition of nectars and fruits consumed by birds and bats in the tropics and subtropics. Biotropica 30:559-586

Biesmeijer J et al. (2006) Parallel declines in pollinators and insect-pollinated plants in Britain and the Netherlands. Science 313:351-354

Branquart E, Hemptinne JL (2000a) Development of ovaries, allometry of reproductive traits and fecundity of *Episyrphus balteatus* (Diptera: Syrphidae). European Journal of Entomology 97:165-170

Branquart E, Hemptinne JL (2000b) Selectivity in the exploitation of floral resources by hoverflies (Diptera: Syrphinae). Ecography 23:732-742

Cameron SA et al. (2011) Patterns of widespread decline in North American bumble bees. Proceedings of the National Academy of Sciences 108:662-667

Cane JH, Gardner DR, Harrison PA (2011) Nectar and pollen sugars constituting larval provisions of the alfalfa leaf-cutting bee *(Megachile rotundata)* (Hymenoptera: Apiformes: Megachilidae). Apidologie 42:401-408

Chalcoff VR, Aizen MA, Galetto L (2006) Nectar concentration and composition of 26 species from the temperate forest of South America. Annals of Botany 97:413-421

Commission E (1992) Council Directive 92/43/EEC of 21 May 1992 on the conservation of natural habitats and of wild fauna and flora. Official Journal L 206:07

Cook SM, Awmack CS, Murray DA, Williams IH (2003) Are honey bees' foraging preferences affected by pollen amino acid composition? Ecological Entomology 28:622-627

Crane E (1990) Bees and beekeeping: science, practice and world resources. Cornell University Press, New York

Curtis T (1991) The flora and vegetation of sand dunes in Ireland. In: Quigley MB (ed) A guide to the sand dunes of Ireland. European Union for Dune Conservation and Coastal Management, Dublin, pp 42-46

de Groot AP (1953) Protein and amino acid requirements of the honeybee (*Apis mellifica* L.). Physiologia Comparata et Oecologia 3:197-285

de Jong TJ (2011) Pollination crisis, plant sex systems, and predicting evolutionary trends in attractiveness. In: Patiny S (ed) Evolution of Plant-Pollinator Relationships. Cambridge University Press, U.K., pp 28-43

de Jong TJ, Batenburg JC, Klinkhamer PG (2005) Distance-dependent pollen limitation of seed set in some insect-pollinated dioecious plants. Acta Oecologica 28:331-335

Diacon-Bolli J, Dalang T, Holderegger R, Bürgi M (2012) Heterogeneity fosters biodiversity: linking history and ecology of dry calcareous grasslands. Basic and Applied Ecology 13:641–653

Dicks L, Corbet S, Pywell R (2002) Compartmentalization in plant–insect flower visitor webs. Journal of Animal Ecology 71:32-43

Dötterl S, Füssel U, Jürgens A, Aas G (2005) 1, 4-Dimethoxybenzene, a floral scent compound in willows that attracts an oligolectic bee. Journal of chemical ecology 31:2993-2998

Ebeling A, Klein A-M, Weisser WW, Tscharntke T (2012) Multitrophic effects of experimental changes in plant diversity on cavity-nesting bees, wasps, and their parasitoids. Oecologia 169:453-465

Ebeling A, Klein AM, Schumacher J, Weisser WW, Tscharntke T (2008) How does plant richness affect pollinator richness and temporal stability of flower visits? Oikos 117:1808-1815

Everard M, Jones L, Watts B (2010) Have we neglected the societal importance of sand dunes? An ecosystem services perspective. Aquatic Conservation: Marine and Freshwater Ecosystems 20:476-487

Forster P (1989) Pollination of *Marsdenia fraseri* (Asclepiadaceae) by *Metriorrhynchus lateralis* (Coleoptera: Lycidae). Coleopterists Bulletin 43:311-312

Fox JF (1992) Pollen limitation of reproductive effort in willows. Oecologia 90:283-287

Füssel U (2007) Floral scent in Salix L. and the role of olfactory and visual cues for pollinator attraction of *Salix caprea* L. PhD thesis, Universitat Bayreuth, Germany

Fussell M, Corbet SA (1992) Flower usage by bumble-bees: a basis for forage plant management. Journal of Applied Ecology 29:451-465

Garibaldi LA et al. (2013) Wild pollinators enhance fruit set of crops regardless of honey bee abundance. Science 339 1608-1611

Giannini TC, Acosta AL, Garófalo CA, Saraiva AM, Alves-dos-Santos I, Imperatriz-Fonseca VL (2012) Pollination services at risk: bee habitats will decrease owing to climate change in Brazil. Ecological Modelling 244:127-131

Goulson D (2010) Bumblebees: behaviour, ecology, and conservation, Second edn. Oxford University Press, UK

Hanley ME, Franco M, Pichon S, Darvill B, Goulson D (2008) Breeding system, pollinator choice and variation in pollen quality in British herbaceous plants. Functional Ecology 22:592-598

Heinrich B (1975) Energetics of pollination. Annual Review of Ecology and Systematics 6:139-170

Herbert E, Bickley W, Shimanuki H (1970) The brood-rearing capability of caged honey bees fed dandelion and mixed pollen diets. Journal of Economic Entomology 63:215-218

Herrera CM, Pérez R, Alonso C (2006) Extreme intraplant variation in nectar sugar composition in an insect-pollinated perennial herb. American Journal of Botany 93:575-581

Herrera J (1988) Pollination relationships in southern Spanish Mediterranean shrublands. The Journal of Ecology 76:274-287

Holzschuh A, Steffan-Dewenter I, Kleijn D, Tscharntke T (2007) Diversity of flower-visiting bees in cereal fields: effects of farming system, landscape composition and regional context. Journal of Applied Ecology 44:41-49

Hornoy B, Tarayre M, Hervé M, Gigord L, Atlan A (2011) Invasive plants and enemy release: evolution of trait means and trait correlations in Ulex europaeus. PLoS ONE 6:e26275

Howe M, Knight G, Clee C (2010) The importance of coastal sand dunes for terrestrial invertebrates in Wales and the UK, with particular reference to aculeate Hymenoptera (bees, wasps & ants). Journal of Coastal Conservation 14:91-102

Hsu JC (1996) Multiple comparisons: theory and methods. Chapman & Hall, Florida, USA

Inouye DW, Pyke GH (1988) Pollination biology in the snowy mountains of Australia: comparisons with montane Colorado, USA. Australian Journal of Ecology 13:191-205

Jander G et al. (2004) Application of a high-throughput HPLC-MS/MS assay to Arabidopsis mutant screening; evidence that threonine aldolase plays a role in seed nutritional quality. The Plant Journal 39:465-475

Junker RR, Blüthgen N (2010) Floral scents repel facultative flower visitors, but attract obligate ones. Annals of Botany 105:777-782

Kearns CA, Inouye DW, Waser NM (1998) Endangered mutualisms: the conservation of plant-pollinator interactions. Annual Review of Ecology and Systematics 29:83-112

Kennedy CM et al. (2013) A global quantitative synthesis of local and landscape effects on wild bee pollinators in agroecosystems. Ecology Letters 16:584-599

Klein AM, Brittain C, Hendrix SD, Thorp R, Williams N, Kremen C (2012) Wild pollination services to California almond rely on semi-natural habitat. Journal of Applied Ecology 49:723-732

Klein AM, Steffan–Dewenter I, Tscharntke T (2003) Fruit set of highland coffee increases with the diversity of pollinating bees. Proceedings of the Royal Society of London. Series B: Biological Sciences 270:955-961

Kohler F, Verhulst J, Van Klink R, Kleijn D (2008) At what spatial scale do high-quality habitats enhance the diversity of forbs and pollinators in intensively farmed landscapes? Journal of Applied Ecology 45:753-762

Krauss J et al. (2010) Habitat fragmentation causes immediate and time-delayed biodiversity loss at different trophic levels. Ecology Letters 13:597-605

Krauss J, Steffan-Dewenter I, Tscharntke T (2003) How does landscape context contribute to effects of habitat fragmentation on diversity and population density of butterflies? Journal of Biogeography 30:889-900

Kremen C, Ricketts T (2000) Global perspectives on pollination disruptions. Conservation Biology 14:1226-1228

Krömer T, Kessler M, Lohaus G, Schmidt-Lebuhn A (2008) Nectar sugar composition and concentration in relation to pollination syndromes in Bromeliaceae. Plant Biology 10:502-511

LeBuhn G et al. (2003) A standardized method for monitoring bee populations–the bee inventory (BI) plot, [online], http://online.sfsu.edu/beeplot, Accessed 01/01/2014

Loper GM, Cohen AC (1987) Amino acid content of dandelion pollen, a honey bee (Hymenoptera: Apidae) nutritional evaluation. Journal of Economic Entomology 80:14-17

Mayer C et al. (2011) Pollination ecology in the 21st Century: key questions for future research. Journal of Pollination Ecology 3:8-23

McCaughey W, Gilliam, M, and Standifier, LN (1980) Amino acids and protein adequacy for honey bees of pollens from desert plants and other floral sources. Apidologie 11:75-86

Meikle RD (1984) Willows and poplars of Great Britain and Ireland. BSBI handbook

Mommaerts V, Wäckers F, Smagghe G (2013) Assessment of gustatory responses to different sugars in harnessed and free-moving bumblebee workers (*Bombus terrestris*). Chemical senses 38:399-407

Munyuli MT, Nyeko P, Potts S, Atkinson P, Pomeroy D, Vickery J (2013) Patterns of bee diversity in mosaic agricultural landscapes of central Uganda: implication of pollination services conservation for food security. Journal of Insect Conservation 17:79-93

Murray TE, Fitzpatrick Ú, Byrne A, Fealy R, Brown MJF, Paxton RJ (2012) Local-scale factors structure wild bee communities in protected areas. Journal of Applied Ecology 49:998-1008

Nocentini D, Pacini E, Guarnieri M, Nepi M (2012) Flower morphology, nectar traits and pollinators of *Cerinthe major* (Boraginaceae-Lithospermeae). Flora-Morphology, Distribution, Functional Ecology of Plants 207:186–196

Öckinger E, Smith HG (2007) Semi-natural grasslands as population sources for pollinating insects in agricultural landscapes. Journal of Applied Ecology 44:50-59

Pascarella JB, Waddington KD, Neal PR (1999) The bee fauna (Hymenoptera: Apoidea) of Everglades National Park, Florida and adjacent areas: distribution, phenology, and biogeography. Journal of the Kansas Entomological Society 72:32-45

Perret M, Chautems A, Spichiger R, Peixoto M, Savolainen V (2001) Nectar sugar composition in relation to pollination syndromes in Sinningieae (Gesneriaceae). Annals of Botany 87:267-273

Potts SG et al. (2011) Developing European conservation and mitigation tools for pollination services: approaches of the STEP (Status and Trends of European Pollinators) project. Journal of Apicultural Research 50:152

Potts SG, Biesmeijer JC, Kremen C, Neumann P, Schweiger O, Kunin WE (2010) Global pollinator declines: trends, impacts and drivers. Trends in Ecology & Evolution 25:345-353

Primack RB (1983) Insect pollination in the New Zealand mountain flora. New Zealand Journal of Botany 21:317-333

Redpath N (2010) Restoration and management of wildflower-rich machair for the conservation of bumblebees. Journal of Insect Conservation 17:491-502

Redpath N, Osgathorpe LM, Park K, Goulson D (2010) Crofting and bumblebee conservation: the impact of land management practices on bumblebee populations in northwest Scotland. Biological Conservation 143:492-500

Roubik DW, Yanega D, Aluja, Buchmann SL, Inouye DW (1995) On optimal nectar foraging by some tropical bees (Hymenoptera: Apidae). Apidologie 26:197-211

Roulston TH, Cane JH (2000) Pollen nutritional content and digestibility for animals. Plant Systematics and Evolution 222:187-209

Roulston TH, Cane JH (2002) The effect of pollen protein concentration on body size in the sweat bee *Lasioglossum zephyrum* (Hymenoptera: Apiformes). Evolutionary Ecology 16:49-65

Salonen A, Ollikka T, Grönlund E, Ruottinen L, Julkunen-Tiitto R (2009) Pollen analyses of honey from Finland. Grana 48:281-289

Sladen FWL (1912) The humble-bee: its life-history and how to domesticate it, with descriptions of all the British species of *Bombus* and *Psithyrus*. Macmillan, UK

Sokal RR, Rohlf FJ (2005) Biometry: the principles and practices of statistics in biological research, Third edn. WH Freeman, New York

Sommerville AH (1992) Willows in the environment. Proceedings of the Royal Society of Edinburgh. Section B. Biological Sciences 98:215-224

Southwick EE, Loper GM, Sadwick SE (1981) Nectar production, composition, energetics and pollinator attractiveness in spring flowers of western New York. American Journal of Botany 68:994-1002

Stace C (2010) New flora of the British Isles, Third edn. Cambridge University Press, UK

Stelfox A (1927) The wild bees of spring. The Irish Naturalists' Journal 1:189-192

Tepedino VJ, Parker FD (1982) Interspecific differences in the relative importance of pollen and nectar to bee species foraging on sunflowers. Environmental Entomology 11:246-250

Teräs I (1985) Food plants and flower visits of bumblebees (*Bombus*: Hymenoptera, Apidae) in southern Finland. Acta Zoologica Fennica 179:1-120

Thomson JD, Wilson P, Valenzuela M, Malzone M (2000) Pollen presentation and pollination syndromes, with special reference to Penstemon. Plant Species Biology 15:11-29

Totland Ø, Sottocornola M (2001) Pollen limitation of reproductive success in two sympatric alpine willows (Salicaceae) with contrasting pollination strategies. American Journal of Botany 88:1011-1015

Tscharntke T, Brandl R (2004) Plant-insect interactions in fragmented landscapes. Annual Reviews in Entomology 49:405-430

Tscharntke T, Gathmann A, Steffan-Dewenter I (1998) Bioindication using trap-nesting bees and wasps and their natural enemies: community structure and interactions. Journal of Applied Ecology 35:708-719

Tscharntke T, Steffan-Dewenter I, Kruess A, Thies C (2002a) Contribution of small habitat fragments to conservation of insect communities of grassland-cropland landscapes. Ecological Applications 12:354-363

Tscharntke T, Steffan-Dewenter I, Kruess A, Thies C (2002b) Characteristics of insect populations on habitat fragments: a mini review. Ecological Research 17:229-239

Underwood AJ (1997) Experiments in ecology: their logical design and interpretation using analysis of variance. Cambridge University Press, UK

Wacht S, Lunau K, Hansen K (2000) Chemosensory control of pollen ingestion in the hoverfly *Eristalis tenax* by labellar taste hairs. Journal of Comparative Physiology A 186:193-203

Weiner CN, Hilpert A, Werner M, Linsenmair KE, Blüthgen N (2010) Pollen amino acids and flower specialisation in solitary bees. Apidologie 41:476-487

Westphal C et al. (2008) Measuring bee diversity in different European habitats and biogeographical regions. Ecological Monographs 78:653-671

Wróblewska A, Stawiarz E (2004) Pollen spectrum of some honeys from Opatów vicinity. Journal of Apicultural Research 48:23-33

Yuan L-C, Luo Y-B, Thien LB, Fan J-H, Xu H-L, Chen Z-D (2007) Pollination of *Schisandra henryi* (Schisandraceae) by female, pollen-eating *Megommata species* (Cecidomyiidae, Diptera) in South-central China. Annals of Botany 99:451-460

5

BEE-MEDIATED POLLEN TRANSFER IN TWO POPULATIONS OF *CYPRIPEDIUM MONTANUM* DOUGLAS EX LINDLEY

Peter Bernhardt*[1], Retha Edens-Meier[2], Eric Westhus[3], Nan Vance[4]

[1]*Department of Biology, Saint Louis University, St. Louis, MO 63103, USA*
[2]*Department of Educational Studies, Saint Louis University, St. Louis, MO 63103, USA*
[3]*Center for Outcomes Research, Saint Louis University, St. Louis, MO 63103, USA*
[4]*P.O. Box 282, Kooskia, ID 83539, USA*

Abstract—The conversion rate of flowers into fruit in *C. montanum* at two sites over four seasons was 52-85%, unusually high for a food mimic orchid. Comparative measurements of the trap-like labellum of *C. montanum* showed it was intermediate in size compared to measurements of six other *Cypripedium* spp. found in North America and China. While visitors to flowers of *C. montanum* represented three insect orders, at two sites, over four seasons only small- to medium-sized, solitary bees (5-10 mm in length) carried the pollen massulae. Bee-visitation occurred at both sites and began within 24-48 hours following labellum expansion. Female bees in the genus *Lasioglossum* (Halictidae) were the most common carriers of massulae. However, species of visiting bees differed between sites and years. At both sites the majority of bees entered and escaped from the labellum in less than 180 seconds and there was no significant difference between the times bees spent in the flowers at both sites. At the site on the Eastside Cascades of Central Oregon, there was no correlation between the length and width of a bee and the time it spent escaping from the basal openings. There was no correlation between bee size and whether the bee carried massulae. Depending on site and year 41-58% of the bees exiting the orchids carried the orchid's pollen. Depending on site and year 75-100% of bees collected exiting the orchids via the basal openings also carried the pollen of at least one other co-blooming species.

Keywords: Basal openings, bees, Cypripedium, *labellum, pollen masses (massulae), pollination, staminodium*

INTRODUCTION

When pollen vectors are absent or infrequent during the flowering period of an out-breeding population this usually results in lower fruit and/or seed production. When this occurs consistently the plant population is usually categorized as pollinator-limited (*sensu* Committee on the Status of Pollinators in North America 2007). Low fruit and seed set are recorded often in orchid species that produce flowers lacking any edible or scent rewards for their pollinator(s). In fact, Tremblay et al. (2005) showed that orchid flowers with food deceptive modes of presentation (*sensu* Ackerman 1986) showed lower rates of fruit set than orchids with pseudocopulatory flowers (*sensu* Dafni & Bernhardt 1990). As floral mimesis dominates orchid speciation, the correlation between low fecundity and the absence of rewards remains of primary concern to conservationists Edens-Meier et al. 2014). Any attempt to protect dwindling populations of orchid species and/or reintroduction following regional extinctions requires the establishment and/or reestablishment of pollinator populations (Dixon 2009).

Cypripedium species (*sensu* Cribb 1999) produce flowers interpreted usually as food deceptive. With important exceptions (e.g. *C. passerinum*; see Catling 1990), mechanical self-pollination in the absence of pollinators is uncommon within this genus. There are no modern references confirming any observations of flower-visiting insects consuming edible rewards and/or collecting nesting materials inside the inflated labellum (Bernhardt & Edens-Meier 2010; Argue 2011). Fragrance analyses of flowers of *Cypripedium* spp. show biochemical variation at the interspecific level (Barkman et al. 1997) but none of their insect pollinators, to date, have been observed collecting their scents unlike the male bees (Euglossini) associated with so many monandrous species of the Neotropics (Dressler 1968, 1981). None of the floral volatiles produced by *Cypripedium* spp. (Barkman et al. 1997) are unique to the genus. They are also produced by flowers of other angiosperm species, more likely to offer nectar and/or pollen as rewards.

It appears that the inflated labellum and diandrous column of at least 17 *Cypripedium* spp. function in the same way. In each case, the potential pollinator enters the labellum, passes under the receptive stigma, and then exits the flower via one of two exit openings. One anther is located above each exit. The anther contacts the escaping insect leaving a dorsal, often irregular, deposit of massulate pollen on the thorax and/or head. However, the diversity and density of pollinators in each *Cypripedium* species may vary broadly according to the combined elements of floral presentation including floral size, internal architecture,

*Corresponding author: bernhap2@slu.edu

colour patterns, epidermal sculptures, scent, massulae morphology etc. (Bernhardt & Edens-Meier 2010; Edens-Meier et al. 2011, 2014).

Therefore, if all cross pollinated *Cypripedium* spp. are food mimics, the natural rates of pollinated pistils, and the conversion rates of pistils into fruit (see Edens-Meier et al. 2010, 2014), should be low (Tremblay et al. 2005). This is not always the case. Bernhardt & Edens-Meier, (2010) reviewed the literature on insect pollination in 15 *Cypripedium* spp. at 29 different sites. They noted that the conversion of flowers into fruit changed at the intraspecific level from year to year and from site to site. Many species suffered seasons in which the conversion rate of flowers into fruit was < 20% (see Primack & Stacy 1998; Lipow et al. 2002; Sugiura et al. 2002; Li et al. 2006; Bänziger 2008; Kull 2008; Sun et al. 2009). It was also true that some species including *C. calceolus* (Kull 2008), *C. fasciculatum* (Lipow et al. 2002), and *C. plectrochilum* (Li et al. 2008) bloomed at sites in which the conversion of flowers into fruit was 33-69%. These rates are entirely competitive with other obligately outcrossing herbs that offer edible pollen and/or nectar as rewards to pollen vectors (see Bernhardt & Montalvo 1979; and Bernhardt & Dafni 2000). Depending on the year, the conversion rate of flowers into fruit can be from 0-25% in a population of *C. acaule* (Primack & Stacy 1998), 33-57% in a population of *C. calceolus* (Kull 2008), and 9-26% in *C. tibeticum* (Li et al. 2006). Likewise, fruit production could vary tremendously between sites over the natural distribution of a single species. This included 0-25% between isolated populations of *C. acaule* (Primack & Stacy 1998) and 10-57% in *C. calceolus* (Kull 2008).

Some *Cypripedium* spp. experience higher fruit set ratios at specific locations in certain years. By following Tremblay et al. (2005) we should conclude that some insects are more common in some sites, over some flowering seasons, and are more likely to be deceived by the same food mimic more than once affecting cross-pollination. However, Bernhardt and Edens-Meier (2010) also noted that, "blaming low fruit set on the performance of a discrete lineage of anthophilous insects as poor pollinia vectors must be approached with caution." The most fecund species including *C. calceolus*, *C. fasciculatum*, and *C. plectrochilum* depended on very different pollinators. *Cypripedium calceolus s.s.* (Kull 2008) was pollinated by a range of small bees in several families. In contrast, *C. plectrochilum* (Li et al. 2008) was pollinated exclusively by a few species in the genus *Lasioglossum* (Halictidae). *Cypripedium fasciculatum* was pollinated by female, parasitoid wasps in the genus *Cinetus* (Ferguson & Donham 2005). Likewise, when *C. acaule* (Primack & Stacy 1998) and *C. tibeticum* (Li et al. 2006) had an infrequent, "successful," season of fruit set (25%) they were pollinated exclusively by gynes of a few *Bombus* spp.

In fact, there appears to be a *Cypripedium* species with pollination rates that may, at certain sites and years, be higher than any of the species discussed above. *Cypripedium montanum* is restricted to the Pacific and interior northwestern North America where it is found in mesic to dry forest clearings and slopes (Sheviak 2002; Vance 2007). Edens-Meier et al. (2011) showed that 94.0% of the pistils, collected at random from withering flowers at two sites in Oregon, over two seasons ($N = 16$), contained germinating pollen on their stigmas and/or pollen tubes penetrating their styles. However, other studies showed that, as usual, the conversion rate of flowers into fruits varies within the same site over different years. A monitoring study on six plots of *C. montanum* in the Bighorn Mountains of Wyoming reported that in 2004 and 2005 average fruit set ($N = 5$, one plot lacked flowering stems) was 41% (range 0-67%) and 38% (range 4-77%) respectively (see Vance 2007). In contrast, Coleman (1995) followed fruit production in California populations over a four-year period and found that the average rate of fruit set was 61.0% over all sites. Huber (unpublished) monitored the rates of natural fructification at his wildflower reserve (GROWISER) in 2003 and again in 2004. He found that 75% stems ($N = 50$) produced at least one mature and dehiscent capsule in 2003 and 85% ($N = 50$) produced fruit in 2004.

Information on pollinators of *C. montanum* remained anecdotal through the 20th century. Luer (1975) observed that a *Bombus* sp. was unable to enter the labellum but small, unidentified, black bees were able to successfully enter through the large, dorsal opening on the labellum and exit via the basal openings (see above).

Therefore, this paper attempts to address five interrelated questions regarding insect-flower interactions in the pollination dynamics of two populations of *C. montanum*. These are the same populations that served previously as sources for observations on natural rates of pollination *in situ* (Vance 2007; Edens-Meier et al. 2010, 2014). First, how do labellum "trap" dimensions in *C. montanum* compare with the same dimensions in flowers of congeners measured in past publications? Specifically, how do labellum sac dimensions in *C. montanum* correlate with pollinator dimensions as in other congeners with different pollinators (Banziger et al. 2005; Li et al. 2006, 2008ab; Banziger et al. 2008; Edens-Meier et al. 2011, 2014)? Second, is this species pollinator-limited at either site or can it rely on mechanical self-pollination (see above) in the absence of pollinators? Third, which insect taxa are the most frequent vectors of pollen masses over different seasons and sites? Fourth, does the taxonomic composition of prospective vectors of pollen masses vary between seasons and sites? Fifth, if *C. montanum* lacks edible rewards, as in all other *Cypripedium* spp., which co-blooming species provide insects with edible rewards at different sites?

MATERIALS AND METHODS

Study sites and field states

The first site used from 6/4-6/16/03 and from 5/22-6/17/04 was in the Blue Mountains of Eastern Oregon (BMEO) at the GROWISER Reserve, Summerville. This site is found primarily on north and east facing slopes at 1050 m elevation with an annual precipitation of 50 cm, half of which falls as snow. The soils are deep volcanic ash with a woodland canopy of *Pseudostuga menziesii*. There are approximately 700 stems of *C. montanum*, at various stages of maturity, growing in woodland gaps and glades but less than 400 produce flowering stems annually (Huber, pers. obs.). While a portion of the property is developed for an *in situ* seeding program (Huber 2002) we performed no observations in these areas and did not use flowers from this program for measurement. In 6/2004, 6/30/2005, 7/06/2005 and from 6/5/2006-6/14/2006 we combined observations on two populations on the Eastside Cascades of central Oregon (ECCO) within the Deschutes National Forest Sisters Ranges District in Jefferson County, Oregon. Within the site the two colonies grow along or near Forest Services Road (FS1190) and are separated from each other by approximately 4/5 km. The site is located within a forest that burned in a 10-hectare wildfire in 2002. The soils derived from weathered tuff and andenite with shallow to deep sandy-gravelly loams formed from volcanic ash over colluvium and renduum. The total number of flowering stems in the two colonies totaled 139 in 2006.

Number of flowers and fruits per scape

The number of flowers on scapes of *C. montanum* appears to be dependent on the physical age of the plant (Huber 2002). As annual variation in flower numbers within the same population may influence visitation rates of potential pollinators we selected 40 flowering stems at random each year and counted the number of flowers/stalk at the BMEO site (2003, 2004). At ECCO we counted 103-206 flowering scapes each season (2004-2006) recoding the total number of flowers produced. We then returned to record the number of maturing capsules at ECCO.

Self-pollination in the absence and presence of pollinators

Flowers of *C. montanum* from both populations were found to be self-compatible (Edens-Meier et al. 2010). To determine whether pollen masses contacted viable stigmas in the absence of pollen vectors in 2004 and 2005, we tagged and covered eight mature buds on eight flowering stalks in tulle bags one or two days before the sepals and lateral petals released the labellum. The flower was considered open after the labellum expanded and opened its dorsal entrance. The flowers bloomed and withered under the tulle bags and we checked the stalks for fruits in July 2004, 2005. During the 2004 season at the BMEO site we also tagged an additional five flower buds on five stalks. The labellum on each bud was removed with cuticle scissors before it expanded to determine if self- or cross-pollination could occur in the absence of visiting insects entering the labellum. These flowers were also checked for evidence of fruit set in July 2004.

Labellum measurements

All terminology for floral dimensions (Fig. 1) and all labellum measurements taken at the BMEO and ECCO sites (2006) follow Li et al. (2006, 2008ab) with two exceptions. We did not record the length and width of the two, basal openings (rear orifices) as in the studies discussed in the Introduction (above) because our primary interest was in the "fit" of insects within the sac (see Luer 1975). We were not able to record the distance in mms of the receptive surface of the stigma to the base of the labellum as in Li et al. (2006) as this would have meant splitting labella open longitudinally thereby reducing further floral presentation of the entire population of a conserved species. We used electronic digital calipers (Fisher Scientific Model 14-648-17) to record floral measurements. Flowers that were trampled, or eaten partially by unknown foragers, or had their labella punctured and deflated by leafcutter bees (*Osmia* spp., Megachilidae, see below) were excluded from all measurements. The dimensions of the *C. montanum* labella at BMEO ($N = 18$, 2004) and the ECCO sites ($N = 43$, 2006) were compared to measurements of six additional *Cypripedium* spp. in North America and China (Li et al. 2006, 2008; Banziger et al. 2008; Edens-Meier et al. 2008; Zheng et al. 2011; Ren unpublished).

Figure 1. Flower of *C. montanum*: an = anther, bo = basal opening, dol = dorsal opening of labellum (note interior purple stripes), ds = dorsal sepal, l = labellum, lp = lateral petal, ov = ovary (Photograph by Nan Vance).

Observations and timing of insect visitors

We observed the behaviour of insects on and in all flowers of *C. montanum* on sunny days at both sites (BMEO 2003, 2004; ECCO, 2004, 2006) totalling approximately 110 field hours. Nocturnal visits by the first author were discontinued after an absence of activity following the setting of the sun. Wearing 3× magnification visors, we observed how and when insects entered the labellum through the central, dorsal entrance and whether they first landed on the labellum or on the contrastingly coloured staminode (see Chi et al. 2008). When an insect entered the labellum we observed whether it exited the flower through one of the two rear orifices (see Li et al. 2006, 2008ab). When insects escaped from a flower we recorded whether they flew away and left the site or whether they were observed to visit a second flower on the same inflorescence, or a flower on another inflorescence in the same colony (see Li et al. 2008a). In 2004 we removed the labellum from five opening buds in the BMEO site before the labellum expanded (see above) to see if insects were attracted to the column of the flower in the absence of the labellum and whether they could remove pollen masses in the absence of the labellum.

We used stopwatches (Sports Time II Chronograph) to record how long it took individual insects to escape from a *C. montanum* flower via the rear exit openings at both sites. We started timing after we saw the insect fall into or fly into the labellum. Timing was stopped after we saw the same insect emerge completely from a rear exit.

Collection and processing of insect-visitor, cross-referencing their pollen loads and measuring the insect specimens

Insect visitors were collected as they visited flowers of *C. montanum* at both sites. Each specimen was killed separately in killing jars containing fumes of ethyl acetate. The specimen was then removed with forceps, placed on a glass slide and bathed in 2-3 drops of ethyl acetate washing pollen grains from the body and/or we scraped the body with the tip of a metal probe to dislodge sticky pollen masses. The solvent on the slide was allowed to evaporate and then the pollen residue was stained with Calberla's solution (Ogden et al. 1974) for five minutes before covering it with a glass cover slip. The dried insect was pinned and labelled. The glass slide was given the same label code as the insect specimen to co-reference insect and pollen identification. The slide was allowed to dry for a minimum of 24 hours before viewing and identifying contents under a light microscope. To facilitate identification of pollen, other than the grains of *C. montanum*, we made a pollen library of grains derived from anthers of co-blooming species flowering at and adjacent to the same study sites. As more than one insect was euthanized in the same jar on the same day, the pollen of a particular species was only scored as present on an insect when we counted >24 grains on the same slide that had the same shape, size, number of apertures and exine sculpturing (see Bernhardt & Weston 1996). There were four insect collection categories.

1) Insects Caught Outside the Flower. At the BMEO site in 2003 we noticed that a large number of insects either perched on the floral organs or hovered a few centimeters around a fully opened *C. montanum* flower, but were never observed actually entering the labellum. These insects were caught to determine whether any carried pollen masses of *C. montanum* during a previous but unobserved visit.

2) Dead, Dying, or Struggling Insects. At the BMEO site in 2003 and the ECCO site in 2006 we examined labella before 10:00 AM each morning and removed the dead and/or inert corpses of insects. The corpses were checked for the presence of *C. montanum* pollen.

3) Potential Pollinators. At the BMEO (2003, 2004) and ECCO (2004, 2006) sites we collected insects observed to enter the labellum through the large, dorsal entrance and leave via one of the two basal openings. As specimens were collected only after they emerged from a basal opening, some were timed (see above) and their entrance-exit times were cross-referenced with their entomological identification, physical dimensions and pollen load analysis.

4) Visitors to *Fragaria vesca* var. *bracteata.* At the BMEO site on 6/6/03 we noted that small bees appeared to alternate their visits to the flowers of *C. montanum* with visits to clumped populations of *F. vesca* var. *bracteata.* We collected insects on flowers of *F. vesca* from 6/6/03 – 6/10/03 to determine how many specimens also carried pollen masses of *C. montanum.*

We measured some insect specimens caught at the ECCO site in 2006 using the same digital calipers used to measure floral architecture. This included the length of the insect from between its frons to the cercus (terminus) of its abdomen (mouth parts were not measured). As insects must squeeze through the opening of one of two basal openings, we also measured the width of the insect by measuring its widest part. In some specimens, we recorded the width of the widest segment at the base of its abdomen (e.g. *Lasioglossum*) while in others we recorded the width of the head or thorax (e.g. *Osmia*). All insect specimens were sent to C.D. Michener for identification and deposition in the Snow Entomological Museum at the U. of Kansas (Lawrence, Kansas).

Statistics

The mean number of flowers per scape at BMEO was compared between 2003 and 2004 using a t-test. The mean number of flowers and fruits (capsules) per scape at the ECCO site in 2004 and 2005 were compared in the same way. The same tests were used to compare the time bees spent inside *C. montanum* flowers at the BMEO site (2004) vs. the ECCO site (2006). Welch's variant was used to account for differences in the variance among sites, and two t-tests were run, one with an anomalously long data point at the BMEO site, and one with that point excluded.

Linear regression was used to determine if there was a relationship between the insect measurements and the time spent in flowers. T-tests were used to determine if there were relationships between the presence of orchid massulae (irregular smears or globs of pollen) on the body and bee length, and the presence of orchid massulae on the body and bee width. Bee lengths and widths were log transformed to follow a normal distribution.

RESULTS

Number of flowers/stem and reproductive success

In 2003 and 2004 stems produced one to three flower buds at the BMEO site. In 2003 the average number of flowers/scape was 2.0 ± 0.57 (mean ± sd, $N = 40$). In 2004 ($N = 40$) there was an average of 2.02 ± 0.53 flowers/stem. No difference was detected between the mean number of flowers/stem for two seasons ($t = -0.81$, $df = 77$, $P = 0.43$). Likewise, the mean number of flowers/scape at the ECCO site varied from as low as 1.46 in 2005 to as

Table 1. Census of flowering and fruiting scapes at ECCO (2004 – 2006).

Year	Scapes	Flowers	Flowers/Scape	Capsules	Capsules/Flower
2004	103	168	1.63	87	0.52
2005	206	301	1.46	164	0.54
2006	139	216	1.55	142	0.66

high as 1.63 in 2004 (Tab. 1). The conversion rate of flowers into fertilized capsules over three seasons (Tab. 1) ranged from 0.52-0.66.

Floral presentation

Flower buds on the same scape opened acropetally or subsynchronously at both sites. Flower buds nodded but, as the dorsal sepal unclasped itself from the labellum and became increasingly erect, the pedicel also bent upwards until the open flower was horizontal to its scape (Fig. 1). The transition from the full bud (labellum visible but un-inflated; sepals clasp the labellum) to the open flower phase occurred over a period of one to four days (24-96 hours) at both sites. We observed small bees and syrphid flies hovering around the flowers during this period but few actually landed on the flower and, of those that did, none were observed to crawl under the partially separated dorsal sepal to find the dorsal surface of the labellum. During this period, anthers examined under 3× were indehiscent.

In open flowers, at both sites, the sepals and lateral petals were greenish-yellowish brown. The staminode was usually yellow with clusters of red-brownish spots. The inflated labellum was white on the outside. Four to nine purple veins ran centrally and longitudinally along the inner surface of the floor of each labellum (Fig. 1). A strong, pleasantly fruity aroma was discernible from flowers on warm days while they stood in light gaps for 20-30 minutes. It was during this period we witnessed the earliest arrival of foraging insects including those caught outside the flower and those that entered the labellum. Nectar glands and nectar secretions were not found in the labellum or on any other floral organ during examination under 3× magnification. Anthers examined under 3× were dehiscent often extruding greasy, hanging pollen masses.

Comparative labellum morphometrics

See Tab. 2 for measurements of the fully expanded labellum ($N = 18$ flowers from 18 scapes) at the BMEO site and at the ECCO site ($N = 43$ flowers from 43 scapes). These measurements suggested that the population of *C. montanum* was intermediate in size compared to the larger rounded sacs of *C. tibeticum* (Li et al. 2006) and *C. reginae* (Edens-Meier et al. 2011) and the smaller, keeled sac of *C. plectrochilum* (Li et al. 2008a). Labellum measurements in *C. montanum* were closer to Chinese species pollinated by small-medium sized bees (Banziger et al. 2008).

Self-pollination rates

None of the bagged flowers set fruit. None of the flowers set fruit if the labellum was removed (see below).

Insects caught outside the flowers

At the BMEO site in 2003 we caught 15 insects either perched on open flowers of *C. montanum* or hovering one or two centimeters around the flower (Tab. 3). The small,

Table 2. Comparative of the labellum in *Cypripedium* spp. (mean and standard deviation, in mms).

Species	*N*	La Length	La Width	La Depth	Entrance Length or Diameter	Entrance Width	Pollinators[1]	References
C. fargesii	20	NA	NA	4.72 (0.80)	6.16 (0.68)[2]	NG[3]	LS	Ren (unpublished)
C. flavum	30	NA	NA	4.77 (0.00)	8.56 (0.01)	NG[3]	BW, Ca	Zheng et al. (2011)
C. flavum	NA	35.0-55.0	30-35.0	NA	10.0-13.0	5.0-9.0	AL	Banziger et al. (2008)
C. montanum (BMEO)	18	26.06 (2.81)	16.13 (2.81)	12.58 (0.97)	8.72 (1.30)	5.89 (0.85)	SMB	
C. montanum (ECCO)	43	25.54 (3.68)	15.87 (2.28)	13.33 (1.65)	7.91 (1.20)	5.89 (0.81)	SMB	
C. plectrochilum	50	NA	NA	1.22 (0.00)	2.80 (0.00)	NG[3]	La	Li et al. (2008)
C. reginae	44	40.79 (3.30)	29.62 (2.92)	24.98 (3.28)	11.96 (2.04)	16.19 (2.92)	LMB	Edens-Meier et al. (2011)
C. tibeticum	20	49.67 (0.04)	37.01 (0.06)	30.37 (0.02)	12.87 (0.01)	NG[3]	BG	Li et al. (2006)
C. yunnanense	NA	35.0- 50.0	25.0-32.0	20	9.0-13.0	6.0–10.0	La	Banziger et al., (2008)

[1]Pollinators = AL (*Andrena* spp.; large body), BG (*Bombus* gynes), BW (*Bombus* workers), Ca (*Calliphora vomitoria*; Calliphoridae), La (*Lasioglossum* spp.), LMB (Large, mixed bees), LS (*Lucilla* sp., Syrphidae), SMB (Small-medium sized bees); [2]*C. fargesii* Entrance Length: $N = 32$; [3]NG = Not given as Labellum entrance measurement presented in preceding column as a diameter

Table 3. Pollen load analyses of insects collected on or in *Cypripedium montanum* at the BMEO site in 2003.

		Pollen Types*						
Insect Taxon	*N*	Aster	Cl	Cm	Pi	Ros	UM	No Pollen
Insects Caught Outside Flowers of *C. montanum*								
Coleoptera								
Anthaxia aenogaster	1	0	0	0	0	0	0	1
Hymenoptera (Apoideae)								
Andrena prunorum	1	0	0	0	0	1	0	0
Eumeninae (wasp)	1	0	0	0	0	0	0	1
Halictus tripartitus	1	1	0	0	1	0	0	0
Hylaeus ellipticus	1	0	0	0	0	0	0	1
Lasioglossum (Evylaeus) sp. a	2	0	0	1	2	0	0	1
L. (Evylaeus) sp. b	1	1	1	1	0	1	1	0
L. olympiae	3	3	2	1	0	3	1	0
Nomada sp.	1	0	0	0	0	0	0	1
Osmia sp.	1	0	0	0	0	0	0	1
Panurginus ineptus	1	0	0	0	0	1	0	0
Totals	14	5	3	3	3	8	2	7
Dead, Dying, or Struggling Insects								
Diptera								
Brachyopa SP.	1	0	0	0	0	1	0	0
Eulonchus sp.	1	0	0	0	0	1	0	0
Psilota sp.	2	0	0	0	0	1	0	1
Hymenoptera								
Osmia simillina	1	0	0	0	0	0	0	1
Totals	5	0	0	0	0	3	0	1
Potential Pollinators (entered labellum dorsal crater and exited via rear orifice)								
Hymenoptera (Apoideae)								
Andrena (Micrandrena) sp.	1	0	0	0	0	0	0	1
A. sp. b	2	0	0	1	0	1	0	1
Halictus rubicundus	1	0	0	0	0	1	0	0
Lasioglossum (Dialictus) sp.	1	0	0	1	0	0	0	0
L. (Evylaeus) sp. a	3	0	1	1	0	2	0	0
L. (*Lasioglossum*) sp	1	0	0	1	0	0	0	0
Lasioglossum olympiae	4	0	1	4	0	4	0	0
Panurginus ineptus	4	0	0	0	0	4	0	0
Totals	17	0	2	8	0	12	0	2

*Aster = Asteraceae; Cl = *Clarkia*, Cm = *Cypripedium* montanum; Pi = *Pinus*, Ros = Mixed unidentified Rosaceae (including *Fragaria*, *Physocarpus capitatus*, *Potentilla*, *Rosa*, *Rubus parviflorus*, UM = Unidentified monocot (probably *Smilacina racemosa*)

black beetle, *Anthaxia aenogaster* (Buprestidae), was extremely common but was collected only once for identification purposes. They were often observed *in copula*, but we did not observe them foraging on floral organs. Bees were collected on the dorsal sepal or lateral petals and were often observed cleaning their legs. Only three bees captured in this way carried massulae of *C. montanum* and all three were bees in the genus *Lasioglossum s.l.* (Halictidae). These three bees also carried the pollen of other co-blooming species. A single specimen of *Lasioglossum* (*Evylaeus*) sp. carried the pollinia of *C. montanum* mixed with the grains of five other pollen forms.

Dead, dying, or struggling insects

A total of nine dead insects, representing three Orders (Coleoptera, Diptera and Hymenoptera) were collected at

Table 4. Pollen load analyses of insects collected in the labella of *Cypripedium montanum* or as they exited via the rear orifices at the BMEO site in 2004.

		Pollen Type*							
Insect Taxon	*N*	Aster	Cal	Cm	Ger	Ros	Ss	UEU	No Pollen
Dead, Struggling Insects									
Coleoptera									
Anthaxia aeongaster	3	0	0	0	0	0	0	0	3
Hymenoptera (Apoideae)									
Bombus fervidus	1	0	0	0	0	1	1	0	0
Eucera (Synhalonia) sp.	1	0	0	0	0	0	0	0	1
Totals	5	0	0	0	0	1	1	0	4
Potential Pollinators (entered labellum dorsal crater and exited via rear orifice)									
Hymenoptera (Apoideae)									
Andrena (Micrandrena) sp	1	0	0	1	0	1	1	0	0
Andrena sp. 1	1	0	0	1	0	0	0	1	0
Andrena sp. 4	1	0	0	1	0	1	0	0	0
Ceratina acantha	8	0	1	3	0	2	1	0	3
Halictus confusus	2	0	0	0	0	2	0	0	0
H. tripartius	7	4	0	3	1	6	1	0	0
Lasioglossum (Dialictus) sp.	3	0	0	1	0	0	0	0	2
L. (Evylaeus) sp	11	2	0	4	0	8	0	0	2
L. tegulariforme	1	0	0	1	0	0	1	0	0
Osmia (Melanomia) sp	1	0	0	0	-	1	0	0	0
Totals	36	6	1	15	1	21	4	1	7

*Aster = Asteraceae, Cal = *Calochortus elegans*, Cm = *Cypripedium montanum*, Ger = *Geranium*, Ros = Mixed unidentified Roaceae, Ss = *Smilacina stellata*. UEU = Unidentified eudicotyledon

BMEO in 2003 and 2004 (Tables 3 and 4) while an additional 25 dead/inert specimens were removed from labella at the ECCO site in 2006 (Tab. 4). As above, the beetle, *Anthaxia aeongaster* (Buprestidae) was observed in great numbers at BMEO and collections presented here do not represent beetle density while the orchid population flowered. These beetles crawled inside the labellum through the dorsal opening and spent hours in the labellum. They were never observed to emerge from the flower via the two basal openings and never carried massulae. Pollen load analyses of inert or dying specimens showed they carried the grains of other co-blooming species at the site especially *Pinus* spp. True flies found dead in the labellum (Tab. 3) at BMEO and the ECCO site (Tab. 5) represented members of the flower-visiting family, Syrphidae, but they did not carry the massulae of the host flower.

On 6/12/2003 at the BMEO site we observed an 11 mm long female of *Osmia simillina* (Megachilidae) enter a labellum. After > six minutes the insect used its mandibles to tear a hole in the rear of the labellum. We caught it in a killing jar as it exited. It didn't carry massulae of *C. montanum.* The following day, this damaged labellum deflated. A 12 mm long male of *Eucera* (*Synhalonia*) sp. (Apidae) was collected at the BMEO site on 5/25/2004 while it was struggling within a labellum. It did not carry massulae. A dormant (legs still twitching), 6 mm long worker of *B. fervidus* (Apidae) was found in a labellum at BMEO on 5/25/2004 and it did not carry massulae. In 2003 and 2004 we observed that a few *Lasioglossum* spp. escaped from the labellum in a novel way. The bee grabbed the tip of the staminode, which protrudes downwards partway into the labellum. Using their first pair of legs to clutch the staminode, the insect climbed out of the labellum, thus avoiding any contact with the stigma or anthers. These bees immediately vacated the site.

On 5/18/2004 we found a withered flower at the ECCO site that contained a dead, female, bee identified as a *Lasioglossum* sp. The bee's head and thorax protruded from one of the flower's basal openings suggesting it died because it could not extricate its abdomen. Both the bee and the massulae it carried were infested with unidentified ascomycetes. The 19 dead, dying, or struggling insects collected at the ECCO site in 2006 (Tab. 5) were bees representing four families; Andrenidae (*Andrena*), Apidae (*Bombus*), Halictidae (*Lasioglossum*) and Megachilidae (*Megachile* and *Osmia*). All specimens were found dead in the flower except for the gyne of *Bombus bifarius.* This 14.32 x 2.54 mm insect landed on the labella of four flowers

Table 5. Pollen load analyses of insects collected at the ECCO site combining dead-struggling specimens with specimens that exited through the basal openings.

Insect Taxon	*N*	Pollen Types*							
		Aster	Cean	Cm	Pi	PL	UM	RM	No Pollen
Dead, Struggling Insects									
Coleoptera									
Anthaxia aenogaster	5	1	3	0	1	0	1	1	1
Diptera									
Unidentified syrphid	1	0	0	0	1	0	0	1	0
Hymenoptera (Apoideae)									
Andrena sp. 1	3	1	3	0	3	1	0	2	0
Bombus bifarius	1	1	1	0	0	0	1	0	0
Lasioglossum athabascense	2	2	2	0	1	0	0	2	0
L. (Evylaeus) sp.	2	1	2	0	1	0	0	1	0
L.sysmbrii	2	1	1	0	1	0	1	1	0
Megachile giliae M	1	1	1	0	0	0	0	0	0
Osmia albolateralis	1	0	1	0	1	0	0	0	0
O ednae	1	0	1	0	1	1	0	0	0
O. malina	1	1	1	0	1	1	0	0	0
Totals	19	9	16	0	11	3	3	8	1
Insects that Exited Via Basal Openings									
Hymenoptera (Apoideae)									
Andrena sp. 1	2	0	2	1	0	0	1	0	0
Lasioglossum athabascense	4	2	4	1	1	1	1	3	0
L. (Dialictus) sp.	7	1	3	2	0	0	0	2	3
L. (Evylaeus) sp.	16	0	14	14	2	0	0	6	0
Osmia ednae	1	0	0	0	0	0	1	0	0
O. (Melanosmia) sp.	1	1	0	0	1	0	0	0	0
Totals	31	4	23	18	3	1	3	11	3

*Aster = unidentified members of the Asteraceae, Cean = *Ceanothus velutinus,* Cm = *C. montanum,* Pi = *Pinus,* PL = papilionoid legume (probably *Lupinus sericeus)* UM = Unidentified monocotyledon, RM = Mixed ticolporate grains (including *Aquilegia formosa, Fragaria vesca, Potentilla gracilis, Rosa gymnocarpa, Ranunculus* sp., *Rubus parviflora*).

on four plants before it squeezed itself into the labellum of the fourth flower and was unable to exit. All specimens collected were female excluding the male, *Megachile gilae.* While all the specimens carried the pollen of co-blooming plants at the Deschutes site none carried the massulae of *C. montanum* suggesting this was the first and last time most of these bees visited the labellum of this orchid.

Potential pollinators

Insects were observed entering the labellum of fully opened (first day) flowers of *C montanum* at both sites over three seasons (Fig. 2 and 3). Bees were observed entering labella on warm, sunny, often cloudless days but only after the flowering stem bearing the flowers stood in a light gap for 20-60 minutes and produced discernible scent. We did not see insects enter a labellum while the flower stood in shade at either site. Depending on the site, and the amount of canopy cover, a labellum of *C. montanum* could be entered by bees as early as 10:24 AM and as late as 4:30 PM. Clumped flowering stems remaining open over a seven-day period at BMEO received daily visits from bees entering labella over periods from one to five hours. Bee visitations to labella of plants at the ECCO site began as early as 11:25 AM and generally ceased prior to 4:30 PM.

The first time we observed an insect entering the large, dorsal entrance on a labellum and exiting via one of the two rear orifices was at BMEO on 6/6/2003. This bee carried a dorsal deposition of pollen mass on its thorax as it exited the flower (Fig. 3, 4). The specimen was later identified as a female *Lasioglossum* (*Dialictus*) sp. (Tab. 3). Combining four seasons of observations at both sites we observed this process >100 times. Bee visits were so numerous at the ECCO site between 10:24 AM-3:10 PM from 6/13-

Figure 2. Unidentified bee inside inflated labellum. Note that the bee's head is not visible as it is obscured by the staminode (Photograph by Nan Vance).

Figure 3. The same bee as in Fig. I emerges from the basal opening. Note the yellow pollen smear deposited on the thorax. Also, note that the bee escapes the interior of the labellum while clutching the lateral petal (Photograph by Nan Vance).

Figure 4. Pinned specimen of *Lasioglossum* sp (collected in 2003) wearing the entire contents of an anther cell on the dorsum of its thorax (Photograph by Andrew Huber).

6/15/2003, we commonly observed two bees in the same labellum at the same time. In all cases of "double" visits observed, the two bees exited the flower via the rear, basal openings. However, these bees left one-by-one as the escape path under the stigma did not appear to accommodate two bees at the same time.

Bees at the BMEO site entered the labellum and exited through the basal openings of the flower within an average of 222.84 ± 297.83 seconds (N = 38) when an abnormally long point (1945 seconds) was included and 176.30 ± 80.98 (N = 37) when that point was excluded. The mean time bees spent in flowers at the ECCO site in 2006 was 179.25 ± 133.17 seconds (N = 28). However, there was no difference in the time spent in flowers between these two sites whether that solitary and abnormally long point at BMEO was included (Welch's t-test, t = -0.8, df = 54, P = 0.4271) or excluded (t = 0.10, df = 41, P = 0.9179).

The behaviour of bees entering the labellum via the dorsal entrance and escaping via one of the two basal openings varied at inter- and intra-specific levels at both sites but we did not observe bees landing directly on the staminode, at either site, if the flower's labellum was intact (see below). The shorter bees (<7 mm in length) later identified as members of the Andrenidae (*Andrena*, and *Panurginus* spp.), Apidae (*Ceratina acanthi*) and Halictidae (*Halictus*; and some *Lasioglossum* [*Evylaeus*] spp.) were more likely to enter the labellum by flying directly into the large, dorsal entrance and landing on the purple veins of the labellum floor (Figures 1, 2). Longer bees (7-10 mm) mostly *Lasioglossum* spp. and *Osmia* spp. (Megachilidae), were more likely to land on the white outer surface of the labellum or, less frequently, on lateral petals before crawling onto the labellum. As these longer bees crawled towards the rim of the dorsal entrance the more likely they would fall into the labellum sac. In both shorter and longer bees, though, the added weight of the insect inside the labellum usually caused the flower to nod on its pedicel and the bee lost contact with the floral epidermis and rolled to the anterior (toe) of the labellum. Each bee usually turned around and crawled upwards. When it passed under the stigma it was lost from view but a bee could slide backwards into the labellum toe again. Every time a bee started over it had to pass under the stigma. When the largest bees (9-10 mm in length) crawled up through a narrow, hair-lined, exit canal (formed by the staminode and the incurved labellum) and attempted to push through one of the rear exit holes, they produced a distinctive buzz, reminiscent of the whining sound made when a female bee applies thoracic vibration to a cluster of porose-poricidal anthers (see Bernhardt 1996). On 6/7/2003 at the BMEO site, a bee later identified as a female of *Halictus triparitus* crawled under the stigma 12 times and poked its head through the basal holes four times before it finally escaped from the flower nine minutes later. In most cases observed, pollen was not deposited on the bee until the dorsal surface of the insect's thorax (Fig. 3 and 4) contacted the dehiscent anther while the insect extricated itself from a basal opening. Some bees <7 mm in length had pollen smeared on their heads and compound eyes instead of on the dorsum of the thorax. Using 3× magnification we noted that anthers of each orchid flower appeared to be emptied of pollen masses within the first two days of bee visitation at both sites. Examination of specimens of long (9-10 mm long) specimens in the genus, *Lasioglossum*, showed that some bees carried off the entire contents of an anther loculus following passive contact with the dehiscent anther as they attempted to escape from either of the basal openings (Fig. 4).

Two separate events could occur after the bee had fully extricated itself from the basal opening. In most cases the bee crawled onto one of the lateral petals or the dorsal sepal. We observed the insect making cleaning motions to its head, abdomen and first pair of legs before flying away. In a few cases the bee fell to the ground upon escape or fell onto one of the lower leaves of the orchid. In this case, the insect often remained motionless for 30 seconds to several minutes before it cleaned itself and flew away. After three seasons of fieldwork we have only one observation of a bee visiting and exiting more than one *C. montanum* flower in the same clump during the same visitation bout. This occurred at BMEO on 6/24/2004 from 11:24-11:42 AM. The unidentified and uncollected bee visited two flowers on the same inflorescence. All other bees, following their exit from the flower, flew away from the flowering scape until lost from view.

During all seasons we observed bee visitations, we never saw a bee land on the contrastingly coloured staminode before it entered the labellum. We only observed bees landing directly on the staminode *after* the labellum was removed by hand. These bees probed the upper surface of the staminode with their mouthparts and then crawled around the column often contacting the exposed stigma. No transfer of pollen masses to these stigmas was ever observed. Female bees did not appear to recognize the dehiscent anthers as a potential source of pollen.

Mean bee lengths and widths

The mean length and width of bees (N = 31) of potential pollinators to *C. montanum* at the ECCO site was 6.85 (sd = 0.89) and 2.57 (sd = 0.49) respectively. For statistical analyses, bee lengths and widths at the Deschutes site (2006) were log transformed to follow a normal distribution. There was no relationship between insect size (length, width and length×width) and the time spent in flowers (linear regression, $F_{3,24}$ = 1.62, P = 0.2111, R = 0.17).

Potential pollinators, bee diversity and pollen load analyses

A total of 84 potential pollinators were collected over four seasons at three sites (Tables 3-5) and 51 specimens (>0.60) belonged to the genus *Lasioglossum* sl. However, bee diversity varied at the BMEO site over two seasons and between the BMEO (2003 and 2004) and ECCO sites (Tables 3-5). *Panurginus ineptus* (all males), *Halictus rubicundus* (female) and *Lasioglossum olympiae* (all females) collected in 2003 were not found in 2004. Likewise, *Ceratina acanthi* (females), *Lasioglossum tegulariforme* (female) and *Halictus confusus* (females) collected in 2004 were not caught in 2003. While collections of *L. athabascense,* exiting flowers of *C. montanum* were made only at the ECCO site, in 2006 we failed to catch any members of the genera *Ceratina, Halictus, Panurginus* spp. at this site (Tab. 3). The ratio of potential pollinators carrying pollen masses of the host flower on their thoraces or heads was slightly higher at the ECCO site in 2006 (0.58) than at BMEO in 2003 (0.47) or 2004 (0.41). The presence of orchid massulae on a bee and the bee's length (t = -0.99, df = 30, P = 0.3313), or width (t = -1.45, df = 30, P = 0.1584) were not related at the ECCO site.

Of the potential pollinators the majority carried the pollen of at least one other co-blooming taxon that offered floral nectar and/or granular pollen (Tables 2-4; Fig. 5); 0.75 (BMEO, 2003), 0.93 (BMEO, 2004) and 1.0 (ECCO, 2006). Pooling results at BMEO in 2003 and 2004 we noted that >0.62 (N = 53) of the bees visited the pollen and/or nectar secreting taxa of at least one member of the Family, Rosaceae (*Fragaria vesca* var. *bracteata, Physocarpus capitatus, Potentilla glandulosa, Rosa* sp., *Rubus parviflorus*) and *Smilacina stellata* (Asparagaceae) before they began visiting flowers of *C. montanum.* On 5/25/04 we observed a *Lasioglossum*-sized bee foraging on flowers of *S. stellata* over a 20-minute period. It interrupted foraging on this species four times to land on labella of adjacent flowers of *C. montanum* but it never entered the labellum. In contrast, pollen load analyses of *C. montanum* collected at the ECCO site indicated that >0.74 of the potential pollinators to *C. montanum* also visited the flowers of the nectar-secreting shrub, *Ceanothus velutinus* (Rhamnaceae; Tab. 5).

Figure 5. Unidentified bee foraging in flower of *F. vesca* after it exited a flower of *C. montanum* approximately two meters away. Note the smear on the dorsum of the thorax (Photograph by Nan Vance).

Bee diversity on Fragaria vesca var. bracteata

Bees foraged on flowers of *Fragaria vesca* var. *bracteata* from 9:15 AM-3:50 PM depending on the time of day in which clumps stood in light gaps. A total of 30 bees were caught on flowers of *F. vesca* var. *bracteata* at BMEO in 2003 (Tab. 5). Seven bees (0.23) carried pollen masses of *C. montanum* (Tab. 6) including two females of *Nomada* sp. (Apidae), three males of *P. ineptus* (Andrenidae) and one female specimen of *Andrena* (*Microandrena*). While these three bee taxa were potential pollinators of *C. montanum* (Tab. 6) at the same site and season, specimens of the same species collected after they exited the orchid flowers failed to carry pollen masses.

Discussion

Labellum of C. montanum *vs. other* Cypripedium *spp.*

Based on earlier and ongoing studies, labellum dimensions in the genus *Cypripedium* appear to correlate with the dimensions of their respective pollen vectors. In particular, they may reflect the canalization of some flower-insect interactions. While the big labellum of *C. tibeticum* accommodates *Bombus* gynes, and some of their smaller workers, the gynes appear to be the only pollinators of this species (Li et al. 2006). Likewise, *C. plectrochilum* has a much smaller and keeled labellum. It is well visited by insects but the only pollen vectors collected, to date, are a few of the smaller species in the genus *Lasioglossum* (Li et al. 2008a). *Apis cerana,* a large bee, couldn't enter the flower of *C.*

Bee Taxon	N	Pollen Types* Cm	Frg	MRos	No Pollen
Andrena sp. a (f)	2	0	2	1	0
Andrena (Microandrena) sp. (f)	3	1	3	3	0
Ceratina acantha (m)	1	1	1	1	0
Nomada sp. (f)	10	2	8	5	2
Osmia sp. (m)	2	0	2	2	0
Panurginus ineptus (m)	12	3	11	11	1
Totals	30	7	27	23	3

Table 6. Bees collected on *Fragaria vesca* var. *bracteata* at the BMEO site in 2003.

*Cm = *C. montanum*, Frg = *F. vesca* var. *bracteata*, MRos = Mixed Rosalean/Ranalaean pollen (but no *Fragaria*); (m) = male, (f) = female

plectrochilum while smaller, slenderer insects (e.g. ants) exited via the rear openings of the flower without contacting the dehiscent anthers. Labellum dimensions of *C. reginae* fall in between *C. tibeticum* and *C. plectrochilum* but are conspicuously larger and deeper than *C. montanum.* Only medium-large bees in the families Apidae and Megachilidae carried the pollen of *Cypripedium reginae.* When small-bodied, halictid bees exited *C. reginae* they always failed to contact the dehiscent anthers while the huge gynes of *Bombus* spp. couldn't fit through the rear exits and never contacted the dehiscent anthers either (Edens-Meier et al. 2011). Therefore, it's important to note that the dimensions of *C. montanum* did not permit the physically largest specimens of *Bombus, Osmia* and *Eucera* species to exit the flower. *Cypripedium montanum* is more likely to accommodate small-medium sized bees as in *C. yunnanense* (Banziger 2008).

Was C. montanum *pollinator-limited at either site?*

Results indicated that neither population was pollinator-limited over the seasons studied, even though a number of bees of appropriate size appeared to escape from the labellum sac at the ECCO site (see above). As *C. montanum* could not self-pollinate in the absence of insects we must conclude that the previous 94% rate of pollen deposited on stigmas reported by Edens-Meier et al. (2010), from both sites, was based on insect-mediated pollination. Of greater importance, as almost all bees left the site after freeing themselves via the basal openings we should conclude that the majority of bee-mediated pollinations at both sites were probably cross-pollinations (xenogamy) instead of vector-mediated, self-pollinations (to getionogamy or autogamy). Small- to medium-sized female bees (5-10 mm in length) with polylectic and/or polyphagic foraging behaviour were the dominant pollen vectors at both sites. At both sites these massulae-carrying bees shared similar time periods for escaping from the rear of the flower although these bees represented a broad range of sizes and taxa. This appears comparable to a review of bee pollination in the North American species complex, *C. parviflorum* (see Argue 2011), closely allied to *C. montanum s.s.* (Li et al. 2011).

Were all visitors to C. montanum *prospective pollen vectors?*

At both sites, flowers of *C. montanum* attracted insects from as many as three insect orders. However, small beetles, flies and some Hymenoptera obviously either lacked the size, and/or behavioural patterns, and/or physical strength to exit the flower via the basal openings. Some bees (e.g. *Bombus, Eucera* and most *Osmia* spp.) were too large to make a legitimate, rear exit escape (see above). Other taxa (e.g. *Andrena prunorum,* eumenid wasp, *Hylaeus ellipticus* etc.), were infrequent visitors that rarely, if ever, entered labella. While the floral architecture of *C. montanum* accommodated an unusually wide variety of small-to medium-sized bees some species alternated between merely perching on the flower and actually entering and exiting the labellum (e.g. *Lasioglossum* spp., *Nomada* sp. and *Panurginus ineptus*). It's likely that some bees did not enter these flowers a second time due to the absence of rewards and any negative stimuli encountered during the escape period.

In this respect, the pollination ecology of *C. montanum* paralleled that of *C. plectrochilum* (Li et al. 2008a). While an unusually diverse range of insects visited both *Cypripedium* spp., all members of the Diptera, Lepidoptera and some Hymenoptera (ants, large bees) lacked appropriate physical dimensions and/or behavioural patterns and did not carry the orchid's pollen either. As *C. montanum* is bee-pollinated it is not surprising that the labellum sac is a death trap for some flies (Tab. 1). However, while the labellum dimensions of *C. montanum* accommodated an unusually broad diversity of small- to medium-sized bees, entrapment proved fatal to bees of varying sizes especially at the ECCO site.

Leaving a *C. montanum* flower via the basal opening did not guarantee successful transference of pollen masses to the bee each time. The flowers appeared to run out of massulae within one to two days after opening even though an individual flower could live from seven to 21 days depending on the site (Edens-Meier et al. 2010). As we could not measure bee depth vs. the distance between the receptive stigma and the floor of the labellum without destroying flowers of a protected species it was not possible to determine which bee species were more likely to leave pollen on the receptive stigma as they crawled under it.

Our observations of bees landing on the staminode, only after the labellum was excised, tended to confirm the theory of Edens-Meier et al. (2014) that the pigmentation pattern on the interlocking staminode-labellum mechanism may represent part of a super-normal stimulus in some *Cypripedium* spp. That is, while the two, differently coloured patterns on the staminode both contrast with this white labellum, the bee probably sees both patterns as contiguous. Together, staminode and labellum floor patterns grade together forming an irregular blotch (*sensu* Kevan & Dafni 1996) often associated with flowers with bilateral symmetry. The pattern canalizes the bee's movements and it lands on, or near, the labellum floor dependably until the labellum is removed. Only then did the bee land on the colour pattern on the staminode as the now missing sac changed the floral symmetry to radial. We do not suggest that this interpretation fits all *Cypripedium* spp. as pigmentation patterns on the staminode and in the labellum vary broadly at inter- and intraspecific levels (e.g. Li et al. 2006, 2008ab and see colour plates in Edens-Meier et al. 2011)

Did pollen vector diversity vary between seasons and sites?

Female bees in the genus, *Lasioglossum* s.l. (Halictidae) carried pollen masses at both sites over three seasons and appeared to be the dominant, but not exclusive, dispersal agents of orchid pollen. Unfortunately, this genus consists of over 1100 species worldwide and it was not possible to identify each specimen to species (C.D. Michener, pers. comm.). We did find, though, that the *Lasioglossum* specimens, captured after they left the basal openings, represented at least three subgenera. Bees in the genus *Halictus* (Halictidae) and in the families Apidae and Andrenidae also carried pollen masses so *C. montanum* exploits both long tongue (Apidae) and short-tongue (Andrenidae, Halictidae) bees, of similar sizes, in the absence of floral nectar. It is clear, though, that the diversity of legitimate pollen carriers varied between 2003 and 2004 at the BMEO site and the diversity of pollen carriers at BMEO differed from the 2006 collection at ECCO. At BMEO, *C. acanthi, L. olympiae, L. tegulariforme*, and *P. ineptus* were collected exiting the flowers for only one season each. *Halictus tripartius* was collected once outside the flowers of *C. montanum* in 2003, and it carried no pollen masses. However, seven specimens of *H. tripartius* were caught exiting the flowers in 2004 and three carried the orchid's pollen. Likewise, members of the genera *Halictus, Nomada* and *Panurginus* were never caught at ECCO in 2006. At that site, four specimens of *L. athabascense* made their only appearance as orchid pollen vectors while some *Osmia* spp. were physically small enough to pass through the rear exit of the flowers. The collection of bees on co-blooming *Fragaria vesca* showed that interpreting the role of a bee as a carrier of *Cypripedium* pollen should include specimens taken from co-blooming flora. It's possible to miss less frequent visitors to the orchid that can carry the pollen masses, when they visit, but prefer to forage on co-blooming species offering nectar and/or pollen. Note, for example, that in 2003 three bee taxa (see above), caught while exiting *C. montanum*, did not carry the orchid's pollen but specimens of the same taxa, caught on *Fragaria vesca* did carry pollen of *C. montanum*.

The exploitation of a wider variety of small-to medium-sized bees must contribute to the increased frequency of pollination in *C. montanum* and its high fruit set assessed by Huber (unpublished) at BMEO. In contrast, we note that *C. plectrochilum* (small labellum), *C. henryi* (Li et al. 2008b; mid-sized labellum) and *C. yunnanse* (Bänziger et al. 2008; mid-sized) were pollinated exclusively by a few *Lasioglossum* spp. (Bänziger et al. 2008; Li et al. 2008b) and their conversion of ovaries into fruits peaked at 45%, 22% and 21%, respectively. Likewise, one population of *C. flavum* was pollinated only by a few *Andrena* spp. and its highest fruit set ratio was only 9.20% (Bänziger et al. 2008). In China, montane populations of *Cypripedium* spp. often show broadly overlapping distributions and flowering periods (Singchi et al. 1999; Perner & Luo 2007). While a narrower spectrum of pollen vectors may lower chances of interspecific hybridization (Bänziger et al. 2008), it might also depress reproductive success within sympatric *Cypripedium* spp. On temperate Chinese mountains most of the potential pollinators are either not attracted to the colours and odours of one *Cypripedium* species and/or are unable to fit into and/or navigate their floral interiors. As in other angiosperms, a *Cypripedium* species should be in danger of becoming pollinator-limited if resident populations of its few pollinator species decline and/or insect foraging seasons change (Committee on the Status of Pollinators in North America 2007).

What factors support a well-visited food mimic system?

Therefore, the high rate of visitation by prospective vectors of massulae in *C. montanum* was supported by two interlocking factors. First, as described above, the sheer number of flowering stems in bloom combined with their shared modes of floral presentation and architectural dimensions exploited a broad and variable diversity of polylectic/polyphagic foragers at both sites over several seasons. Exploitation of resident, small-to medium- sized bee faunas in *C. montanum* parallels results obtained from multiple sites and seasons in the pollination ecology of Eurasian, *C. calceolus* (Nillson 1979; Kull 1999; Bernhardt & Edens-Meier 2010).

Second, to exploit the broadest diversity of polylectic female bees *C. montanum* must appear to offer nectar and/or pollen as in the majority of non-specific (generalist) frauds (Ackerman 1986; Dafni & Bernhardt 1990). It's unlikely that successful pollination continues throughout the comparatively long floral lifespans of *C. montanum* in the absence of a dependable, co-blooming flora for generalist bees. We suspect that our frequent observations of potential pollinators at both sites were due, at least in part, to the diversity and density of co-blooming species. Flowers offering only pollen (Bernhardt 1996) and/or pollen and nectar rewards were always available at both sites, over the flowering seasons of *C. montanum*, even though floral diversity differed between sites. We also speculate that habitats rich in food sources for bee offspring could encourage more nesting and increased populations of multi-voltine, bee taxa and one nectariferous species may be sufficient to support adult nutrition. Over 79% of the potential pollinators of *C. montanum* at ECCO collected pollen of *Ceanothus velutinus* (a mass flowering, nectar-secreting shrub). Different *Fragaria* spp. follow the distribution and overlapping flowering periods of some Chinese *Cypripedium* spp. (Li et al. 2008a) serving as food sources for orchid pollinators. Perhaps extensive populations of *Fragaria* spp. should be observed and regarded as insect collection sites more often and used as potential sources of pollinators of some *Cypripedium* spp. when their distributions and flowering periods overlap (see above).

However, we do not suggest that *Cypripedium* spp. with only a few specialist pollinators are always pollinator-limited and typified by low fruit set ratios. *Cypripedium fasciculatum* does not self-pollinate and is dependent on only a few *Cinetus* spp. (Diapriiadae; Ferguson & Donham 1999). Lipow et al. (2002) compared natural rates of pollination (stigmas and pistils containing pollen tubes) in *C. fasciculatum*, at three disjunctive sites, over their respective flowering seasons. At one site in Oregon, the frequency of pollinated pistils was 69.2%. This, we argue, was and remains competitive with any *Cypripedium* spp. pollinated by many species of small-to medium-sized bees (Bernhardt & Meier 2010).

Variation of fruit set rates in C. montanum

We do not suggest that populations of *C. montanum* must always enjoy high rates of fruit set when their patchy populations co-occur with a diverse bee fauna and a co-blooming, flora offering nectar and/or pollen. The range of this species is from 0 – 2400 m throughout the Pacific Northwest and interior of Alaska, British Columbia and the continental United States (Sheviak 2002). The reproductive ecology of isolated populations is expected to vary between differing microhabitats and years. Nilsson (1979) and Kull (1999) found that, *C. calceolus* was also pollinated by a diverse assemblage of small-to medium-sized bees. The conversion ratio of pistils into capsules in *C. calceolus* was varied from 4-57% according to year and site (see review by Bernhardt & Edens-Meier 2010).

Previous results obtained by Edens-Meier et al. (2010) and Lipow et al. (2002) also suggest that natural rates of insect-mediated pollination in populations of *C. montanum* and *C. fasciculatum* may be higher than rates of matured fruit at the same site and year. There are several reasons why this occurs in *C. montanum, C. fasciculatum* and flowering populations of other *Cypripedium* species. First, some flowers in a population of *C. reginae* never set fruit because their maturing buds were ruined by late freezes (Edens-Meier et al. 2011). Second, the majority of orchid species postpone ovule fertilization. Megasporogenesis won't occur unless the flower is pollinated first and fertilization followed by fruit maturation and dehiscence usually takes weeks or months (Arditti 1992). This subjects slowly maturing ovaries to sudden fluctuations in climate, predation and human impact over extended periods. During that maturation period some ovaries of *C. reginae* are eaten by larvae of geometrid moths (Edens-Meier et al. 2011). Luo Yi-bo (pers. comm.) observed domesticated yaks trampling *C. tibeticum, C. flavum, C. guttatum and C. yunnanensis* at higher elevations in Yunnan from 2003-2006. Browsing and trampling by North American ungulates (Karow, pers. comm. 2005) and cattle (Vance unpublished 2007) may also be severe on remaining populations of *C. montanum*.

Finally, we do not argue with the exhaustive review by Tremblay et al. (2005) that pollinator visits to sexual mimics surpass pollinator visits to food mimics resulting in higher rates of fruit set. We do suggest that floral mimicry in an orchid species needs to be evaluated on a species-by-species and population-by-population basis over several seasons before conservation policies can be established (Bernhardt & Edens-Meier 2010). *Cypripedium montanum* is a food mimic but has a conversion rate of flowers into fruits competitive or even surpassing some sexual mimics (Tremblay et al. 2005) within two sites over its natural range. Fruit set in *C. montanum* appears competitive with some other obligate, outcrossing, angiosperms that offer edible rewards (see review in Bernhardt & Meier 2010). Understanding the fine points of why this occurs (pollinator diversity, co-blooming nectar and pollen flora, interactions between pollinator dimensions and floral architecture, fruit predators, topography, and prevailing climate, etc.) should help us conserve and increase fecundity in remaining populations of this threatened species.

ACKNOWLEDGEMENTS

We thank Dr. C.D. Michener (Snow Entomological Museum) for identifying specimens and for sending specimens to other authorities when he could not identify to species. Andrew Huber deserves special thanks for recording fruit set in July and giving us access to the GROWISER lodge allowing us to stay *in situ* to monitor flowering periods. This research was supported in part by funding from the US Department of Agriculture Forest Service, Pacific Northwest Research Station.

REFERENCES

Arditti JD (1992) Fundamentals of orchid biology. John Wiley and Sons Incorporated, New York.

Ackerman JD (1986) Mechanisms and evolution of food deceptive pollination systems in orchids. Lindleyana 1:108-113.

Argue CL (2011) The pollination biology of North American orchids: Volume 1: North of Florida and Mexico. Springer, New York.

Bänziger J, Sun H, Luo YB (2005) Pollination of a slippery lady slipper orchid in south-west China: *Cypripedium guttatum* (Orchidaceae). Botanical Journal of the Linnean Society 148:251-264.

Bänziger J, Sun H, Luo YB (2008) Pollination of wild lady slipper orchids *Cypripedium yunnanense* and *Cypripedium flavum* (Orchidaceae) in south-west China: Why are there no hybrids? Botanical Journal of the Linnean Society 156:51-64.

Barkman TJ, Beaman JH, Gage DA (1997) Floral fragrance variation in *Cypripedium*: Implications for evolutionary and ecological studies. Phytochemistry 44: 875-882.

Bernhardt P (1996) Anther adaptations in animal pollination. In: D'Arcy WG, Keating, WG (eds) Anther adaptations in animal pollination. Cambridge University Press, Cambridge, pp 192-221.

Bernhardt P, Edens-Meier R (2010) What we think we know *vs.* what we need to know about orchid pollination and conservation: *Cypripedium* as a model lineage. Botanical Review 76:203-219.

Bernhardt P, Weston P (1996) The pollination ecology of *Persoonia* (Proteaceae) in eastern Australia. Telopea 6:775-804.

Catling PM (1990) Auto-pollination in the Orchidaceae. In: Arditti, VJ (ed) Orchid biology: reviews and perspectives. Timber, Portland, pp 121-138.

Chi J, Luo, YB, Bernhardt P, Ran JC, Liu ZJ, Zhou Q (2008) Pollination by deceit in *Paphiopedilum barbigerum* (Orchidaceae): A staminode exploits the innate colour preferences of hoverflies (Syrphidae). Plant Biology 11:17-28.

Coleman RA (1995) The wild orchids of California. Comstock, Ithaca, New York.

Committee on the Status of Pollinators in North America (2007) Status of Pollinators in North America. National Research Council. The National Academies Press, Washington, D.C., www,nap,edu.

Cribb P (1999) Cypripedioideae. In: Pridgeon A, Cribb PJ, Chase MW, Rasmussen FN (eds) Genera Orchidacearum Volume I. Oxford University Press, United Kingdom, pp 105-164.

Dafni A, Bernhardt P (1990) Pollination of terrestrial orchids of southern Australia and the Mediterranean Region: Systematic, ecological and evolutionary implications. In: Hecht M, Wallace B, Macintyre RJ (eds). Evolutionary Biology: 24. Plenum Publishing Corporation, New York, pp. 193-252.

Darwin C (1877) The various contrivances by which orchids are fertilised by insects. 2nd ed. John Murray, London.

Dixon KW (2009) Pollination and restoration. Science 325:571-573.

Dressler RL (1968) Pollination by euglossine bees. Evolution 22:202-210.

Dressler R (1981) The orchids: Natural history and classification. Harvard University Press, Cambridge, Massachusetts.

Dressler R (1993) Phylogeny and classification of the orchid family. Dioscorides Press, Portland, Oregon.

Edens-Meier RM, Arduser M, Westhus E, Bernhardt P (2011) Pollination ecology of *Cypripedium reginae* Walter (Orchidaceae): Size matters. Telopea 13:327-340.

Edens-Meier RM, Vance N, Luo YB, Li P, Westhus E, Bernhardt P (2010) Pollen-pistil interactions in North American and Chinese *Cypripedium* L. (Orchidaceae). International Journal of Plant Sciences 171:370-381.

Edens-Meier RM, Luo YB, Pemberton R, Bernhardt P (2014) Pollination and floral evolution of slippper orchids. In: Edens-Meier R, Bernhardt P (eds) Darwin's Orchids: Then and Now. University of Chicago Press, Chicago, Illinois, pp

Ferguson CS, Donham K (1999) Pollinator of clustered lady's slipper orchid *Cypripedium fasciculatum* (Orchidaceae) in Oregon. North American Native Orchid Journal 5:180-183.

Ferguson CS, Donham K, Brown JL (2005) *Cypripedium fasciculatum* (Orchidaceae) Anthesis and fruit set in relationship to Diapriid activity. Selbyana 26:3-113.

Huber AG (2002) Mountain lady's slipper (*Cypripedium montanum*): Establishment from seeds in forest openings. Native Plants Journal 3:151-154.

Kevan P, Dafni A (1996) Floral symmetry and nectar guides: Ontogenetic constraints from floral development, color pattern rules and functional significance. Botanical Journal of the Linnean Society 120: 371-377.

Kull T (1999) *Cypripedium calceolus* L. Journal of Ecology 87:913-924.

Li JH, Liu ZJ, Salazar G, Bernhardt P, Perner H, Tomohisa Y, Jin XH, Chung SW, Luo YB (2011) Molecular phylogeny of *Cypripedium* (Orchidaceae: Cypripedioideae) inferred from multiple nuclear and chloroplast regions. Molecular Phylogenetics and Evolution 61:308-320.

Li P, Luo YB, Bernhardt P, Tang XQ, Kou Y (2006) Deceptive pollination of the lady's slipper *Cypripedium tibeticum* (Orchidaceae). Plant Systematics & Evolution 262:53-63.

Li P, Luo YB, Bernhardt P, Kou Y, Perner H (2008a) Pollination of *Cypripedium plectrochilum* (Orchidaceae) by *Lasioglossum* spp. (Halictidae); the roles of generalist attractants *versus* restrictive floral architecture. Plant Biology 10:220-230.

Li P, Luo YB, Deng Y, Kou Y (2008b). Pollination of the lady's slipper *Cypripedium henryi* Rolfe (Orchidaceae). Botanical Journal of the Linnean Society 156:491-499.

Lipow SR, Bernhardt P, Vance NC (2002) Comparative rates of pollination and fruit set in widely separated populations of a rare orchid (*Cypripedium fasciculatum*). International Journal of Plant Sciences 163:775-782.

Luer CA (1975). The native orchids of the United States and Canada excluding Florida. The New York Botanical Garden, W.S. Cowell Ltd., Ipswich, England.

Nilsson LA (1979) Anthecological studies on the Lady's Slipper, *Cypripedium calceolus* (Orchidaceae). Botaniska Notiser 132:329-347.

Ogden EC, Raynor GS, Hayers JV, Lewis DM (1974) Manual of sampling airborne pollen. Hafner Press, London.

Perner H, Luo YB (2007) Orchids of Huanglong. Huanglong National Park and the Sichuan Publishing Group, Sichuan, China.

Sheviak, CJ (2002) *Cypripedium.* In: Flora of North America, Magnoliophyta: Liliidae: Liliales and Orchidales. Volume 26. Oxford University Press Inc., New York, New York. pp 499-507.

Singchi C, Zhanvyo T, Luo YB (1999) Native Orchids of China In Colour. Science Press, Beijing, China.

Tremblay RL, Ackerman JD, Zimmerman JK, and Calvo RC (2005) Variation in sexual reproduction in orchids and its evolutionary consequences: a spasmodic journey to diversification. Biological Journal of the Linnean Society 84:1-54.

Vance N (2007) *Cypripedium montanum* Douglas ex Lindley (mountain lady's slipper): A technical conservation assessment. Species conservation project, USDA Forest Service, Rocky Mountain Region, Golden CO, http://www.fs/usda.gov/detail/r2/landmanagement/, accessed 02/01/2014.

6

A CASUAL CANTHAROPHILY: THE MEETING BETWEEN *ASTYLUS VARIEGATUS* (COLEOPTERA: MYLERIDAE) AND *OXYPETALUM BANKSII* (APOCYNACEAE: ASCLEPIADOIDEAE)

Milene Faria Vieira* and Rúbia Santos Fonseca

Departamento de Biologia Vegetal, Universidade Federal de Viçosa, 36570-000 Viçosa, Minas Gerais, Brazil

Abstract—Cantharophily is reported for the first time in a Brazilian asclepiad, involving the mylerid *Astylus variegatus* and the nectariferous flowers of *Oxypetalum banksii*, a plant mainly pollinated by wasps. The use of nectar as food by *A. variegatus*, considered pollinivorous and granivorous, is also novel. The mutual interaction described here is an example of a plant-pollinator interaction with generalist insects visiting a plant with a specialized pollination system. It's also temporary and occasional and, therefore, is often overlooked in studies of plant-pollinator interactions. In this study, we found that the casual meeting between *O. banksii* and *A. variegatus* was a key event for the reproduction of both.

Key words: *Asclepiad, Atlantic Forest, Brazil, beetles, specialized pollination, wasps*

The asclepiads (Apocynaceae: Asclepiadoideae, *sensu* Endress & Bruyns 2000) have highly elaborate and complicated flowers. There is an unusual synorganization between parts and also between organs of different categories (Endress 1994). This has led to the evolutionary origin of new organs that are not present in other angiosperm groups in this combination: corona (more or less well developed outgrowths of the corolla and/or stamens), gynostegium (congenitally fused and highly synorganized gynoecium and androecium) and pollinaria (pollen aggregated into masses (pollinia) with mechanical clips (translators) which attach them to pollinators).

The pollination in these plants is a result of pollinaria removal (five per flower) and insertion of pollinia in stigmatic chambers (five chambers per flower - more details in Wyatt 1976; Kunze 1991). In Brazil, a country rich in asclepiads and many of them endemic and rare (Rapini et al. 2002), studies on interactions between asclepiads and their pollinators are restricted to that of Vieira and Shepherd (1999a) and Medeiros et al. (2008). According to Ollerton and Liede (1997), pollinators of these plants are diverse, including Diptera (flies), Hymenoptera (bees and wasps) and Lepidoptera (butterflies and moths). These authors commented that specialized beetle pollination is rare in the asclepiads.

In fact, only three recent studies have demonstrated cantharophily (pollination by beetles) in asclepiads. Ollerton et al. (2003) documented for the first time in South Africa, a specialized pollination system that involves asclepiads (*Asclepias woodii*, *Sisyranthus trichostomus* and *Xysmalobium involucratum*) and chafer beetles (Scarabaeidae: Cetoniini). Shuttleworth and Johnson (2008) described, also in South Africa, a bimodal pollination system in the asclepiad *Xysmalobium undulatum*; its pollination is performed by two unrelated species (or functional groups) of pollinators: chafer beetles and pompilid wasps (Hymenoptera: Pompilidae). These authors, in 2009 and at the same study site, described four specialized pollination systems in asclepiads, one of them involving three asclepiad species of *Pachycarpus* and chafer beetles (Shuttleworth & Johnson 2009).

Other studies have reported beetles as flower visitors of asclepiads, but did not confirm their role as pollinators. Woodell (1979) recorded cetoniid chafers (*Mausoleopsis aldabrensis*) visiting flowers of three asclepiads on Aldabra. Forster (1989) documented what may be specialized beetle pollination in *Marsdenia fraseri*, involving the lycid *Metriorrhynchus lateralis* (Coleoptera: Lycidae) in Australia. Hawkeswood (1994) observed, also in Australia, *Micraspis frenata* beetles (Coleoptera: Coccinellidae) on flowers of *Asclepias curassavica* and *Gomphocarpus physocarpus*. The author recorded the novelty of the use of nectar as food for these beetles. Vieira and Shepherd (1999a) studied a floral visitor guild of seven species of *Oxypetalum* in Brazil and found an unidentified species of cantharid (Coleoptera: Cantharidae) and *Enoclerus* sp. (Coleoptera: Cleridae) on flowers of *Oxypetalum banksii* feeding on nectar; they were not seen carrying pollinaria.

Oxypetalum banksii Schult. is an endemic vine from Brazil, inhabiting areas of the Cerrado and Atlantic Forest (Rapini et al. 2010). In the municipality of Coimbra (20°50'07"S and 42°49'52"W), Zona da Mata of Minas Gerais State, southeastern Brazil and site of the present study,

*Corresponding author; email: mfvieira@ufv.br

this plant inhabits areas with highly modified vegetation, such as pastures and roadsides. Individuals are found isolated and flowering occurs throughout the year (Vieira & Sherpherd 1999a). Their flowers are nectariferous (Vieira & Shepherd 2002a) and pollinated by wasps (mainly Vespidae), which carry pollinaria on the mouthparts (Vieira & Sherpherd 1999a). *Oxypetalum banksii* is self-compatible, but depends on pollinators due to herkogamy, which prevents spontaneous self-pollination (Vieira & Shepherd 1999b).

The beetle *Astylus variegatus* (Germar, 1824) (Coleoptera: Myleridae) occurs in several states of Brazil, including Minas Gerais, as well as Paraguay and Argentina (Souza & Carvalho 1994). Adults are seen in flowers of native or cultivated plants, feeding on pollen (Schlottfeldt 1944; Costa Lima 1953; Silva et al. 1968; Matioli & Figueira 1988). Rossetto and Rossetto (1976) observed for the first time, in southeastern Brazil, a large number of adults of *A. variegatus* feeding in panicles of sorghum (*Sorghum bicolor* (L.) Moench) and damaging seeds. These authors commented that the population of beetles disappeared abruptly; one day after a finding of infestation, no insect was found in the field or nearby. Their larvae, which live in soil, are considered agricultural pests because they feed on corn seeds, in newly-sown crops (Matioli et al. 1990).

In March 2011, a month in which *A. variegatus* adults are commonly observed in southeastern Brazil (Rossetto & Rossetto 1976; Matioli & Figueira 1988; Matioli et al. 1990), 300 individuals of this species were recorded in a flowering plant of *O. banksii*. Given this fact, the objectives of this study were to describe the visiting behaviour of *A. variegatus* on flowers of *O. banksii*, in order to understand its role in the plant reproduction. For this, we carried out observations and quantified pollinaria removal and insertion of pollinia in 136 flowers. In addition, photographic records were made and 82 beetles were collected to confirm pollinaria attached to their bodies, the number of beetles with pollinaria and the number of pollinaria carried by insects.

In the plant, the ratio of individuals of *A. variegatus* and flowers was balanced (about 1:1). No other insect was observed, except a few individuals (about six) of an unidentified species of cantharid beetle. In a flower, up to seven individuals of *A. variegatus* were observed, including couples in copulation (Fig. 1).

Astylus variegatus remained for long periods in a single flower and visited several flowers of the plant without damaging the flowers. During copulation and flower visitation, the female takes the nectar as food, the only resource available, while the male is loaded in her back (Fig. 1b, c). Males were also observed feeding on nectar. The use of this floral resource by *A. variegatus*, considered pollinivorous (Matioli & Figueira 1988; Matioli et al. 1990) and granivorous (Rossetto & Rossetto 1976) is a new record.

Beetles introduced the mouthparts in the corolla tube to access the nectar, placing the head between the corona and the gynostegium (Fig. 1a), in a behaviour similar to wasps (see Vieira & Shepherd 1999a). Individuals of *A. variegatus* were also observed in panicles of the grass *Brachiaria decumbens* Stapf, close to individual of *O. banksii*. These beetles probably fed on their seeds, similar to that observed by Rossetto and Rossetto (1976).

We observed flowers with up to five pollinaria removed, the average being 2.04 per flower (Tab. 1). This value is higher than that recorded by Vieira and Shepherd (2002b), of 1.45 pollinaria removed per flower by wasps. According to these authors, in a study with seven species of *Oxypetalum*, *O. banksii* showed one of the highest values of pollinaria removal and one the lowest values of pollinia insertions (0.02 pollinia per flower). This tendency of higher pollinaria removal and low pollinia insertions was maintained because

FIG. 1. *Astylus variegates* on flowers of *Oxypetalum banksii*. (a) Three beetles on the flower, one with the mouthparts inserted in the corolla tube between the corona and gynostegium. (b) Couple in copulation on flowers, female with two pollinaria and male with one pollinarium on the mouthparts. (c) Couple in copulation, the female with the mouthparts inserted in the corolla tube. c, corona; g, gynostegium; p, petal. Bar = 5 mm.

Flowers examined (no.)	Pollinaria removed (no. flower / % flower)						Pollinaria removed per flower (mean)
	0	1	2	3	4	5	
136	13	38	36	33	11	5	2.04
	9.56	27.94	26.47	24.26	8.09	3.68	

TAB. I. Pollinarium removal in flowers of *Oxypetalum banksii.*

only two flowers were pollinated, each flower with one pollinium inserted (0.01 pollinia per flower). Low insertion of pollinia seems to be the reason for the low natural fruit set in *O. banksii* (0.8%; Vieira & Shepherd 1999b) which has also been commonly recorded in other asclepiads (see Wyatt & Broyles 1990).

Our findings included zero (69.50%), one (26.80%), two (2.44%) and three pollinaria (1.22%) on beetles' mouthparts. Vieira and Shepherd (1999a) observed one or two pollinaria on wasps' mouthparts. Among the 25 insects (of 82 in total) seen with pollinaria, 11 were males and 14 were females.

The visiting behaviour of females and males of *A. variegatus* on flowers of *O. banksii*, their ability to remove pollinaria and most likely insert pollinia, are factors that favour their role as pollinators. It is the first record of cantharophily in a Brazilian asclepiad and of mylerid beetles acting as pollinators of this group of plants.

Another factor that favours the role of *A. variegatus* as pollinator of this asclepiad is its self-compatibility (Vieira & Shepherd 1999b), therefore independent of pollen flow between plants for fruiting. Moreover, the flowers of *O. banksii* have a peculiar morphological feature described by Vieira and Shepherd (2002a), which allows a production of two fruits (twin follicles) after one inserted pollinium in a flower. The other asclepiads, including other *Oxypetalum* species, after the insertion of one pollinium, produce only one follicle per flower (more details in Sage et al. 1990; Vieira & Shepherd 2002a). Twin follicles result in higher production of seeds and thus there seems to be a compensation for their low reproductive success. To illustrate, after the insertion of one pollinium per flower, *Oxypetalum appendicultum* Mart. produces a single follicle with up to 166 seeds, while the twin follicles of *O. banksii* produce up to 282 seeds (Vieira & Shepherd 1999b). Thus, the two pollinated flowers observed in this study could produce up to 564 seeds.

Pollination of *O. banksii* by *A. variegatus* is dependent on an opportunistic meeting between the plant and beetle. It is an example of a plant-pollinator interaction with generalist insects visiting plants with a specialized pollination system. This interaction favours the reproduction of *O. banksii*, especially considering its unstable habitat (pastures and roadsides), where the presence of pollinators is uncertain. The mutualistic interaction described here is temporary and occasional and, therefore, is often overlooked in studies on plant-pollinator interactions. The present study, however, shows that the casual meeting between *O. banksii* and *A. variegatus* is a key event for the reproduction of both.

ACKNOWLEDGMENTS

We thank Paulo Sérgio Fiúza Ferreira for the identification of beetles and Jeff Ollerton for comments on the earlier version of the manuscript. Financial support was provided by CNPq and Fapemig (PPM III).

REFERENCES

Costa Lima A (1953) Insetos do Brasil. Coleopteros. Escola Nacional de Agronomia, Rio de Janeiro.

Endress ME, Bruyns PV (2000) A revised classification of the Apocynaceae s.l. Botanical Review 66:1–56.

Endress PK (1994) Diversity and evolutionary biology of tropical flowers. Cambridge University Press, Cambridge.

Forster PI (1989) Pollination of *Marsdenia fraseri* (Asclepiadaceae) by *Metriorrhynchus lateralis* (Coleoptera: Lycidae). Coleopterists Bulletin 43:311–312.

Kunze H (1991) Structure and function in asclepiad pollination. Plant Systematics and Evolution 176:227–253.

Hawkeswood TJ (1994) Notes on the Australian ladybird beetle *Micraspis frenata* (Erichson) (Coleoptera: Coccinellidae) feeding on nectar from *Asclepias* and *Gomphocarpus* flowers (Asclepiadaceae). Giornale Italiano di Entomologia 7:67–71.

Matioli JC, Figueira AR (1988) Dinâmica populacional e efeitos da temperatura ambiental e precipitação pluviométrica sobre *Astylus variegatus* (Germar, 1824) e *A. sexmaculatus* (Perty, 1830) (Coleoptera; Dasytidae). Anais da Escola Superior de Agricultura Luiz de Queiroz 45:125–142.

Matioli JC, Rossi MM, Carvalho CF (1990) Ocorrência e distribuição mensal de *Astylus variegatus* (Germar, 1824) e *A. sexmaculatus* (Perty, 1830) (Coleoptera: Dasytidae) em alguns municípios do Estado de Minas Gerais. Anais da Sociedade Entomológica do Brasil 19:373–382.

Medeiros JF, Rapini A, Barbosa UC, Py-Daniel V, Braga PIS (2008) Primeiro registro de Simuliidae (Diptera) com polinários de Asclepiadoideae (Apocynaceae). Neotropical Entomology 37:338–341.

Ollerton J, Liede S (1997) Pollination systems in the Asclepiadaceae: a survey and preliminary analysis. Biological Journal of the Linnean Society 62:593–610.

Ollerton J, Johnson SD, Cranmer L, Kellie S (2003) The pollination ecology of an assemblage of grassland asclepiads in South Africa. Annals of Botany 92:807–834.

Rapini A, Mello-Silva R, Kawasaki ML (2002) Richness and endemism in Asclepiadoideae (Apocynaceae) from the Espinhaço range of Minas Gerais, Brazil – a conservationist view. Biodiversity and Conservation 11:1733–1746.

Rapini A, Koch I, Kinoshita LS, Simões AO, Spina AP (2010) Apocynaceae. In: Forzza RC et al. (ed.) Lista de Espécies da Flora do Brasil. Jardim Botânico do Rio de Janeiro, Rio de Janeiro. [on line] URL: http://floradobrasil.jbrj.gov.br/2010/FB004747.

Rossetto CJ, Rossetto D (1976) *Astylus variegatus* (Germar, 1824) (Coleoptera Dasytidae) danificando sorgo. Bragantia 35:131–132.

Sage TL, Broyles SB, Wyatt R (1990) The relationship between the five stigmatic chambers and two ovaries of milkweeds flowers: a three dimensional assessment. Israel Journal of Botany 39:187–196.

Schlottfeldt CS (1944) Insetos encontrados em plantas cultivadas e comuns - Viçosa, Minas Gerais. Revista Ceres 6:52–65.

Shuttleworth A, Johnson SD (2008) Bimodal pollination by wasps and beetles in the African milkweed *Xysmalobium undulatum*. Biotropica 40:568–574.

Shuttleworth A, Johnson SD (2009) New records of insect pollinators for South African asclepiads (Apocynaceae: Asclepiadoideae). South African Journal of Botany 75:689–698.

Silva AGA, Gonçalves CR, Galvão DM, Gonçalves AJL, Gomes J, Silva MN, Simoni L (1968) Quarto Catálogo dos Insetos que Vivem nas Plantas do Brasil. Seus Parasitos e Predadores. Parte II – 1° Tomo. Ministério da Agricultura, Rio de Janeiro.

Souza B, Carvalho CF (1994) Aspectos morfológicos do adulto de *Astylus variegatus* (Germar, 1824) (Coleoptera, Melyridae). Pesquisa Agropecuária Brasileira 29:689–694.

Vieira MF, Shepherd GJ (1999a) Pollinators of *Oxypetalum* (Asclepiadaceae) in southeastern Brazil. Revista Brasileira de Biologia 59:693–704.

Vieira MF, Shepherd GJ (1999b) Sistema reprodutivo de *Oxypetalum appendiculatum* Mart. e *Oxypetalum banksii* Roem. & Schult. subsp. *banksii* (Asclepiadaceae). Acta Botanica Brasilica 13:237–241.

Vieira MF, Shepherd GJ (2002a) *Oxypetalum banksii* subsp. *banksii*: a taxon of Asclepiadaceae with an extragynoecial compitum. Plant Systematics and Evolution 233:199–206.

Vieira MF, Shepherd GJ (2002b) Removal and insertion of pollinia in flowers of *Oxypetalum* (Asclepiadaceae) in southeastern Brazil. Revista de Biología Tropical 50:37–43.

Woodell SRJ (1979) The role of unspecialized pollinators in the reproductive success of Aldabran plants. Philosophical Transactions of the Royal Society B 286:99–108.

Wyatt R (1976) Pollination and fruit set in *Asclepias* - a reappraisal. American Journal of Botany 63:845–851.

Wyatt R, Broyles SB (1990) Reproductive biology of milkweeds (*Asclepias*): recent advances. In: Kawano S (ed) Biological approaches and evolutionary trends in plants. Academic Press, San Diego, pp 255–272.

7

AGGRESSION BETWEEN FLORAL SPECIALIST BEES ENHANCES POLLINATION OF *HIBISCUS* (SECTION *TRIONUM*: MALVACEAE)

B. J. Sampson1, C. T. Pounders, C. T. Werle, T. R. Mallette, D. Larsen, L. Chatelain and K. C. Lee

USDA-ARS Thad Cochran Southern Horticultural Laboratory, Poplarville, MS 39470 USA

Abstract—Specialist oligoleges are bees with narrow dietary niche breadths often touted to have foraging specializations that enhance their pollination efficiency above that of co-foraging generalists (polyleges) such as honeybees *Apis mellifera* L. This study and many comparative pollinator efficacy studies in crops show that oligoleges on a per bee basis are not always the most efficient pollinators. Percentage of stigmatic contact visits by populations of oligolectic *Ptilothrix bombiformis* (Cresson) in Maryland and Mississippi show that adult bees contact stigmas and transfer 70+ pollen grains during 2% - 3% of floral visits. As low as these percentages seem, pollination efficiency of this bee varies from negative values due to the removal of stigmatic pollen during some visits to increases of 30% as *Hibiscus* petals close and spikes of 300% - 1000% as *Ptilothrix* adults display aggression inside flowers. Aggressive *P. bombiformis* tussle with each other, often grappling with, lunging at, and biting conspecifics. Opponents will often lose their balance, tumble around inside flowers and, in the process, more efficiently pollinate host blooms. Such aggressive interactions constituted 5% of visitation bouts to *Hibiscus* flowers, yet accounted for ~20% of contact visits that transferred 10 or more pollen grains per stigma. Tussles therefore represent brief agonistic entanglements that can enhance the pollination efficiency of solitary bees at host plants with large herkogamous blooms. More complex behavioural interactions between different sexes and species of foraging bees may explain the importance of greater bee diversity to overall pollinator effectiveness.

Keywords: aggression, courtship, effectiveness, mutualism, pollinators, seed set

INTRODUCTION

Pollen flow in most flowering plant taxa depends on numerous pollinator species as well as on male and female bees to varying degrees. Bee species that are specialists (oligoleges) gather pollen and nectar from a narrow range of related floral hosts, others are generalists (polyleges) visiting upwards of 100 - 200 plant species for food. Presumably among these, oligoleges are the most efficient pollinators due to close evolutionary relationships with their preferred floral hosts and associated foraging specializations. However, in some bee-plant mutualisms of economic importance, individual adult bees appear equitably efficient at pollinating their respective floral hosts despite difference in gender and level of sociality. For instance, male bees are often considered trivial pollinators because they gather no pollen and do not care for their offspring and so tend to be rare floral visitors. Male bees of some oligolectic species may be passive yet effective pollinators due to their hairiness and sheer abundance. In fields of summer squash, for instance, males and females of the oligolege *Peponapis pruinosa* (Say) are both efficient pollinators, their visits producing seeded *Cucurbita pepo* fruits of similar size. A 9:1 male-biased foraging density of *Peponapis pruinosa,* perhaps due to the males' permanent residence in a field, seemingly render male *Peponapis* as effective as or more effective than females (Cane et al. 2011). Polyleges such as honeybees can be efficient pollinators too, even those species that resort to flower robbery to obtain nectar. At rabbiteye blueberry blooms, *Vaccinium virgatum* Aiton, legitimate floral visits by oligolectic *Habropoda laboriosa* Fab., *Osmia ribifloris* (Cockerell), and polylectic honeybees *Apis mellifera* L. set similar fruit sets of ~40% (Sampson & Cane, 2000). Polylectic male carpenter bees (*Xylocopa* spp.) and honeybees were believed to circumvent pollination by robbing blueberry flowers of nectar through holes cut into the bases of corollas. Yet, floral robbers turned out to be efficient pollinators of blueberry. Two robber visits transferred as much compatible pollen as did single legitimate visits by oligolectic *H. laboriosa* (Sampson et al. 2004). Even exotic bees are effective pollinators of plants with which they share little or no evolutionary history. For example, an exotic generalist from Mexico, the orchid bee *Euglossa dilemma* Friese, is now naturalised in Florida and can outperform indigenous bee species as pollinators of invasive *Solanum* (Liu & Pemberton 2009). In the mutualism studied here, polylectic bumblebees (*Bombus* spp.) are ~5 time more efficient at contacting *Hibiscus* stigmas than are the plant's own oligoleges, mallow bees *Ptilothrix bombiformis* (Cresson) (Apidae: Emphorini), a difference likely due to the larger body size of *Bombus* workers (Willmer & Finlayson 2014).

From these examples, we posit similarly sized bees at least on an individual basis deposit similar amounts of pollen and set similar fruit and seed sets despite differences in gender, level of sociality, degree of floral host specialization, and geographic origin (Greenleaf & Kremen 2006; Matsuki

*Corresponding author: blair.sampson@ars.usda.gov

et al. 2008). However, the pollination effectiveness of any bee species is determined by a bee's average pollination efficiency, and more importantly, by overall foraging abundance (Winfree et al. 2014). Oligoleges in fact are often the most effective pollinators due to high foraging abundances maintained by a strong innate fidelity for host blooms. Males are often less fidelic than females are; however, they remain close to host blooms with the expectation of finding mates. At host blooms, male bees behave quite differently than females do. Therefore, their foraging movements, amorous approaches to females, and aggressive displays could have unintended consequences for pollination efficiency and rates of host plant outcrossing: negative or positive (Spira et al. 1992; Nagamitsu & Inouye 1997; Santa Anna-Aguayo et al. 2014).

The large herkogamous blooms of *Hibiscus* have a wide gap between their anthers and stigmas that large, hairy, and fast moving bees such as *P. bombiformis* periodically cross (Baker & Hurd 1968). Higher bee abundance could increase pollination of herkogamous flowers in two fundamental ways: first, through the simultaneous increase in floral traffic and stigmatic contact and second, and more intriguingly, by aggressive intrafloral interactions, which may stir greater pollinator movement. In effect, abundance and efficiency, the two factors of a pollinator's effectiveness, are not mutually exclusive. Pollinator abundance could enhance the pollination efficiencies of bee species by increasing incidences of agonistic interactions inside host flowers (Cane et al. 2011; Santa Anna-Aguayo et al. 2014). Therefore, we observed tussles between aggressive bees of a solitary oligolege *Ptilothrix bombiformis* and given the enormous amount of pollen they discharge, we posit that such pollinator aggression increases pollen transfer and seed set of their *Hibiscus* hosts.

Materials and Methods

We studied the *Ptilothrix-Hibiscus* mutualism because *P. bombiformis* are large abundant solitary bees (13 - 17 mm) and oligoleges that vector pure loads of pollen gathered from related species of Malvaceae, principally from wild and cultivated *Hibiscus* (Rust 1980). In the field, *Ptilothrix* males (Fig. 1A) are identifiable by distinctive patrol flights above flowers, which often involve darting in and out of blooms in search of receptive females. *Ptilothrix* females were identified by more methodical flight paths between *Hibiscus* flowers and by dense and often pollen-laden pollen brushes (scopae) on their hind legs (Fig. 1B). *Ptilothrix* at Mississippi *Hibiscus* came from natural nesting populations at two semi-urban/rural sites: Poplarville in 2009 and McNeill in 2014.

Hibiscus is an excellent species to study bee-plant interactions because plants produce daily 1 to 5 large, short-lived (1-day-old), herkogamous flowers that present bees with enormous spiny pollen grains (~150 - 200 µm) that are easily counted on stigmas with a low-power magnifying lens. Since counts of stigmatic pollen can be performed on intact pistils, we can draw a direct relationship between bee pollination efficiency (stigmatic pollen loads) and host reproductive success (seed set). Chosen for these studies were

FIGURE 1. Intrafloral movements of *Ptilothrix bombiformis.* (A) A tussle between two male bees. Audible buzzes and grappling quite pronounced. Note the field of scattered pollen grains on the petals left by the two squabbling males and the bees close proximity to the stigma. (B) Co-visitation by two *P. bombiformis* females without significant interaction.

three *Hibiscus* species, each grown from wild-collected seed: *H. laevis* All., *H. coccineus* (Mendik.) Walter, and *H. moscheutos* sbsp. *incanus* (J.C. Wendl.) Ahles. Plants of the first two species, *Hibiscus laevis* and *H. coccineus,* were irrigated daily in 12 L pots at the Poplarville site in 2009. Seeds of *Hibiscus moscheutos incanus* were sown along 75-m of unirrigated row at the McNeill site in 2014. Diploid *H. laevis* and *H. coccineus* (2n = 38 chromosomes, Group II *Hibiscus* sect. *Trionum*) were grown together on the same container pad because both species freely intercross and should not affect seed set of the other. Although, diploid *H. moscheutos incanus* (Group I, *Hibiscus* sect. *Trionum,* 2n = 38) has the same number of chromosomes as *H. laevis* and *H. coccineus,* the former's putative genetic incompatibility with Group II species required us to plant *H. moscheutos* elsewhere, so as to ensure optimal seed set in open pollination controls (Wise & Menzel 1971).

Lumite® mesh cages (1.5 m^3) excluded bees from flowers on four plants of each *Hibiscus* species. Cages were removed for 15 - 60 min between 0800 and 1015 h to expose newly open flowers to *P. bombiformis,* a species responsible for 98 - 99% of floral visits at both the Poplarville and McNeill study sites. Observations of floral visitors lasted 10 d in 2009 (21 - 31 Aug) and 1 month in 2014 (16 July - 14 August). Eleven levels of floral visitation were recognized in *Ptilothrix*: (1) no visits, (2) single male visits (3) single female visits, (4) two non-interacting males, (5) two non-interacting females, (6) a non-interacting male and female, (7) two interacting males, (8) two interacting females, (9) an interacting male and female, 10) near simultaneous visits by 3-5 (3+) bees without interaction, and (11) open pollination. Aggressive bees that come into physical contact often grappled, bit, and buzzed one another, behaviours hereafter referred to as "tussles". Non-interacting bees generally avoided each other in blooms and never made physical contact or acted in a belligerent fashion. After each *Ptilothrix* floral visit or visitation bout, a hands-free jeweller's loupe aided in counting pollen grains that accrued on each of *Hibiscus*' five or six stigmatic lobes. In an attempt to increase sample size, grass stems were tied around flaring petals of select *H. moscheutos incanus* flowers (petal angle: $\theta' = 17° - 23°$) to shape blooms into a narrower tube (petal angle: $\theta' = 0° - 10°$) that might bring *Ptilothrix* bees into closer contact with stigmas and incite agonistic intrafloral behaviour. Petal angle in degrees (θ') was calculated for radially symmetrical flowers of *Hibiscus* using the equation:

$$\theta' = tan^{-1}\left(\frac{Dc - Db}{2h}\right) \qquad 0° \leq \theta' \leq 90°$$

Where *Dc* is flower diameter at its aperture, *Db* is calyx diameter, and *h* is floral depth. θ' is derived from an imaginary right triangle with x = *h* where x extends perpendicular from the outer edge of the calyx to the flower aperture, and y = ½(*Dc* – *Db*) is the distance between the petal's outer edge and the point where x intersects y at a right angle.

After a single visit by a bee to a *H. laevis* or a *H. coccineus* bloom, or after 10 to 20 visits to each *H. moscheutos incanus* flower, paper price tags were tied to ovaries to flag seed capsules for harvest 7 - 30 days later. Cages were then replaced. A high proportion of seedless capsules in our *H. coccineus* population required the manual cross-pollination of additional flowers to determine full seed set in to rule out sterility in this species. Optimal seed set in the field for the other two species was considered to be analogous to sets acquired by open pollination whereby flowers during their entire life were freely accessible to pollinators. Likewise, optimal seed set in *H. moscheutos* could be estimated from a plot of cumulative seed set versus successive bee visitation. Field experiments were completely randomized (CRD) with bee visitation bouts and seed capsules serving as replicates. To deal with skewed data distributions containing a high proportion of zeros, non-parametric Kruskal-Wallis ANOVAs and multiple Wilcoxon two-sample exact tests were used to identify treatment differences among mean single-visit stigmatic loads in all three *Hibiscus* species and seed set in the type II *Hibiscus* species. Chi-square test identified any non-random patterns in the frequency of contact visits resulting from intrafloral tussles between *P. bombiformis*. Non-linear regression (PROC NLIN in SAS) estimated both *Ptilothrix* pollination efficiency and the least amount of stigmatic pollen needed for full seed set in *H. moscheutos incanus*.

Results and Discussion

Floral oligoleges are often touted as being highly efficient pollinators. Their numerous foraging specializations often favour rapid synchronous flight, copious pollen harvest, and strong host fidelity (Strickler 1979; Liu & Pemberton 2009). Synchronous flight enables bees to vector viable pollen to host stigmas during peak receptivity. Efficient pollen harvests result in the delivery of pure loads of compatible pollen. Strong host fidelities displayed by male and female oligoleges often increase bee foraging abundance, interfloral traffic, and thus pollination rates. Although these three fundamental adaptations of oligoleges theoretically favour high rates of host outcrossing, plants themselves have very different genetic interests than those of their pollinators. Therefore, if reproductively advantageous to do so, plants limit floral access to very specific and highly efficient pollinators by means of such floral adaptations as small stigmas, narrow corollas, long pistils, and changes in floral phenology (Harder and Aizen 2010; Sletvold et al. 2010).

Our studies as well as those reviewed by Larsson (2005) suggest that floral hosts for genetic reasons will evolve floral architectures that limit or promote pollinator efficiency, even the efficiencies of specialized floral oligoleges. As a result, many bees of similar size may be more equally efficient depending on how well they "fit" host floral morphology. For instance, unrelated bee taxa on a per bee basis appear to pollinate diverse plant taxa (e.g., *Vaccinium, Cucurbita, Solanum,* and *Hibiscus*) equally or nearly so (Spira et al. 1992; Sampson et al. 2000; 2004; Liu & Pemberton 2009; Cane et al. 2011). Efficiencies in fact may be lower for highly specialized bees that are capable of rapidly harvesting host pollen while minimizing pollen load loss by avoiding female parts of host flowers. This seems true of *Ptilothrix bombiformis*, a bee species that manages to contact host stigmas 2% - 3% of the time (Spira et al. 1992), hardly a proficiency one expects from such large, pilose, and specialised pollinators (Fig. 1). Clearly, the mating system of *Hibiscus* shares with blueberry a wide separation between stamens and pistils, which reduces the likelihood of geitonogamous mating (i.e., self-pollination, Sampson et al. 2013). *Hibiscus'* large 7 - 8 cm wide flowers increases the number of paths a bee can take to exit a flower without ever contacting a stigma. In such a case, a pollinator's size importantly affects pollination efficiency more than its degree of pollen specialisation. Similarly, at other floral hosts, the size of a bumblebee or a syrphid fly determines the ability of each insect to load plant stigmas with pollen. For each millimetre of intergular span, a worker bumblebee loads periwinkle (*Vinca minor*) stigmas with 10 additional pollen grains. Larger bumblebees and hoverflies were 4 to 5 times more efficient pollinators of blueweed (*Echium vulgare*) than were their smallest counterparts (Willmer & Finlayson 2014). In the bee-plant mutualism studied here, larger and more rotund bumblebees, though resembling *Ptilothrix*, transfer ~5 times more *Hibiscus* pollen (Spira et al. 1992). Bumblebees are not as effective as *P. bombiformis*, since they rarely visit *Hibiscus* flowers and then only for nectar.

Hibiscus plants throughout their range seem to limit the individual pollination efficiency of their principal pollinator, *P. bombiformis.* Using the pollination threshold provided by Spira et al. (1992), Mississippi and Maryland *Ptilothrix* deposit about 70 pollen grains with each successful contact with *Hibiscus* stigmas, which occurs only ~3% of the time. In Mississippi, 21% of contact visits by *P. bombiformis* each transferred ≥10 pollen grains, on average, a satisfactory quantity to accrue enough pollen for seed set and offset any pollen knocked off stigmas by visiting bees. The latter was a surprising finding; we had not expected that 7% of *Ptilothrix* visits each resulted in the removal of 1 - 10 stigmatic pollen grains. We chose to lower our pollination threshold by a factor of seven from that reported by Spira et al. (1992), as 75 - 100 grains per flower were sufficient for full seed set in two of our three floral host species (*H. moscheutos incanus* and *H. laevis*, (Figs 2 - 4). In contrast, optimal seed set in more northern populations of *H. moscheutos* reportedly require many more pollen grains for full seed set, 360 – 700 grains per stigma (Spira et al. 1992). This higher threshold could very well reflect a different phenotypic response to a colder climate, which may affect pollen availability, pollen viability, or ovule fertility (García et al., 2000). Accordingly, our revised pollination threshold for a contact visit at our Mississippi plantings is the transfer of 10+ pollen grains. Since full seed in *Hibiscus* requires multiple visitation, we estimate that *H. laevis* and *H. moscheutos* requires a minimum of 8 to 10 contact visits by *Ptilothrix* (Figs. 3 and 4). Estimating a similar pollination threshold for *Hibiscus coccineus* was difficult because of high degree of sterility in this population. Our *H. coccineus* plants produced very small capsules (~0.35 g) containing ~75% - 80% fewer seeds than the other two species despite our best efforts to manually pollinate stigmas (Fig. 3). Single visits by *Ptilothrix* females did produce seed sets equal to the optimal seed set for *H. coccineus* (i.e., 5 seeds, Fig. 3), confirming the higher pollination efficiencies of pollen-gathering female bees when compared with the efficiencies of male bees. However, possible sterility resulted in very poor seed set in *H. coccineus* and only manual pollination (saturated stigmatic pollen loads) produced capsules with a few small seeds (Fig. 3).

Although floral hosts sometimes limit bee pollination efficiency in order to prevent autogamous pollen transfer, there may be sufficient variability in a bee's pollination efficiency to result in pollen surpluses on stigmas (Spira et al. 1992). At our fields, *Ptilothrix* bees constituted 98 - 99% of floral visitors and were amply abundant to transfer stigmatic pollen loads at or above the 100-grain threshold for full seed set (Figs 2 - 4). Most *P. bombiformis* visits led to zero pollen deposition rates or, what was noteworthy, to the removal of stigmatic pollen grains 7% of the time. Although pollination efficiencies of single visits by both male and female *Ptilothrix* were low and sometimes negative, efficiency was quite a dynamic pollinator trait. For instance, *P. bombiformis* behaviours such as foraging and courtship and to a lesser extent, floral traits such as petal closure and stigmatic area, increased rates of host pollen deposition by 30% - 1000%. In solid stands of our *H. laevis* and *H. coccineus* for instance, 23% - 29% of visits transfer 10+

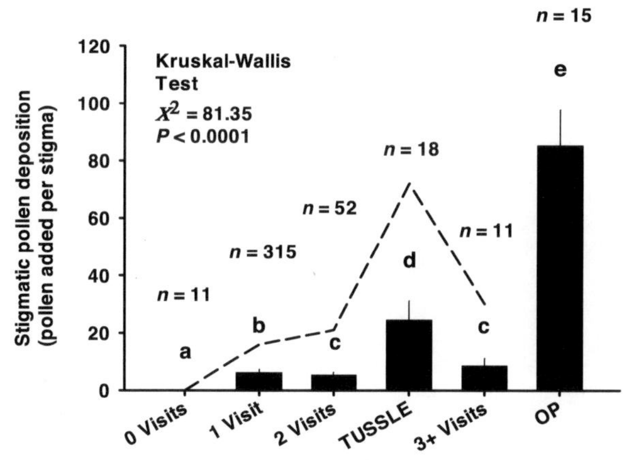

FIGURE 2. Net stigmatic pollen loads transferred to three species of *Hibiscus.* Flowers after tussles between 2 bees, no visits, single visits, double visits, triple, or near-simultaneous visits, and open pollination (OP) by *Ptilothrix bombiformis.* Black bars represent mean stigmatic pollen deposition including 1±SE. Sample sizes are shown above bars and mean differences denoted by different letters according to multiple Wilcoxon 2-sample exact tests. The dashed line above the first 4 bars indicates the percentage of visits that transferred 10+ pollen grains onto *Hibiscus* stigmas ("contact visits").

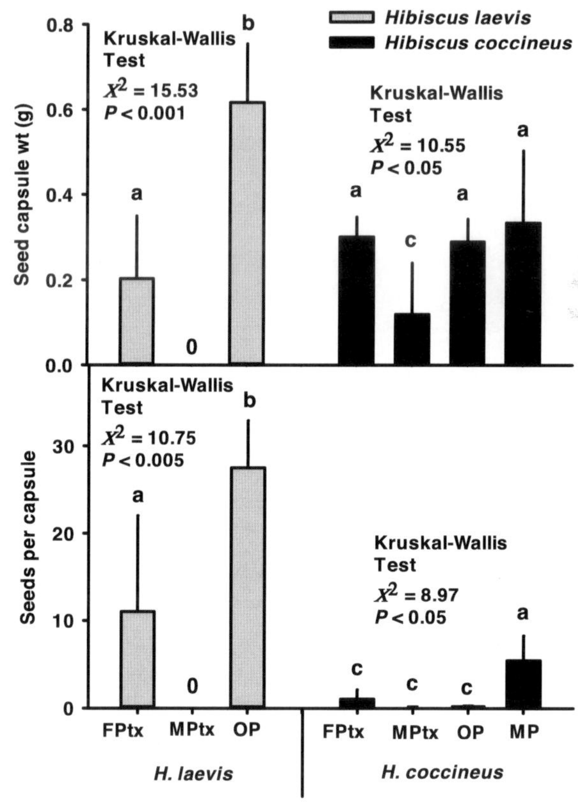

FIGURE 3. Comparative pollination efficiencies of single contact visits by male *Ptilothrix bombiformis* (MPtx) and females (FPtx) in the 2009 study based on seed capsule weight and seed set (seeds per capsule) of *Hibiscus laevis* and *H. coccineus.* For comparison, some flowers were openly pollinated (OP) and manually pollinated (MP) with stigmas saturated with compatible pollen (*H. coccineus* only). Bars (+SE) denote mean stigmatic pollen loading per visit. Different letters above the bars indicate means that were different according to multiple comparisons of rank scores using Wilcoxon two-sample exact tests across the two Type II species of *Hibiscus*: *H. laevis* and *H. coccineus.*

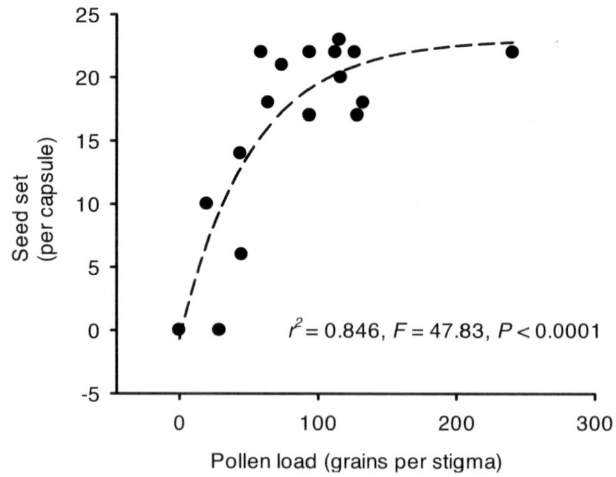

FIGURE 4. Effect of *Ptilothrix bombiformis* pollination efficiency on *H. moscheutos incanus* seed set in the 2014 study. Line shows a strong curvilinear response of host reproductive output to stigmatic pollen loading by visiting *P. bombiformis*.

pollen grains to host stigmas. These two species also have flowers bearing larger upturned stigmas. In contrast, 19% of visits effect significant pollen transfer (10+ grains) in a floral host such as *Hibiscus moscheutos incanus* that has comparatively straighter pistils and smaller stigmatic lobes (Fig. 1). Aging *Hibiscus* flowers also furl petals, which bring bees into greater contact with stigmas. A reduction in petal angle ($\Delta\theta'$ = -13° to -17°) enhanced *P. bombiformis* efficiency 30% from 5.39 ± 0.75 grains per visit (N = 253 bee visits) to 7.75 ± 1.68 grains per visit (N = 47 bee visits, $\chi^2 = 5.24$, df = 1, $P < 0.05$). Such a link between host floral morphology, pollination efficiency, and large pollen loads might be advantageous to plants such as *Hibiscus* whose vigorous offspring are products of intense pollen-tube competition (Spira et al. 1992). Thus, selection favouring dynamic floral traits may have encouraged rarer bee visits that efficiently and simultaneously transfer substantial quantities of multiple pollen genotypes.

Although, single floral visits by *Ptilothrix* appear rather inefficient, when two bees enter flowers and interact physically, the scenario changes entirely. Two *P. bombiformis* by interacting inside blooms mutually increase their pollination efficiency by an astonishing 300% – 1000%. Hence, bee intrafloral behaviour shaped by host floral morphology exerted the greatest influence on this bee's pollination efficiency. As expected, single visits by pollen-laden females loaded stigmas with twice as much pollen as male visits had (Table 1); a sufficient quantity to initiate seed set in fertile *H. laevis* and largely infertile *H. coccineus* (Fig. 3). However, sequential visitation by *Ptilothrix* bees does not guarantee the accrual of stigmatic pollen. Field observations in Mississippi concur with those of Spira et al (1992) in Maryland, 1 to 5 sequential visits by *P. bombiformis* to *H. moscheutos* flowers were not additive in their ability to accrue pollen onto stigmas (Fig. 2). Bees must strike the stigma to effect pollination, a rather rare event. Unexpectedly however, 7 in 10 floral visits involving two tussling *P. bombiformis* contacted stigmas and transferred compatible pollen loads. Tussles between *Ptilothrix* bees of either gender involve bumping, lunging, biting, grappling, and tumbling around in host flowers (Fig. 1A), agonistic behaviours that increased rates of stigmatic contact from 20% – 30% to 70% (Fig. 2) and transfer rates 11-fold from 3 to 34 pollen grains per bee visit (Table 1). Aggressive male bees often tussle with rival males in flowers or struggle with unreceptive pollen-foraging females during copulation attempts (Santa Anna-Aguayo et al. 2014). Females accosted by males in this way often emit audible distress buzzes from inside flowers. Such tussles while quite rare (5% of floral visits) accounted for 18% of contact visits ($\chi^2 = 94.72$, df = 2, $P < 0.0001$, Table 1). Copulation attempts by bees inside flowers resulted in a doubling of stigmatic pollen transfer. In

TABLE 1. Mean amount of pollen transferred per visit (Mean ± SE) to *Hibiscus* stigmas by *Ptilothrix bombiformis* that interact (contact) or do not (no contact). Recognized visitation interactions include 1) no visits, virgin flowers, 2) single visits by male bees, 3) single visits by female bees, 4) almost simultaneous visits by two males, male and female, or two females, and 5) almost simultaneous visits by two males, two females, or a male and a female that led to a tussle inside a flower. Different letters indicate significantly different means according to multiple Wilcoxon 2-sample exact tests.

Mean ± SE stigmatic pollen loads per bout Co-visitor	% all visits or bouts (% all contact visits, ≥ 10 pollen grains per bout)	Co-visitor None	 Male	 Female
None	81% (67%)	0 ± 0a (n = 11)	3.23 ± 0.91b (n = 90)	7.37 ± 1.59c (n = 225)
Male (no contact)	14% (15%)		3.00 ± 1.24b (n = 10)	5.93 ± 2.24b (n = 15)
Female (no contact)				5.89 ± 1.52b (n = 27)
Male (contact, tussle)	5% (18%)		34.5 ± 18.50* (n = 2)	13.3 ± 3.10d (n = 8)
Female (contact, tussle)				33.3 ± 13.6d (n = 8)

fact, *Ptilothrix* females on average transfer 5 times more pollen after tussling with males and other females. Tussling bees improve pollen loading in three ways. First, a withdrawing bee, once disturbed, will often take a rather chaotic spiral flight path out of a flower, which guides it closer to the stigma (Boyadzhiev 1999). Second, two grappling bees often lose traction and become a tumbling ball, which effectively doubles pollinator size and hence the probability of stigmatic contact. Third, pollen grains ejected from the bodies of tussling bees will settle onto host stigmas (Fig. 1A). These three outcomes together may lead to huge surpluses of stigmatic pollen, which in non-pollen-limiting species of *Hibiscus*, intensifies pollen-tube competition, leads to non-random mating, and thereby enhances paternal reproductive success (Spira et al. 1992).

Higher abundances of both male and female oligoleges enhance pollination efficiencies in unexpected ways. Aggressive bees interacting with one another inside flowers during foraging trips and courtship can transfer more host pollen to stigmas. While *Ptilothrix* males pollinate *Hibiscus* less efficiently than females do, they make up for a low individual efficiency with a greater collective efficiency. Male antagonism toward rival males and attempted copula with females can incite enough chaos inside flowers to triple stigmatic pollen transfer and contribute enough pollen grains to sire one-fifth of a host plant's seeds. Moreover, the furling petals of aging *Hibiscus* blooms may further enhance pollination efficiency by coaxing *Ptilothrix* to fly into stigmas. Greater bee abundance and higher species diversity may collectively enhance host pollination through higher incidences of aggression between floral visitors. Therefore, the gestalt concept of "the whole is greater than the sum of its parts" also seems applicable to bee pollination. Pollination is not simply the gradual linear accrual of pollen, but is punctuated with intense pollen loading wrought by pollinators interacting inside flowers, sometimes violently.

Acknowledgements

We thank Reggie Davis, Robin Hayes, Timothy Swain, and Joseph Anderson for their field assistance. We are especially grateful to the United States Department of Agriculture Pathways Program for providing student training and assistance.

References

Baker HG, Hurd PD (1968) Intrafloral ecology. Annual Review of Entomology 13:385-414.

Boyadzhiev KN (1999) Spirals and conchospirals in flight of insects. College Math Journal 30:23-31.

Cane JH, Sampson BJ, Miller SA (2011) Pollination value of male bees: the specialist bee *Peponapis pruinosa* (Apidae) at summer squash (*Cucurbita pepo*). Environmental Entomology 40:614-620.

García D, Zamora R, Gómez JM, Jordano P, Hódar JA (2000) Geographical variation in seed production, predation, and abortion in *Juniperus communis* throughout its range in Europe. Journal of Ecology 88 436-446.

Greenleaf SS, Kremen C (2006) Wild bees enhance honeybees' pollination of hybrid sunflower. Proceedings of the National Academy of Sciences 103:13890-13895.

Harder LD, Aizen MA (2010) Floral adaptation and diversification under pollen limitation. Philosophical transactions B DOI: 10.1098/rstb.2009.0226.

Larsson M (2005) Higher pollinator effectiveness by specialist than generalist flower-visitors of unspecialized *Knautia arvensis* (Dipsacaceae). Oecologia 146:394-403.

Liu H, Pemberton RW (2009) Solitary invasive orchid bee outperforms co-occurring native bees to promote fruit set of on invasive *Solanum*. Oecologia 159:515-525.

Matsuki,Y, Tateno R, Shibata M, Isagi Y (2008) Pollination efficiencies of flower-visiting insects as determined by direct genetic analysis of pollen origin. American Journal of Botany 95:925-930.

Nagamitsu T, Inoue T (1997) Aggressive foraging of social bees as a mechanism of floral resource partitioning in an Asian tropical rainforest. Oecologia 110:432-439.

Rust RW (1980) The biology of *Ptilothrix bombiformis* (Hymenoptera: Anthophoridae). Journal of the Kansas Entomological Society 53:427-436.

Sampson BJ, Cane JH (2000) Pollination efficiencies of three bee (Hymenoptera: Apoidea) species visiting rabbiteye blueberry. Journal of Economic Entomology 93:1726-1731.

Sampson BJ, Danka RG, Stringer SJ (2004) Nectar robbery by bees *Xylocopa virginica* and *Apis mellifera* contributes to the pollination of rabbiteye blueberry. Journal of Economic Entomology 97:735-740.

Sampson BJ, Stringer SJ, Marshall DA (2013) Blueberry floral attributes and their effect on the pollination efficiency of an oligolectic bee, *Osmia ribifloris* Cockerell (Megachilidae: Apoidea). HortScience 48:136-142.

Santa Anna-Aguayo AI, Schaffner CM, Golubov J, Garcia-Franco J, Martínez AJ (2014) Behavioral repertoires and interactions between *Apis mellifera* (Hymenoptera: Apidae) and the native bee *Lithurgus littoralis* (Hymenoptera: Megachilidae) in flowers of *Opuntia huajuapensis* (Caryophyllales: Cactaceae) in the Tehuacan Desert. Florida Entomologist manuscript *in press.*

Sletvold N, Grindeland M, Ågren J (2010) Pollinator-mediated selection on floral display, spur length and flowering phenology in the deceptive orchid *Dactylorhiza lapponica*. New Phytologist 188:385-392.

Spira, TP, Snow AA, Whigham DF, Leak J (1992) Flower visitation, pollen deposition, and pollen-tube competition in *Hibiscus moscheutos* (Malvaceae). American Journal of Botany 79:428-433.

Strickler K (1979) Specialization and foraging efficiencies of solitary bees. Ecology 60:998-1009.

Vásquez DP, Aizen MA (2004) Asymmetric specialization: a pervasive feature of plant-pollinator interactions. Ecology 85:1251-1257.

Willmer PG, Finlayson K (2014) Big bees do a better job: interspecific size variation influences pollination effectiveness. Journal of Pollination Ecology 14:244-254.

Winfree R, Williams NM, Dushoff J, Kremen C (2014) Species abundance, not diet breadth, drives the persistence of the most linked pollinators as plant-pollinator networks Disassemble. American Naturalist 185:600-611.

Wise DA, Menzel MY (1971) Genetic affinities of the North American species of *Hibiscus* Sect. *Trionum*. Brittonia 23:425-437

8

FLORAL SCENT AND POLLINATORS OF THE HOLOPARASITE *PILOSTYLES THURBERI* (APODANTHACEAE)

Sedonia D. Sipes*[1], Kara E. Huff Hartz[2], Hardik Amin[2], Aldwin Anterola[1], and Daniel L. Nickrent[1]

[1]*Department of Plant Biology Southern Illinois University Carbondale Illinois 62901-6509*
[2]*Department of Chemistry and Biochemistry Southern Illinois University Carbondale Illinois 62901-4409*

Abstract—Floral scent is likely important to the pollination of parasitic plants, despite that it has not been well-studied. We studied the pollination ecology of the North American stem holoparasite *Pilostyles thurberi* (Apodanthaceae) at two field sites in Texas. To identify effective pollinators, we collected floral visitors to *P. thurberi* flowers, observed their foraging behavior, and looked for *P. thurberi* pollen on their bodies. *Augochloropsis metallica* bees (Halictidae) and eumenine potter wasps (Vespidae) were pollinators. *P. thurberi* flowers are visually inconspicuous but produce a strong fruity fragrance. GC/MS analysis of whole floral extracts and dynamic headspace samples revealed the fragrance to be an unusually simple bouquet of raspberry ketone and several eugenols. Comparison of scent profiles to those from uninfected host plants (*Dalea formosa*) allowed putative separation of parasite and host volatiles. This is the first report of the constituents of floral fragrance in Apodanthaceae.

Keywords: Floral scent, dioecy, endoparasite, bee pollination, raspberry ketone, eugenol

INTRODUCTION

Darwin (1859) coined the term "tangled bank" to refer to the intricate networks present among flowering plants and other species in their communities, interactions that include both mutualists and antagonists. Owing to their direct physical connection to another plant, parasitic angiosperms present additional life history complexities. Recently, reproductive traits were reviewed across all parasitic plants (Bellot & Renner 2013) and two conclusions emerged: about 10% of parasitic plants are dioecious (higher than the angiosperm average of 6%; Renner and Ricklefs 1995) and the vast majority of parasitic plants are pollinated by animals. Across the 12 clades of parasitic plants the most frequent pollinating agents are insects such as beetles, butterflies, bees, wasps, and flies.

This paper addresses pollinators and floral scent in the holoparasite *Pilostyles* Guillemin (Apodanthaceae), a genus of ca. 14 species with widely disjunct distributions from North to South America, western Asia, and Australia. If one includes *Berlinianche* within *Pilostyles*, the distribution also includes tropical Africa. The holoparasites of Apodanthaceae were formerly included within Rafflesiaceae, which was shown via molecular phylogenetic analyses to be a polyphyletic assemblage of parasites belonging to four different orders (Nickrent et al. 2004). Apodanthaceae was first shown to be a member of Cucurbitales by Nickrent et al. (2004), later confirmed by Filipowicz and Renner (2010). Apodanthaceae are dioecious (and some possibly monoecious) endoparasites whose only externally visible parts are their unisexual flowers.

Flower visitors have been recorded for several species in Apodanthaceae, although in many instances, the efficacy of these as pollinators has not been determined. *Trigona* spp. (stingless honeybees) have been recorded visiting flowers of the neotropical genus *Apodanthes* in Costa Rica (Gomez 1983). In Panama, Croat (1978) reported bees, small butterflies and even mosquitoes visiting *Apodanthes caseariae*. *Pilostyles hamiltonii* in Australia is possibly wasp pollinated (Dell & Burbidge 1981) whereas in the South American *P. berteroi* unidentified solitary bees are reported to be pollinators (Kummerow 1962). Recently, Bellot & Renner (2013) recorded floral visitors on flowers of *Pilostyles haussknechtii* in Iran and *P.* [*Berlinianche*] *aethiopica* in Zimbabwe. In addition to ants, which are unlikely pollinators, calliphorid flies visited *P. aethiopica* and ulidiid flies visited *P. haussknechtii*; the fly taxa were determined to be effective pollinators based on their foraging behavior and pollen carriage.

Despite this diversity of pollinators documented for *Pilostyles* spp., the pollination relationships of these parasitic plants may be constrained by the limited visual cues produced by the plant: the flowers are small (< 2 mm) and inconspicuous due to their placement on interior branches of the host plant. We hypothesize that, for plants such as these, chemical cues may be especially important in attracting pollinators. At the same time, floral chemistry in stem parasites may be constrained by a number of factors related to their parasitic lifestyle, such as influences of host chemistry and overall anatomical and morphological reductions (Kuijt 1969).

Floral scents are chemically complex cues that are involved in pollination (Vainstein et al. 2001, Raguso

*Corresponding Author; email ssipes@plant.siu.edu

2008), and major contributions to pollination ecology have been made with the analysis of floral fragrances using techniques such as dynamic headspace sampling combined with gas chromatography and mass spectrometry (Raguso & Pellmyr 1998, Tholl & Röse 2006). To date, only subjective descriptions (i.e. by the human nose) have been recorded for members of Apodanthaceae, despite that floral scent may be of special importance in these plants. Here we report observations and identities of floral visitors, assessment of their potential as pollinators, and analysis of floral scent chemistry of *P. thurberi* Gray, a parasite of *Dalea* and *Psorothamnus* species in the southwestern United States.

MATERIALS AND METHODS

Study sites

We conducted field studies June 23-26, 2007 at two populations in north central Texas: one near Post in Garza county, and one near Benjamin in Knox county. At these sites, the host plant of *Pilostyles thurberi* is *Dalea formosa* Torr. Both locations were semi-arid rocky escarpments with a plant community dominated by *Juniperus ashei*, *Chrysothamnus nauseosus*, *Prosopis glandulosa*, and *D. formosa*. *P. thurberi* flowers in early summer at these sites, and its flowering does not overlap with that of its host. The Post site was at or near peak flower (most infected plants with ~10-20 open *P. thurberi* flowers), while the Benjamin site was earlier in relative phenology (about half the infected plants had some open *P. thurberi* flowers while the remainder had unopened buds). Population size is difficult to determine precisely because endophytes may be present in host plants but not visible prior to flowering. Based on hosts with visible *P. thurberi* flowers/buds, we estimate that at least 100 infected host plants were present at both sites.

Assessment of floral visitors

We observed and collected floral visitors throughout the day for 1-2 sunny days at each site. During these days, visitor observations and collections were conducted between the hours of 9:00 and 16:00 by 1-3 observers. Observations filled all available time within these hours except the times required to start/stop the scent collections. Observations consisted of at least one person walking slowly among infected host plants looking for visitors, and one person sitting near an infected host plant for ~15-30 minute intervals watching carefully for visitors to *P. thurberi* flowers. To discriminate between pollinators and non-pollinating visitors, the behavior and constancy of insects were noted in the field. Collected insects were examined for placement of *P. thurberi* pollen on the body. These slides were compared to voucher slides of *P. thurberi* pollen. Floral visitors were

FIGURE 1. A) Male flowers of *Pilostyles thurberi* emerging from the stem of *Dalea formosa*. B) Male flower intact (left), flower with petals removed to show central column with anthers dehiscing pollen (center), and female flower in longitudinal section showing ovules (right). C) Dynamic headspace sampling apparatus. D) A male *Augochloropsis metallica* bee (dead) posed on a *Dalea formosa* stem.

identified by SDS and deposited in the Southern Illinois Pollinating Insect Collection, Department of Plant Biology, SIUC.

Analysis of floral scent

Dynamic headspace sampling was used to sample floral scents at the Post and Benjamin study sites, following established protocols for collection of volatiles (Raguso & Pellmyr 1998). We collected scent samples from six male and four female flower clusters, each on a different host individual. Flower clusters were bagged using Reynolds oven bags (a clean, inexpensive plastic that emits minimal volatiles) cut to size and formed into tubes using a food sealer. These tubes were slipped over infected *D. formosa* stems with *P. thurberi* flower clusters and sealed on each end with twist ties (Fig.1C). Leaves of the host *D. formosa* were excluded (trimming when necessary) from the bag to minimize the amount of host volatiles collected. Volatiles were collected for 1 hr using Supelco Personal Air Samplers (PAS-500) calibrated to 200 ml/min flow rate, and scent traps made from 10 mg Super Q adsorbent (Alltech Associates, Deerfield, IL) placed in glass pipettes and sandwiched between small balls of glass wool. Samples were eluted with 300 µl of hexane in the field and stored on ice until they were transported back to SIUC, where they were stored at -80°C until analysis. To help distinguish volatiles produced by *P. thurberi* from those of its host, or other environmental contaminants, we collected two ambient samples as well as scents from leaves of five putatively uninfected *D. formosa* individuals. A *D. formosa* individual was considered putatively uninfected if it had neither *P. thurberi* flowers nor flower scars on its basal stems from previous years' flowering. The flowering time of *D. formosa* does not overlap with that of *P. thurberi*, so we needed only to sample vegetative headspace from the former as a control. We collected and analyzed headspace scent from a total of 10 stems with *P. thurberi* flowers (6 with male flowers and 4 stems with female flowers), 2 uninfected *D. formosa* individuals, and 2 ambient controls.

In addition to the head space samples, whole flower extracts were also analyzed by gas chromatography/mass spectrometry (GC/MS). The whole flower extracts were obtained by sonicating 100 mg of air dried flowers (which were still strongly scented), with 1.0 ml of *n*-hexane for 5 minutes. The extract was analyzed by GC/MS.

In the lab, both the whole flower extracts and the headspace samples were concentrated down to a volume of approximately 100 µl using nitrogen gas, and analyzed using a cylindrical ion trap MS (Saturn 2100T, Varian) equipped with a 3900 GC. 1.0 to 3.0 µl of each scent extract was injected and analyzed using a method adapted from Adams (2007). Samples were injected in splitless mode with an injector temperature of 220°C. The extracts were separated using a capillary column (5% phenyl 95% dimethylpolysiloxane, VF-5MS, Varian Factor Four, 30 m x 0.25 mm x 0.25 µm) at a flow rate of 1.0 ml/min using helium (Airgas Mid America, 99.999%) as carrier gas. The initial oven temperature was 60°C, and the temperature was ramped at a rate of 3°C /min to a final temperature of 246°C, giving a total analysis time of 62 minutes. The analytes were detected in the electron ionization (70 eV) mode scanning from 10-650 m/z. The temperature parameters for the mass spectrometer were as follows: trap temperature: 170°C, manifold temperature: 40°C and transfer line temperature: 240°C. The GCMS data were collected and processed using the Varian Workstation software (ver. 6.9).

To identify peaks originating from *P. thurberi* floral scents, the resulting chromatograms of the extracts were compared to chromatograms of *n*-hexane solvent blanks, ambient samples collected from the field sites, and headspace samples from vegetative branches of uninfected *D. formosa*. Only peaks with a signal-to-noise ratio greater than 3 were considered, and selective ion monitoring was used to reduce the background noise level. The analytes were identified by retention time and mass spectrum in comparison to authentic standards. Because we replicated the analytical conditions of Adams (2007), we used Adams's (2007) retention times and an NIST05 library search of mass spectra when authentic standards were not available.

Several polar compounds that were prominently present in the whole flower extracts (eugenols and raspberry ketone) were not detected in the headspace samples. To confirm the presence of these, we split some of the samples that had sufficient amount left after analysis (1 whole flower extract, 3 headspace samples from male flowers, 3 headspace samples from female flowers, and 2 samples from uninfected *D. formosa*) and derivatized the second fraction in an attempt to improve detection of these polar compounds. Derivatization is routinely used to improve GC/MS detection of alcohols (e.g. Casaña-Giner et al. 2003). The second fraction was treated with N,O-*bis*(trimethylsilyl)trifluoroacetamide (BSTFA) which replaces the –OH moiety with $-OSi(Me)_3$. This derivatization reaction converts phenol and alcohol groups to trimethylsilyl ethers and improves the peak efficiencies and the detection limits of the analytes with an –OH moiety. To derivatize the extracts, we added 125 µl of the pre-dried extract (using anhydrous Na_2SO_4, Sigma) to 100 µl of BSTFA in 1% trimethylchlorosilane (Thermo Pierce) and 15 µl of pyridine (99%, ACS grade, ACROS). The solution was mixed and incubated at 65 °C for 2 hours and analyzed within 24 hours after derivatization. The chromatograms for the silylated headspace and whole flower *P. thurberi* samples were compared to silylated *n*-hexane blanks for identification, as well as a silylated authentic standard of raspberry ketone (Sigma).

RESULTS

Assessment of floral visitors

Both male and female *P. thurberi* flowers were visited by male *Augochloropsis metallica* bees (Halictidae) (Fig. 1D) as well as several unidentified taxa of potter wasps (Eumeninae: Vespidae). We collected four *A. metallica* and two potter wasps at Post, and one *A. metallica* and three potter wasps at Benjamin; however, more individuals were observed than were collected. *A. metallica* especially was difficult to collect

TABLE 1. Floral scent compounds detected in *P. thurberi* whole flower extract. Compounds were identified by comparison to authentic standards.

Retention time, minutes	Chemical Species
21.85	Eugenol
27.93	Methyl isoeugenol
30.10	Raspberry ketone

with nets, given the location of the *P. thurberi* flowers within the host plant branches. The individuals collected were taken using a double armed glass flask with tubing attached to both arms; one tube was used to aspirate the pollinators into the flask. These taxa were observed at both sites collecting nectar from flowers of both sexes and appeared to be foraging with constancy on *P. thurberi* (i.e. visiting numerous *P. thurberi* flowers sequentially). Small amounts of *P. thurberi* pollen were carried on the ventral surface of their bodies, which would potentially contact the stigmatic surface of female flowers as the bees crawled across the flower clusters. Therefore, both taxa are probably serving as pollinators of *P. thurberi*. We did not observe these visitors go to flowers of any other plant species, and we did not detect pollen of other species in the samples taken from these visitors.

Analysis of floral scent

Both fresh and dried flowers of *P. thurberi* had a strong fruity scent similar to raspberries. Constituents of floral scent in *P. thurberi* detected by GC/MS included raspberry ketone (4-(4-hydroxyphenyl)-2-butanone), eugenol, methyl isoeugenol, and isoeugenol isomers; these were all found from both male and female flowers (Tables 1-2, Figs. 2-4). Raspberry ketone and the eugenols were detected by GC/MS analysis in whole-flower extraction but not in the underivatized headspace samples.

Derivatization greatly improved detection of the eugenols in the headspace and whole flower samples. One peak (RT 25.5 min.) was found in each scent trap extract and *Pilostyles* extract. A NIST database search showed a good match to isoeugenol trimethylsilyl ether. However, isoeugenol has two isomers (*cis*- and *trans*-) as well as a structural isomer (eugenol) and two other peaks (27.5 min. and 29.5 min.) with similar mass spectra were observed in some of the other scent trap extracts. Pure authentic standards of isolated eugenol isomers (i.e. eugenol, *cis*-isoeugenol, and *trans*-isoeugenol) were not available, so we were not able to determine by comparison which isomer corresponded to each peak. However, based on the known relative order of retention times of silylated eugenol isomers (Saitta et al. 2009), it is most likely that the first peak is eugenol, the second peak is *cis*-isoeugenol and the third peak is *trans*-isoeugenol.

The trimethylsilyl ether of raspberry ketone, identified by comparison of the retention time and mass spectrum to the derivatized authentic standard, was detected in the whole flower derivatized sample (Table 2, Figs. 3-4), but not in the derivatized (or underivatized) headspace samples. Headspace sampled in the lab from dried whole flowers also lacked raspberry ketone, suggesting that this compound was present in *P. thurberi* flowers but either not captured by or not eluted from our scent traps.

Derivatized samples also contained a peak (RT 36.7 minutes) putatively identified via NIST database match as hydrocinnamic acid *p*-(trimethylsiloxy)-, trimethylsilyl ester, which was not detected in any of the underivatized samples. We did not detect any eugenol-related compounds, raspberry ketone, or hydrocinnamic acid in uninfected *D. formosa* headspace samples, suggesting that these compounds do indeed originated from *P. thurberi*.

Headspace samples also contained a suite of other compounds, including numerous terpenoids (Table 3). Limonene and α-pinene were the most common monoterpenes and these were found in each of the extracts that contained monoterpenes. Camphene, α-thujene, β-pinene, and a monoterpene ester, bornyl acetate, were found in nearly all of the headspace samples of *P. thurberi* and uninfected *D. formosa*. Additionally, 22 sesquiterpenoid peaks were found in the 20-35 minute time frame by their characteristic fragment ions (m/z 105, 161), and molecular weight (204). Authentic standards for all potential NIST05 matches in this time range were not available to confirm their identity. However, we were able to assign putative IDs to these by comparing NIST05 matches, Adams's (2007) retention times, and a quantitative list of compounds found in *Dalea formosa* by Lucero et al. (2005). The monoterpene and sesquiterpene peaks were found in the uninfected *D. formosa* headspace samples as well as some of the *P. thurberi*

TABLE 2. Compounds found in derivatized (silylated) headspace samples and whole flower extract from *P. thurberi*. Letters in parentheses correspond to labeled peaks and mass spectra in Figs. 3 and 4. None of these compounds were detected in the uninfected *D. formosa* vegetative samples. [1]Putative identification by comparison to retention times in Saitta et al 2009. [2]Identification by comparison to authentic standard. [3]Putative identification using NIST05.

Retention Time, minutes	Chemical Species	# Samples Detected
17.6	Unidentified compound MW 192 (A)	3 male headspace, whole flower extract
25.5	Eugenol trimethylsilyl ether[1] (B)	3 male & 3 female headspace, whole flower extract
27.5	*cis*-Isoeugenol trimethylsilyl ether[1] (C)	1 male headspace
29.5	*trans*-Isoeugenol trimethylsilyl ether[1] (D)	2 male & 3 female headspace
31.32	Raspberry ketone trimethylsilyl ether[2] (E)	Whole flower extract only
36.65	Hydrocinnamic acid, *p*-(trimethylsiloxy)-, trimethylsilyl ester[3] (F)	Whole flower extract only

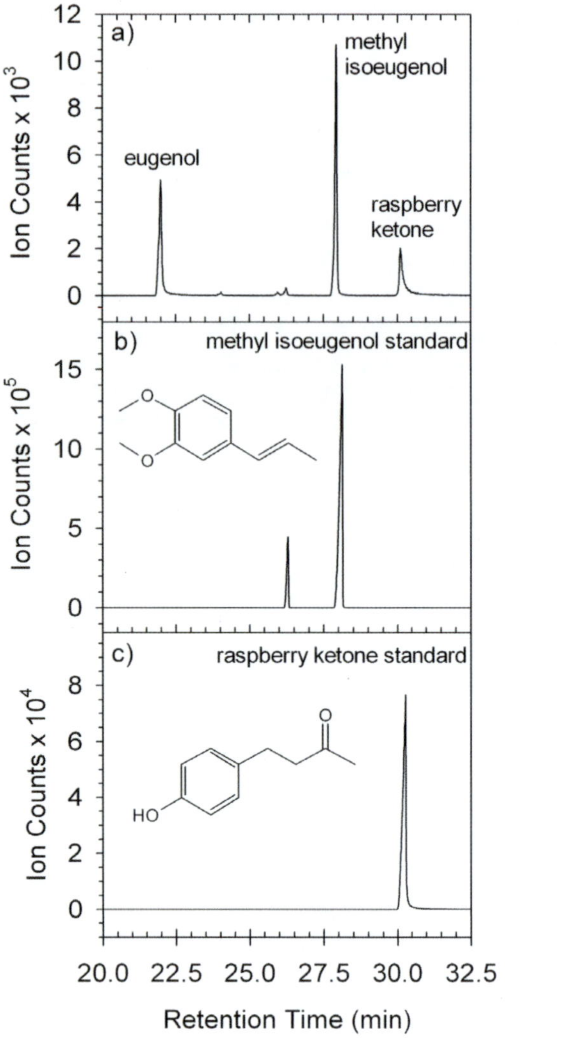

FIGURE 2. Gas chromatogram of a) underivatized whole flower extract of P. thurberi, b) methyl isoeugenol standard, and c) raspberry ketone standard. Chromatogram is filtered to show only 164 and 178 ions in order to remove several artifactual peaks arising from siloxane contaminants.

headspace samples, but were not detected in the *P. thurberi* whole flower extracts.

Discussion

The foraging behavior of both *Augochloropsis metallica* and the eumenine wasps, along with the placement of pollen on their bodies suggests that these visitors are effective pollinators of *P. thurberi.* Additionally, both were foraging with at least temporary constancy on *P. thurberi* during our observation times. Most eumenine wasps are predators of lepidopteran larvae (O'Neill 2001). The eumenines at our study sites appeared to be visiting *P. thurberi* opportunistically while searching the host plant *D. formosa* for prey. If the wasps' preferred prey are host-specific herbivores of *Dalea,* the wasps may be consistent visitors to *P. thurberi* simply because these flowers are convenient nectar sources in close vicinity to their prey. This multi-trophic interaction is worthy of more study: if *P. thurberi* flowers attract predators of *D. formosa* herbivores, infected host plants could actually experience increased fitness, much like plants that attract ant guards by offering extrafloral nectaries (Beattie & Hughes 2002). Under such a scenario, the parasite could potentially function either as an antagonist, commensalist, or mutualist of its host plant, depending on the abundance of herbivores and herbivore predators in the community.

Augochloropsis metallica a polylectic bee, i.e. females collect pollen provisions for larvae from many plant species (Hurd 1979). Because we did not observe or collect any foraging *A. metallica* females, we do not know if *P. thurberi* is used as a pollen host by this bee species. The male bees we observed were collecting nectar. However, even though male bees do not collect pollen for nest provisioning, they are often as good or better pollinators than female bees (e.g. McIntosh 2005). Female bees groom their bodies during foraging, moving pollen from various body parts to the hind leg or abdominal scopae for transport back to the nest or

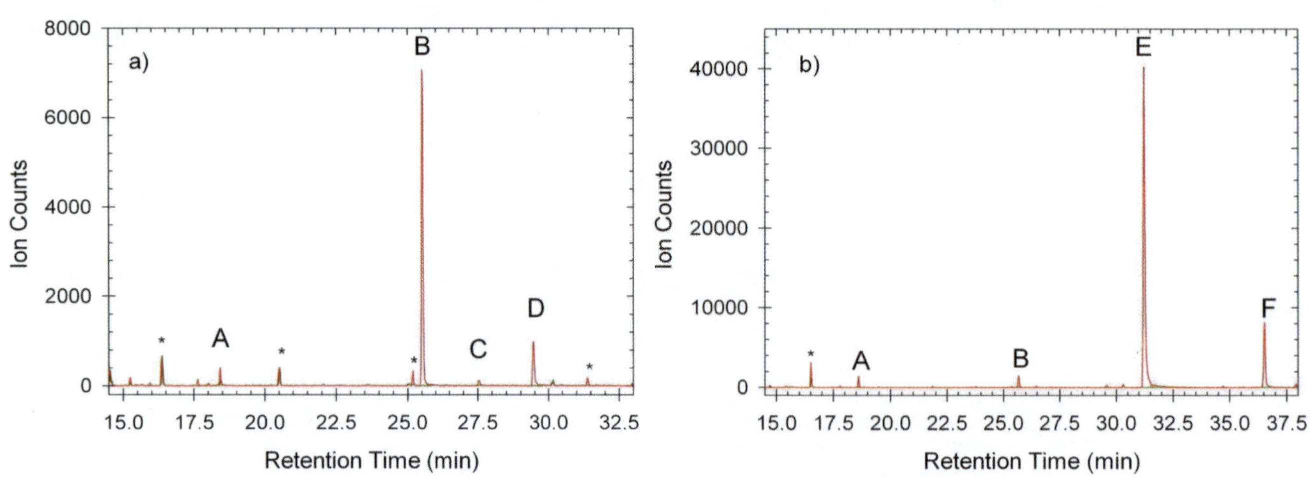

FIGURE 3. Chromatograms from derivatized samples of a) headspace sample from male *P. thurberi*) and whole flower *P. thurberi,* showing ions m/z 179 and 236. Letters correspond to putative identifications listed in Table 2. Floral samples shown in red and derivatized hexane blank in green. Peaks marked by * are contaminants due to the extraction solvent.

colony. Grooming is known to reduce pollen carryover and therefore pollinator efficacy (e.g. Barrett et al. 1994). Male bees, including those observed in this study, do not groom which may account for their high efficacy as pollinators in several studies. *A. metallica* males flew from one *Dalea* plant to the next, diving directly down towards the stems where the *P. thurberi* flowers were located, without spending much time searching.

Peak flowering in *Dalea* is April and May but some flowers can be found during June when peak flowering occurs in *Pilostyles*. It is also possible that some pollinators are active across the flowering times of both plant species. For this reason, it is possible that the two species could share pollinators. *Dalea* species are primarily bee pollinated and, attract a rich fauna of native bees (Epple 2005, Cane 2006, 2012). However, *Dalea* is not among the 75 genera of plants in North America known to be visited by *A. metallica* (Hurd 1979). Moreover, *D. formosa* flowers do not resemble those of *P. thurberi* and do not smell of raspberries or fruit (DLN and SDS, pers. obs.), so it is unlikely that host and parasite consistently share pollinators due to shared floral characteristics.

Our findings, combined with others in the literature, suggest that *Pilostyles* species are pollinated by a variety of insects, including bees, wasps, and flies. Our limited observation period leaves us with the question of how representative our data are across time. However, floral visitors described as "metallic green bees" have been observed visiting *P. thurberi* in several previous years by an amateur naturalist and landowner of the Post site (Z. Kirkpatrick, pers. comm.). *Augochloropsis metallica* is not the only metallic green bee at these field sites (we collected *Agapostemon texana* from contemporaneously-flowering Asteraceae at the Post site), but these observations at least suggest *A. metallica* may be a consistent visitor to *P. thurberi*.

The flowers of *P. thurberi* do not present visually conspicuous floral displays to attract these or other floral visitors (Fig. 1A). Flowers are tiny (< 2 mm) and often occur on stems within the interior of the host plants, where they are largely obscured from view by the host's leafy shoots. It is very likely that floral scent plays an especially important role in the pollination relationships of this plant species. However, unlike most entomophilous angiosperms that produce complex floral bouquets consisting of dozens of compounds, *P. thurberi*'s floral odors are relatively simple,

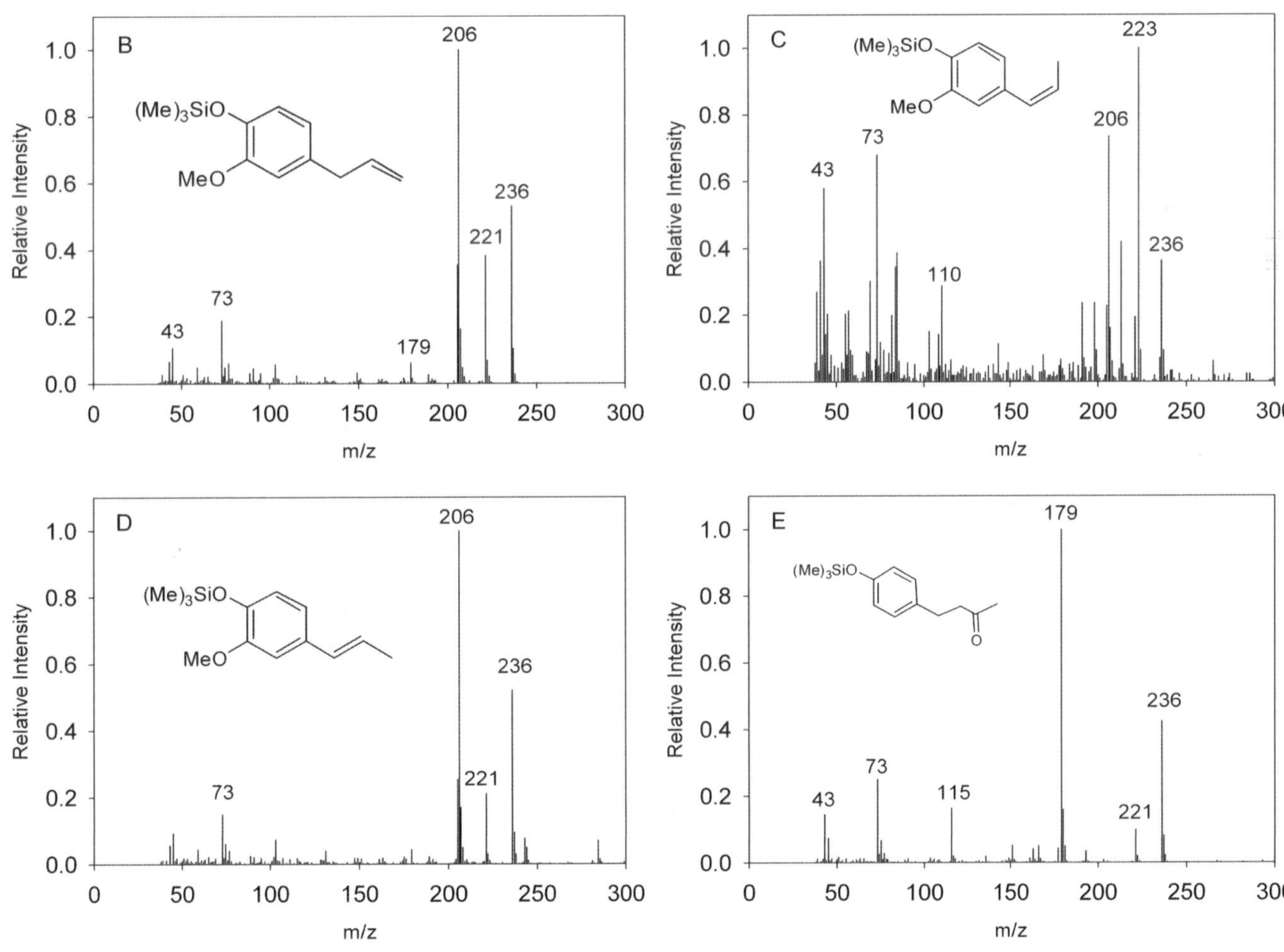

FIGURE 4. Mass spectra of selected peaks from gas chromatograms of derivatized samples; letters refer to peaks labeled in in Fig. 3. Structures of silylated derivatives are shown, where B is eugenol trimethylsilyl ether, C is *cis*-isoeugenol trimethylsilyl ether, D is *trans*-isoeugenol trimethylsilyl ether, and E is raspberry ketone trimethylsilyl ether.

TABLE 3. Compounds detected in underivatized headspace samples of both *P. thurberi* and uninfected *D. formosa*. These all are considered to be originating from *D. formosa*. Unidentified sesquiterpenes are listed with most abundant mass spectral fragments. [1]Putative identification using NIST05, Adams's (2007) retention times, and Lucero et al. (2005). [2]Identification by comparison to authentic standard.

Retention Time, minutes	Chemical Species	# Samples Detected	
		P. thurberi	*D. formosa*
5.15	α-Thujene[1]	5 (3 male & 2 female)	2
5.36	α-Pinene[2]	7 (4 male & 3 female)	2
5.83	Camphene[2]	5 (3 male & 2 female)	2
6.61	β-Pinene[2]	6 (3 male & 2 female)	1
6.83	Myrcene[1]	4 (2 male & 2 female)	1
7.44	α-Phellandarene or 3-Carene	3 (2 male & 1 female)	1
7.78	α-Terpinene[2]	-	1
8.17	Limonene[2]	7 (4 male and 3 female)	2
8.73	*trans*-Ocimene[1]	2 (2 male)	2
9.17	γ-Terpinene[2]	2 (2 male)	2
10.22	Terpinolene[2]	2 (2 male)	2
18.41	Bornyl acetate[2]	6 (4 male & 2 female)	1
19.74	m/z 136, 131, 93	3 (2 male & 1 female)	-
21.01	α-terpinyl acetate[1]	3 (2 male & 1 female)	1
21.85	Unidentified sesquiterpenoid m/z 204, 161, 133, 105	1 female	1
22.121	α-copaene[1]	3 (2 male & 1 female)	1
23.41	Unidentified sesquiterpenoid m/z 204, 189, 161,133,119,105	1 female	1
23.93	β-Caryophyllene[2]	4 (2 male & 2 female)	1
24.37	Unidentified sesquiterpenoid m/z 204, 161, 105	1 female	1
25.41	α-Humulene[2]	1 female	1
26.08	γ-Muurolene[1]	1 female	1
26.22	γ-Curcumene[1]	1 female	1
26.43	Germacrene D[1]	3 (2 male & 1 female)	1
26.91	Unidentified sesquiterpenoid m/z 204, 161, 133, 119, 105, 91	3 (2 male & 1 female)	-
27.08	Unidentified sesquiterpenoid m/z 204, 189, 161, 119, 105, 81	2 (1 male & 1 female)	1
27.18	Unidentified sesquiterpenoid m/z 204, 161, 105	1 female	1
27.34	β-Curcumene[1]	1 female	1
27.74	γ-Cadinene[1]	2 (1 male, 1 female)	1
28.00	Unidentified sesquiterpenoid m/z 204, 161, 13, 119, 105	5 (2 male, 3 female)	1
28.12	Cadina-1(2),4-diene[1]	4 (2 male, 2 female)	1
28.52	Unidentified sesquiterpenoid m/z 204, 161, 119, 105	2 (1 male, 1 female)	1
28.69	Unidentified sesquiterpenoid m/z 204, 161, 133, 105	2 (1 male, 1 female)	1
28.84	Unidentified sesquiterpenoid m/z 204, 161, 122, 107	4 (2 male, 2 female)	1
29.51	Unidentified sesquiterpenoid m/z 204, 161, 133, 121, 105	1 female	1

consisting of raspberry ketone, methyl isoeugenol, and several isomers of eugenol. All of these are known to be general insect attractants and male bee attractants (e.g. Nogueira et al. 2001). The sweet raspberry scent of *P. thurberi* flowers is quite strong to the human nose, and raspberry ketone is detectable by humans at thresholds as low as 1 ppb (Larsen & Poll 1990). The raspberry ketone fragrance remains strong from air-dried flowers, even months or years after collection. Another holoparasitic plant, *Cytinus visseri* (Cytinaceae), has an aliphatic ketone-dominated scent a combination of 3-hexanone and 1-hexen-3-one, that attracts its rodent pollinator, *Elephantulus brachyrhynchus* (Johnson et al. 2010).

Most literature accounts of raspberry ketone as a floral scent constituent deal with the specialized relationships between certain orchids and male tephritid flies, who use both raspberry ketone and methyl isoeugenol collected from flowers as sex pheromones (Tan and Nishida 1995, 2005). Bellot and Renner (2013) did not report the sex of the calliphorid and ulidiid fly pollinators of *P. aethiopica* and *P. haussknechtii*. Although there is nothing in the literature to suggest that male halictid bees display similar chemically-mediated behaviors, similar behaviors are found in males of the distantly-related orchid bees (Euglossini) (Dodson et al. 1969). The pattern is intriguing and worthy of more detailed studies.

The terpenoids detected in the *P. thurberi* headspace samples, all of which were also detected in uninfected *D. formosa* individuals, are most likely being produced by *D. formosa*. All the terpenoids we IDed through comparison with authentic standards have been reported elsewhere in *D. formosa* leaves (Lucero et al. 2005). It seems unlikely that both host and parasite are synthesizing the exact same suite of terpenoids. Moreover, the lack of any terpenoid peaks detected in the whole flower extracts from *P. thurberi* strongly suggests they are not originating from the *P. thurberi* flowers. These compounds could have emitted from the host plant stem, which of necessity was bagged along with the *P. thurberi* flowers during headspace collection. Alternatively, these could be secondary compounds taken up from the host through the parasite's haustorial connection and then emitted by the flowers. Regardless of their origin, they may influence the pollination relationships of *P. thurberi*. Host chemistry has been shown to influence a parasitic plant's interactions with both herbivores (e.g. Marvier 1996) and pollinators (e.g. Troncoso et al. 2010). Troncoso et al (2010) showed that the mistletoe *Tristerix verticillatus* emitted a different suite of volatiles depending upon which of three host plants it parasitized, and that these odor differences affected visitation rates by insects.

We do not know the degree to which *P. thurberi*'s floral scent is different from other plants in its community, nor do we know how it compares to its congeners with different pollination systems. For example, *P. aethiopica* is pollinated by blowflies (Calliphoridae) (Bellot & Renner 2013). Blowfly pollinators typically visit flowers with scents similar to rotting meat or dung, and evolutionary shifts to and from fly pollination have been found to be accompanied by the evolution/loss of floral scents with sulphurous compounds (Shuttleworth & Johnson 2010). How such transitions might play out in a parasitic plant, whose chemical ecology likely involves complex interactions among the parasites, host plants, pollinators, and herbivores, remains a topic for future research.

Acknowledgements

We would like to thank Z. Kirkpatrick for access to the Post field site and field assistance, R. Raguso for guidance in floral scent methodology, and two anonymous reviewers for helpful comments that improved the manuscript.

References

Barrett, SCH, Harder LD, and Cole WW (1994) Effects of flower number and position on self-fertilization in experimental populations of *Eichhornia paniculata* (Pontederiaceae). Functional Ecology 8: 526-535.

Beattie A (1971) A technique for the study of insect-borne pollen. Pan-Pacific Entomologist 47:82.

Beattie AJ, Hughes L (2002) Ant-plant interactions. In: Herrera CM, Pellmyr O (eds) Plant-Animal Interactions: An Evolutionary Approach. Blackwell Publishing, Malden, MA.

Bellot S, Renner SS (2013) Pollination and mating system of Apodanthaceae and the distribution of reproductive traits in parasitic angiosperms. American Journal of Botany 100:1083-1094.

Blarer A, Nickrent DL, Endress PK (2004) Comparative floral structure and systematics in Apodanthaceae (Rafflesiales). Plant Systematics and Evolution 245:119-142.

Cane, JH (2006) An Evaluation of Pollination Mechanisms for Purple Prairie-clover, *Dalea purpurea* (Fabaceae: Amorpheae). American Midland Naturalist 156: 193-197.

Cane JH, Weber M, Miller S (2012) Breeding biologies, pollinators, and seed beetles of two prairie-clovers, *Dalea ornata* and *Dalea searlsiae* (Fabaceae: Amorpheae), from the intermountain west, USA. Western North American Naturalist 72:16-20.

Casaña-Giner V, Oliver JE, Jang EB, Carvalho L (2003) Syntheses and behavioral evaluation of fluorinated and silylated analogs of raspberry ketone as attractants for the melon fly, *Bactrocera cucurbitae* (Coquillett). Journal of Entomological Science 38:111-119.

Croat T (1978) Flora of Barro Colorado Island. Stanford University Press, Stanford CA.

Darwin C (1859) On the origin of species by means of natural selection. London: Murray.

Dell B, Burbidge AH (1981) Notes on the biology of *Pilostyles* (Rafflesiaceae) in Western Australia. W. Austral. Herb. Res. Notes 5:71-79.

Dodson CH, Dressler RL, Hills HG, Adams RM, Williams NH (1969) Biologically active compounds in orchid fragrances. Science 164:1243-1249.

Epple AO (1995) A Field Guide to the Plants of Arizona. Falcon Publishing. Helena, Montana.

Filipowicz N, Renner SS (2010) The worldwide holoparasitic Apodanthaceae confidently placed in the Cucurbitales by nuclear and mitochondrial gene trees. BMC Evolutionary Biology 10:48.

Gomes AL, Fernandes GW (1994) Influence of parasitism by *Pilostyles ingae* (Rafflesiaceae) on its host plant, *Mimosa naguirei* (Leguminosae). Ann. Bot. 74:205-208.

Gomez LD (1983) Rafflesiaceae. In: Burger W (ed) Flora Costaricensis, vol. 13. Fieldiana Botany, Chicago, IL, pp 89-93.

Hurd PDJ (1979) Superfamily Apoidea. In: Krombein KV, Hurd PDJ, Smith DR, Burkers BD (eds) Catalog of Hymenoptera in American North of Mexico. Smithsonian Press, Washington, D.C.

Johnson SD, Burgoyne PM, Harder LD, Dötterl S (2010) Mammal pollinators lured by the scent of a parasitic plant. Proceedings of The Royal Society B 278:2303-2310.

Kummerow J (1962) *Pilostyles berterii* Guill., eine wenig bekannte Rafflesiacee in Mittelchile. Zeits. Bot. 50:321-337.

Larsen M, Poll M (1990) Odour thresholds of some important aroma compounds in raspberries. Zeitschrift für Lebensmittel-Untersuchung und -Forschung 191:129-131.

Lucero ME, Estell RE, Sedillo RL (2005) The composition of *Dalea formosa* oil determined by steam distillation and solid-phase microextraction. Journal of Essential Oil Research 17: 645-647.

Marvier MA (1996) Parasitic plant-host interactions: plant performance and indirect effects on parasite-feeding herbivores. Ecology 77:1398-1409.

McIntosh M (2005) Pollination of two species of *Ferocactus*: interactions between cactus-specialist bees and their host plants.

Functional Ecology 19:727-734.

Nickrent DL, Blarer A, Qiu Y-L, Vidal-Russell R, Anderson FE (2004) Phylogenetic inference in Rafflesiales: the influence of rate heterogeneity and horizontal gene transfer. BMC Evolutionary Biology 4:40.

Nogueira PCdL, Bittrich V, Shepherd GJ, Lopes AV, Marsaioli AJ (2001) The ecological and taxonomic importance of flower volatiles of *Clusia* species (Guttiferae). Phytochemistry 56:443-452.

O'Neill KM (2001) Solitary wasps: behavior and natural history. Cornell University Press, Ithaca, NY.

Raguso RA, Pellmyr O (1998) Dynamic headspace analysis of floral volatiles: A comparison of methods. Oikos 81:238-254.

Raguso RA (2008) Wake up and smell the roses: The ecology and evolution of floral scent. Annual Review of Ecology, Evolution, and Systematics 39:549-569.

Renner SS, Ricklefs RE (1995) Dioecy and its correlates in the flowering plants. American Journal of Botany 82:596–606.

Rutherford RJ (1970) The anatomy and cytology of *Pilostyles thurberi* Gray (Rafflesiaceae). Aliso 7:263-288.

Saitta M, Salvo F, Di Bella G, Dugo G, La Torre GL (2009) Minor compounds in the phenolic fraction of virgin olive oils. Food Chemistry 112:525-532.

Shuttleworth A, Johnson SD (2010) The missing stink: sulphur compounds can mediate a shift between fly and wasp pollination systems. Proceedings of the Royal Society B 277: 2811-2819.

Tan KH, Nishida R (1995) Incorporation of raspberry ketone in the rectal glands of males of the Queensland fruit fly, *Bactrocera tryoni* Froggatt (Diptera: Tephritidae). Applied Entomology and Zoology 30:494-497.

Tan KH, Nishida R (2005) Synomone or kairomone? – *Bulbophyllum apertum* flower releases raspberry ketone to attract *Bactrocera* fruit flies. Journal of Chemical Ecology 31:497-507.

Tholl D, Röse USR (2006) Detection and identification of floral scent compounds. In: Dudareva N, Pichersky E (eds) Biology of Floral Scent. CRC Press, Boca Raton FL, pp 3-26.

Thomson JD (1986) Pollen transport and deposition by bumble bees in *Erythronium*: influences of floral nectar and bee grooming. Journal of Ecology 74:329–341.

Troncoso AJ, Cabezas NJ, Faúndez EH, Urzúa A, Niemeyer HM (2010) Host-mediated volatile polymorphism in a parasitic plant influences its attractiveness to pollinators. Oecologia 162:413-425.

Vainstein A, Lewinsohn E, Pichersky E, Weiss D (2001) Floral fragrance. New inroads into an old commodity. Plant Physiology 127:1383-1389.

9

INHIBITION OF BIOCHEMICAL TERPENE PATHWAYS IN *ACHILLEA MILLEFOLIUM* FLOWERS DIFFERENTLY AFFECTS THE BEHAVIOR OF BUMBLEBEES (*BOMBUS TERRESTRIS*) AND FLIES (*LUCILIA SERICATA*)

Anne-Amélie C. Larue-Kontić & Robert R. Junker*

University Salzburg, Department of Ecology and Evolution, Hellbrunnerstraße 34, 5020 Salzburg, Austria

Abstract—Floral scents serve multiple functions in the interactions with organisms. Flowers of *Achillea millefolium* (Asteraceae) emit scent bouquets dominated by terpenoids. These flowers are mainly visited by flies and beetles, whereas bumblebees, common visitors at other Asteraceae, are absent from *A. millefolium* flowers. In order to test how a reduced mono- and sesquiterpenoid emission affect insect behaviour we inhibited the biochemical pathways towards the production of terpenoids of *A. millefolium* plants and conducted behavioural choice tests. The inhibition resulted in reduced emission rates of most mono- and sesquiterpenes and thus altered the olfactory phenotype of the flowers. In a flight cage, flies usually chose flowers with a natural scent bouquet, bumblebees clearly preferred flowers treated with inhibitors. These findings confirm that floral scents play a pronounced role in foraging decisions of flower visiting insects and support the notion that responses towards scent are animal species-specific emphasising the role of scents as floral filters.

Keywords: flight cage, floral scent bouquet, Fosmidomycin, hydroponic plants, Lovastatin

INTRODUCTION

Plants emit floral scent bouquets to communicate with diverse flower visitors. Some floral volatiles act as attractants for mutualists and simultaneously repel antagonistic insects (Kessler et al. 2008; Junker & Blüthgen 2010; Schiestl 2010). Accordingly, floral scents have been shown to determine the presence, absence and frequency of flower visits of multiple insect species (Junker et al. 2010; Larue et al. 2016). *Achillea millefolium* (Asteraceae) flowers are mostly visited by syrphids, flies and beetles (Larue et al. 2016). Although bumblebees and honeybees are common flower visitors in other Asteraceae e.g. *Cirsium arvense*, they naturally do not visit flowers of *A. millefolium* (Larue et al. 2016). The floral scent bouquet of *A. millefolium* is dominated by mono- and sesquiterpenoids (Larue et al. 2016) that can serve defensive functions in floral scent bouquets (Junker & Blüthgen 2010; Schiestl 2010; Junker et al. 2011). The precursors of mono- and sesquiterpenoids, isopentenyl diphosphate (IPP) and dimethylallyl diphosphate (DMAPP) derive from two pathways. Monoterpenoids are formed in the plastidic methylerythritol phosphate (MEP) pathway whereas sesquiterpenoids are synthesized in the cytosolic mevalonate pathway (MVA) (Dudareva et al. 2005). In a previous study (Larue et al. 2016), we experimentally augmented the scent bouquets of *A. millefolium* with volatiles extracted from a plant species (*C. arvense*) that predominantly emits aromatic compounds. Following this augmentation, *A. millefolium* flowers were visited by honeybees and bumblebees. In order to discriminate between the attractive effect of aromatics and the repellent effect of terpenoids, in this study we inhibited the biochemical pathways of hydroponic *A. millefolium* plants involved in terpene-synthesis. Inhibited and control plants were used in flight cage choice tests with bumblebees (*Bombus terrestris*) and flies (*Lucilia sericata*) to study their behaviour towards terpenes.

MATERIALS AND METHODS

Organisms

Achillea millefolium seeds originating from a single population near Hamburg, Germany (obtained from the Botanical Garden of the University of Hamburg, Germany), were placed on filter papers (Rundfilter, 5.5 cm Ø, Macherey-Nagel, Düren, Germany) in petri dishes and were treated with 250 µl of a watery gibberellic acid solution (1000 ppm, Sigma-Aldrich Handels GmbH, Vienna, Austria). After germination, seedlings were transplanted into 2 ml micro test tubes (Safe lock tubes, Eppendorf AG, Hamburg, Germany) filled with 1.2 ml agar grow medium (1.5 g agar dissolved in 120 ml water, Bacto-Agar, A. Hartenstein GmbH, Würzburg, Germany), which was enriched with a fertilizer combination suited for the nutritional requirements of young plants (180 µl of Grow, Bloom and Micro fertilizer, mixture ratio 2:1:1, Advanced Hydroponics, Vienna, Austria). In order to prevent dehydration, tubes were closed until the roots of the

*Corresponding author: robert.junker@sbg.ac.at

seedlings were strong enough to supply plants with nutrients and water. At that time the lower tip of each of the tubes was removed using pincers and tubes were inserted into 55 L metal water basins filled with deionized water and fertilizer solution with the same fertilizer-water ratio as described above, allowing roots to grow into the solution. Fertilizer solution was changed every two weeks and the fertilizer mixture adjusted to the growth stage of the plants according to the manufacturer's instructions. For ventilation, the water fertilizer solution was aerated *via* a silicon tube leading directly into the water of each basin using an electrical pump (LP-100 air-pump, Resun, Shenzhen Xing Risheng Industrial Co., Ltd.; Shenzhen, China). Room temperature remained constantly at 18 °C and fluorescent tubes (Osram, T8 L 58W/865 LUMILUX Daylight G13, PK Beleuchtungstechnik, Hohenbrunn, Germany) as well as red/blue LED-lamps (LED Pflanzenlicht 14W, esmart, Pinnow, Germany) illuminated the plants for 11 hours a day. *Bombus terrestris* and *Lucilia sericata* used in experiments were purchased (biohelp, Vienna, Austria) and kept and reared in the lab. Insects were fed with sugar solution and flower pollen (Hochland Bio-Blütenpollen, Hoyer, Polling, Germany). The used insect species represented close relatives of common flower visitor species observed on flowers of *A. millefolium* or other Asteraceae in the field (Larue et al. 2016).

Inhibition of biochemical pathways

In order to inhibit the metabolic products of the methylerythritol phosphate (MEP) and mevalonate pathway from *A. millefolium* we used pathway-specific inhibitors. For best results, we used a combination of Fosmidomycin (blocking the MEP pathway) and Lovastatin (blocking the mevalonate pathway) as isopentenyl diphosphate (IPP) is exchanged via/through the plastid membrane between both pathways (Dudareva et al. 2005; Junker et al. 2011). For applying the specific inhibitors to hydroponic plants via/through the roots, inhibitors were dissolved in water. For the treatments, a 10 mM Fosmidomycin stock solution and a 1 mM Lovastatin stock solution were used. 25 mg of Fosmidomycin (Life technologies Austria, Vienna, Austria) were dissolved in 12.2 ml distilled H_2O. To increase the water solubility of Lovastatin (Sigma-Aldrich Handels GmbH, Vienna, Austria) a method analogous to that described in Kita et al. (1980) was implemented. To exclude any effects of solvents and additives (see Kita et al 1980), a control solution was prepared in the same way as done for the Lovastatin solution excluding both inhibitors. All solutions were stored in multiple aliquots at -20 °C until usage. Before starting the inhibition experiments, treatment and control hydroponics were taken out of the metal basins, root tips were cut to improve absorption of inhibitors and plants placed in plastic vases (Erlenmeyer flasks of PP, 100 ml, A. Hartenstein GmbH, Würzburg, Germany) containing one of the following solutions: Control plants received 91.75 ml nutrient solution and 8.25 ml control solution. Plants of the treatment group were supplied with a mixture of 91.75 ml nutrient solution and the pathway-specific inhibitors, i.e. 0.75 ml of the 10 mM Fosmidomycin stock solution and 7.5 ml of the 1 mM Lovastatin stock solution.

Floral volatile collection and analysis

Floral headspace of treatment (N = 10) and control (N= 3) plants were collected using the dynamic headspace sampling (compare Larue et al. 2016). Volatiles were sampled 12 hours after inhibition experiment started, as pre-trials showed that inhibitors worked best after that period of time. Flowers of the hydroponic plants were enclosed into scentless polyethylene tetraphthalate (PET) oven bags (Toppits® Cofresco Fischhalteprodukte GmbH & Co. KG, Minden, Germany), without picking them from the plant for 15 min and the enriched headspace were sampled for 2 min into volatile traps with a flow rate of 200 ml min^{-1}. Volatile traps were filled with a mixture of 1.5 mg Tenax-TA (mesh 60-80; Supelco, Germany) and 1.5 mg Carbotrap B (mesh 20-40; Supelco, Germany). Floral volatiles were analysed by using an automatic thermal desorption system (TD, model TD-20, Shimadzu, Japan) coupled with gas chromatograph/mass spectroscopy (GC-MS, model QP2010 Ultra EI, Shimadzu, Japan). GC was equipped with a ZB-5 column (Zebron ZB-5, 5% phenyl polysiloxane, length 60 m, inner diameter 0.25 mm, film thickness 0.25 μm, Phenomenex, Newport Beach, USA). The carrier gas (helium) had a flow of 1.5 ml min^{-1} and the oven temperature was hold for 1 min at 40 °C before it raised with 6 °C min^{-1} until the maximum of 250 °C. The MS interface was 260 °C and the ion trap worked at 200 °C. Floral compound identification was performed with a GCMSsolutions software (Version 2.72, Shimadzu Corporation), mass spectra of authentic standards as well as spectral libraries (ADAMS, ESSENTIALOILS-23P, FFNSC 2, W9N11) and Kovats indices generated using n-alkanes to identify floral volatiles.

To compare the floral scent bouquets from control and inhibited flowers 12 hours after inhibition, we calculated Euclidean distances and used them for non-metric multidimensional scaling (NMDS).

In order to directly compare the floral scent bouquets of experimental plants, we fitted control and inhibited plants onto an ordination (NMDS) based on Euclidean distances and tested whether the centroids of both groups significantly deviated using the R package vegan (Dixon 2003).

Flight cage choice tests

In order to compare the insects' behaviour towards control flowers of *A. millefolium* and those with an inhibited floral bouquet, bumblebees (N = 17) and flies (N = 11) were allowed to choose between flowers of a control and a treatment plant approximately 12 hours after inhibition. For behavioural bioassays, we used the same plants as for scent sampling. Bumblebees and flies were kept in flight cages in the laboratory at a temperature of 20 ± 2 °C. After capturing them from the flight cages, all insects were directly tested under daylight conditions (Sylvania Grolux T5, 54W, Havells Sylvania Germany GmbH, Vienna, Austria) at a temperature of 20 ± 2 °C and each individual was used only once. Up to three naïve insects per species were tested simultaneously. The flight cage where insects were tested was of pyramidal shape with 50 cm side length and 30 cm height. The upper part consists of mosquito nets and the ground was made from PVC containing two holes 20 cm apart from each other. Flowers were attached to the holes (height from ground about 10 cm) and were supplied with water, nutrients and inhibitors from underneath the flight cage. Insects were released in a flight cage and the observation and recording started after 20 seconds of acclimation. Depending on the insects' activity the total observation time varied between 10 - 40 minutes. During this time, approaches to and landing on flowers were recorded. The number of corymbs per plant and number of flower per corymb was held similar in control and treatment plants. Insects' preferences for inhibited or control plants were expressed as the difference D of the proportional number of insects landing on inhibited flowers minus the proportion of insects landing on control flowers, i.e. positive values indicate a preference for inhibited flowers, negative values for control flowers. To take account for the nested design of our experiment (i.e. up to three insect individuals were included in one trial) we tested whether D significantly deviates from zero (against the null hypothesis ($D = 0$)) using linear mixed effect models (LME) with approaches or landings as dependent variable and experimental trials as a random factor. LME were conducted using the R package *nlme* (Pinheiro et al. 2015). All statistical analyses were performed in R (R Development Core Team 2013).

RESULTS

Floral volatiles emitted from inhibited and control plants

A. millefolium flowers treated with inhibitors 12 hours prior to volatile sampling emitted reduced amounts of individual monoterpenes and a sesquiterpene compared to control plants (Tab. 1). Unexpectedly, flower treated with inhibitors emitted higher amounts of limonene (not significant after Bonferroni correction; Tab. 1). As a consequence, floral scent bouquets from control and treatment plants were well separated in a NMDS based on Euclidean distances (Fig. 1) and centroids of control and inhibited plants deviated significantly (R^2 = 0.56; P = 0.004).

TABLE 1. Scent emission of *A. millefolium* flowers 12 hours after application of inhibitors to hydroponic plants. Shown are mean emission rates and standard deviation (SD) of each compound [ng hr^{-1} dry weight^{-1}] from control and treatment flowers. P-values for comparison of the emission of individual substances in control and treated flowers using Wilcoxon-test are also given. Scent compounds are listed in order of the Kovats' index within each class. Significant deviations after Bonferroni correction are indicated by asterisks.

Compounds [ng hr^{-1} dry weight^{-1}]	RI	Control (N= 3) Mean	SD	Treatment (N= 10) Mean	SD	P-values
Monoterpenes						
α-Thujene	926	193.02	89.39	1.66	2.54	0.005
α-Pinene	938	600.25	191.67	714.76	583.83	0.922
Camphene	956	147.46	77.34	56.93	42.56	0.002*
Sabinene	979	2731.83	491.82	106.05	67.65	0.002*
β-Pinene	982	221.69	83.58	271.34	307.74	0.846
α-Terpinene	1022	127.76	63.41	0.00	0.00	NA
ρ-Cymene	1030	459.98	186.46	99.94	103.79	0.002*
Limonene	1033	98.94	45.12	481.69	389.86	0.027
β-Phellandrene	1036	127.69	45.09	57.22	113.66	0.084
1,8-Cineole	1039	301.45	128.66	156.57	160.08	0.027
(E)-β-Ocimene	1049	82.04	29.50	7.03	9.87	NA
γ-Terpinene	1064	281.79	156.67	13.37	15.76	NA
α-Terpinolene	1095	59.41	43.68	0.00	0.00	NA
Allo-Ocimene	1132	10.62	10.19	0.08	0.28	NA
Camphor	1158	78.31	65.82	10.13	19.49	0.004
sum		349.97	652.53	123.64	281.57	0.025
Sesquiterpenes						
β-Caryophyllene	1445	27.16	31.87	14.68	14.94	0.049
Aromatics						
Benzaldehyde	963	209.07	42.99	192.95	93.28	0.432
Aliphatics						
(Z)-3-Hexen-1-ol	854	99.34	47.84	36.70	22.95	0.002*
(Z)-3-Hexenyl acetate	1005	507.31	168.70	20.96	12.70	0.002*
sum		303.32	238.72	28.83	20.14	0.333

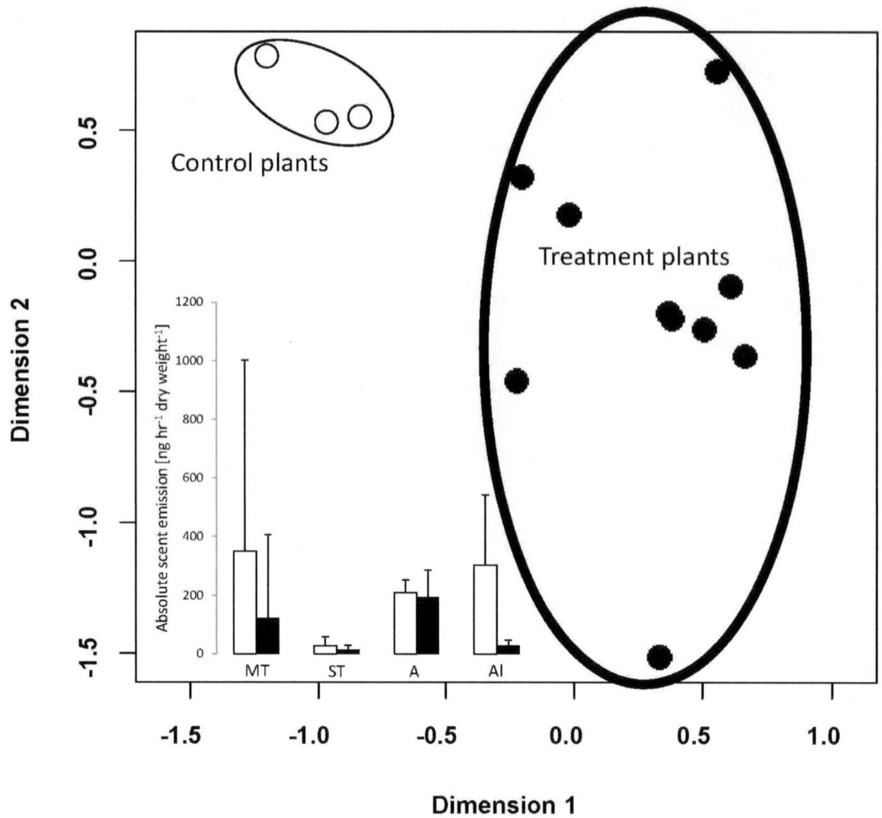

FIGURE 1. Nonmetric multidimensional scaling (NMDS) based on Euclidean distances of floral scent bouquets (proportional emission) from inhibited (treatment, filled circles) and control (open circles) *Achillea millefolium* plants 12 hours after starting the inhibition. The bar graph shows the absolute scent emission (mean ± SD) of monoterpenes (MT), sesquiterpenes (ST), aromatics (A) and aliphatic compounds (Al) from control (white bars) and inhibited (black bars) plants.

Flight cage choice tests

In choice tests where flies and bumblebees were allowed to choose between flowers of inhibited and control *A. millefolium* plants, flies showed a preference for control flowers in landing approaches (lme: $Df = 6$, F-value = 6.99, $P = 0.04$; Fig. 2) and landings (lme: $Df = 6$, F-value = 6.99, $P = 0.04$; Fig. 2). Bumblebees preferred to land on flowers with an inhibited scent bouquet (lme: $Df = 9$, F-value = 6.37, $P = 0.03$, Fig. 2) whereas they were indecisive in landing approaches but with a tendency towards inhibited flowers (lme: $Df = 9$, F-value = 1.50, $P = 0.25$; Fig. 2).

DISCUSSION

Pathway-specific inhibitors of the biochemical pathways towards mono- and sesquiterpenoid production affected the floral scent emission of *Achillea millefolium* plants. Nearly all scent compounds of the targeted biochemical pathways (mono- and sesquiterpenoids) were affected in a way as expected, i.e. their emission rates were reduced in inhibitor-treated plants compared to control plants. Thus, the scent bouquets of control and treatment plants were clearly different. According to the hypothesis that floral scents affect the behaviour of flower-visiting animals, bumblebees and flies discriminated between control and inhibited flowers of *A. millefolium* in choice tests in a flight cage. Bumblebees that usually do not visit *A. millefolium* flowers in the field (Larue et al. 2016) preferred inhibited flowers; flies, common visitors of *A. millefolium* flowers, preferred control flowers emitting a natural scent bouquet. Note that the species used in the experiments are not the same species as observed in the field study but represent close relatives. These results support the notion that floral scents are important traits in controlling the identity and frequency of flower visitors in interactions with plant species (Junker et al. 2010; Larue et al. 2016).

The inhibition of the pathways led to a significant reduction in the emission rates of many mono- and sesquiterpenes, but not to scent bouquets deprived of terpenoids. It is debated whether animals respond to whole bouquets (ratios of the compounds) or to key compounds that mediate specific functions such as attraction or repellence. It has been shown that in specialized and generalized pollination systems either individual key-compounds or a small number of volatiles mediate the attraction to flowers (Wright & Smith 2004; Schäffler et al. 2015). Likewise, floral scent bouquets may have a defensive function due to the emission of individual compounds (Junker et al. 2011). Other examples, however, show that whole bouquets of volatile compounds in specific ratios are used to locate host plants (Bruce et al. 2005). Although we observed clear preferences of both, bumblebees and flies, for treatment or control plants, respectively, our data do not allow to clearly discriminate between the mode of action of floral scent bouquets in our system. Potentially, terpene inhibition reduced the emission rate of a compound that functions as repellent for bumblebees or as attractant for flies. Alternatively, the inhibition clearly altered the ratio of emission rates of the compounds synthesized by *A. millefolium*, which may have resulted in modified behavioural responses. Future studies require additional experiments potentially using synthetic blends of volatiles to

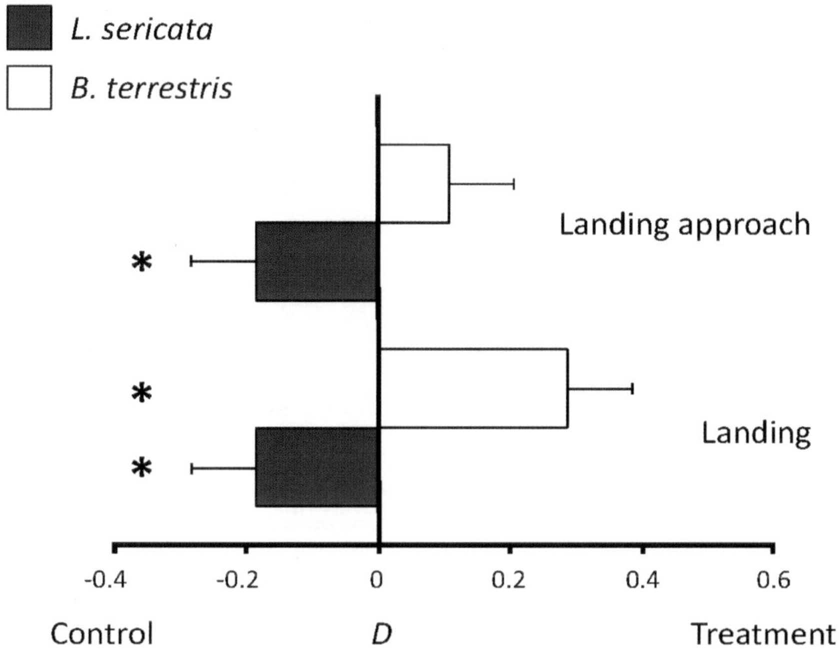

FIGURE 2. Results of flight cage choice tests: Responses of *Lucilia sericata* (dark grey bars, $N = 11$) and *Bombus terrestris* (white bars, $N = 17$, D, mean $\pm$ STD) to flowers of *Achillea millefolium* plants with natural scent bouquets (control) or scent bouquets with inhibited terpene-pathways (treatment). D is defined as the proportional number of insects visiting flowers of treatment plants minus the proportional part of insects visiting control flowers. To test whether responses (D) significantly deviate from zero we calculated linear mixed effect models (LME) and tested against the null hypothesis ($D = 0$). Significant deviations are indicated by asterisks * $P < 0.05$.

fully reveal the mode of action of attraction and repellence in generalized systems.

The experimental manipulation of floral scent bouquets is a powerful tool to study the relevance of volatiles in flower-visitor interactions. Our data confirm that both positive and negative responses by insects to volatiles contribute to the behaviour of flower visitors and thus visitation patterns. This allows flowers that do not have morphological barriers (such as *A. millefolium*) to select flower visitors from the locally available spectrum to facilitate the visitation by some animals but reduce or prevent the visitation of others. Therefore, floral scents are potent floral filters with strong effects on the behaviour of consumers of floral resources.

ACKNOWLEDGEMENTS

The authors are grateful to Susanne Popp for the help in the inhibition experiments and Jessica Bossems for helping to establish the hydroponic system. We also thank Jonas Kuppler for help in the cultivation of plants as well as Stefan Dötterl, Klaus Lunau, and Irmgard Schäffler for providing infrastructure and/or for help with scent analysis. This work was funded by the Deutsche Forschungsgemeinschaft (DFG, JU2856/1-1).

REFERENCES

Adams RP (2007) Identification of essential oil components by gas chromatography / mass spectrometry, 4 edn. Allured Publishing Corporation, Carol Stream, Illinois.

Bruce TJA, Wadhams LJ, Woodcock CM (2005) Insect host location: a volatile situation. Trends in Plant Science 10:269-274.

Dixon P (2003) VEGAN, a package of R functions for community ecology. Journal of Vegetation Science 14:927-930.

Dudareva N, Andersson S, Orlova I, Gatto N, Reichelt M, Rhodes D, Boland W, Gershenzon J (2005) The nonmevalonate pathway supports both monoterpene and sesquiterpene formation in snapdragon flowers. Proceedings of the National Academy of Sciences of the United States of America 102:933-938.

Junker RR, Blüthgen N (2010) Floral scents repel facultative flower visitors, but attract obligate ones. Annals of Botany 105:777-782.

Junker RR, Gershenzon J, Unsicker SB (2011) Floral odour bouquet loses its ant repellent properties after inhibition of terpene biosynthesis. Journal of Chemical Ecology 37:1323–1331.

Junker RR, Höcherl N, Blüthgen N (2010) Responses to olfactory signals reflect network structure of flower-visitor interactions. Journal of Animal Ecology 79:818-823.

Kessler D, Gase K, Baldwin IT (2008) Field experiments with transformed plants reveal the sense of floral scents. Science 321:1200 - 1202.

Kita T, Brown MS, Goldstein JL (1980) Feedback-regulation of 3-hydroxy-3-methyglutaryl coenzyme a reductase in livers of mice treated with mevinolin, a competitive inhibitor of the reductase. Journal of Clinical Investigation 66:1094-1100.

Larue A-AC, Raguso RA, Junker RR (2016) Experimental manipulation of floral scent bouquets restructures flower–visitor interactions in the field. Journal of Animal Ecology 85:396-408.

Pinheiro J, Bates D, DebRoy S, Sarkar D, R Core Team (2015) nlme: Linear and Nonlinear Mixed Effects Models. R package version 3.1-120.http://CRAN.R-project.org/package=nlme.

R Development Core Team (2013) R: a language and environment for statistical computing. R Foundation for Statistical Computing, Vienna.

Schäffler I, Steiner KE, Haid M, van Berkel SS, Gerlach G, Johnson SD, Wessjohann L, Dötterl S (2015) Diacetin, a reliable cue and private communication channel in a specialized pollination system. Scientific Reports 5:12779.

Schiestl FP (2010) The evolution of floral scent and insect chemical communication. Ecology Letters 13:643-656.

Wright GA, Smith BH (2004) Variation in complex olfactory stimuli and its influence on odour recognition. Proceedings of the Royal Society of London Series B-Biological Sciences 271:147-152.

10

INSECT POLLINATION: COMMODITY VALUES, TRADE AND POLICY CONSIDERATIONS USING COFFEE AS AN EXAMPLE

Vernon G. Thomas*[1], Peter G. Kevan[2]

[1]*Department of Integrative Biology, College of Biological Science, University of Guelph, Guelph, Ontario N1G 2W1, Canada*
[2]*School of Environmental Sciences, Ontario Agricultural College, University of Guelph, Guelph, N1G 2W1, Canada*

Abstract—Science has shown the importance of animal pollinators to human food security, economy, and biodiversity conservation. Science continues to identify various factors causing pollinator declines and their implications. However, translation of the understanding of pollinators' roles into current policy and regulation is weak and requires attention, both in developed and developing nations. The national and international trade of commodities generated via insect pollination is large. Trade in those crops could be a means of influencing regulations to promote the local existence of pollinating species, apart from their contributions to biodiversity conservation. This paper, using the example of international coffee production, reviews the value of pollinating species, and relates them to simple economics of commodity production. Recommendations are made that could influence policy and decision-making to promote coffee production, trade, and pollinators' existence. Assumptions and considerations are raised and addressed. Although the role of insect pollinators in promoting fruit set and quality is accepted, implementing pollination conservation in forest habitats may require assured higher prices for coffee, and direct subsidies for forest conservation to prevent conversion to other crop lands. Exporting and importing governments and trade organizations could establish policy that requires insect pollination in the coffee certification process. The European Parliament and the North American Free Trade Agreement could be instrumental in creating policy and regulation that promotes insect pollination services in coffee production. The reciprocity between the services of insect pollinators in certified coffee production and their services in forest biodiversity production should be implicit in future policy negotiations to enhance both systems.

Keywords *animal pollination, trade, commodities, policy, biodiversity, conservation, coffee*

INTRODUCTION

The roles of pollinating species in the generation of human foods and biodiversity production have been well-established (Buchmann & Nabhan 1996; Kevan 1999; Klein et al. 2007; Ollerton et al. 2011). The scientific understanding of the factors underlying declines in pollinating species and negative impacts upon certain pollinators and biodiversity is not so well-established and continues to be researched (Allen-Wardell et al. 1998; Kevan 2001; Biesmeijer et al. 2006; Potts et al. 2010; Brittain & Potts 2011; Mayer et al. 2011). However, the translation of much of what is understood about the importance of pollination into government policy and legislation is weak by comparison. This is often attributed to scientists not wishing to be seen as advocates (Sisk et al. 2011), and the disparate natures of modern science and government policy. Where legislation concerning pollinating insects exists, it is often based on commercial honey bee (*Apis mellifera*) operations managed mainly for honey production, rather than pollination, outside of North America (Tang et al. 2007), and may ignore the diverse array of insects that contribute pollinating services regionally.

There is no specific international treaty that focuses on the conservation of pollinating species and provides the ratified basis of national laws. General policy on pollinator conservation arises from the 1992 United Nations Convention on Biological Diversity. In 1996, the Conservation and Sustainable Use of Agricultural Biodiversity Program was created under the 1992 Biodiversity Convention and used in 1998 to convene the São Paulo Declaration on Pollinators. This declaration led in 2000 to the formation of the International Pollinator Initiative (IPI) (International Pollinator Initiative 1999), and a later (2002) Plan of Action (POA) (Byrne & Fitzpatrick 2009; Williams 2003). The principal aims of the Plan of Action comprise monitoring of pollinators and factors affecting them, upgrading taxonomic information, refining the economic values of pollination services, and promoting the conservation/restoration of pollinator diversity in managed and natural ecosystems. Each aim further involves assessment, adaptive management, capacity building, and mainstreaming. These aims are to be co-ordinated and conducted at a world-wide level (Williams 2003). Although the IPI-POA is only an *agreement* among nations, it is the basis upon which most of the current pollinator conservation and protection research and concern rests. The POA, however, does not mention trade or policy.

The African Pollinator Initiative placed emphasis on mainstreaming pollination science into policy decisions (Eardley et al. 2006). Mayer et al. (2011) identified a series

*Corresponding author; email: vthomas@uoguelph.ca

of questions that could direct future pollination research and touched on the need for policy involvement. This point was developed further by Biesmeijer et al. (2011), who identified the societal and policy aspects of pollination research as vital in furthering the interest of pollination services. In 2009, the International Risk Governance Council identified the need for policy in sustaining pollination services, and defined the risks related to pollinator declines in addition to the deficits in policy, governance and understanding (IRGC 2009).

In this paper, we review evidence for the huge economic value of pollination services and the value of the crop commodities that are traded internationally. Using the example of international production of coffee and the trade that it generates, we indicate how a greater incorporation of pollination by managed and wild pollinators could enhance yields and adjacent biodiversity maintenance. We indicate how the implementation of this ecological idea has to be reconciled with socio-economic issues of commodity trade at local and international scales in order for policy changes to occur. This paper is an initial foray into the process of integrating insect pollination services into policies directing agricultural coffee production. We make suggestions and recommendations to government and non-government agencies that could facilitate this process.

Which Pollination Systems Should Be Focused On?

Flowering plants exhibit enormous variation in their manner of pollination and fruit/seed set. The reliance of world crop production upon animal pollinators was reviewed by Klein et al. (2007), who identified a wide range of dependency, from essential to non-essential. The success of efforts to involve greater pollination from animal pollinators will depend on the degree to which their service is vital to a given commodity's production (Richards 2001) and subsequent trade. Thus most attention should be placed on those crops for which animal pollination is essential, or may increase the efficiency of, fruit/seed set. This systems approach assumes that a regional decline in the abundance of pollinators and that a pollination "deficit" (Steffan-Dewenter et al. 2005; Aizen et al. 2009; Bauer & Wing 2010; Potts et al. 2010; Cameron et al. 2011) exists in that commodity's production, and that intervention can reduce that deficit. However, there is ongoing debate about this issue (Ghazoul 2005a,b; Steffan-Dewenter et al. 2005).

Assessed Global Values of Animal Pollination Services

The assessments provided by the literature are in constant flux because of changing commodity prices on world markets, and especially because of revisions based on a growing awareness of the role of animal pollinators in crop production. The equation – Pollination Service Value = V x D x P (Southwick & Southwick 1992; Drucker 2004) where V is the annual crop value, D is the assumed level of dependence on insect pollination, and P is the proportion of insect pollinators visiting the crop that are honey bees, is not the most accurate model for measuring service value because of the inherent limitations of the assumptions and the overt bias towards honey bees' services (see Allsopp et al. 2008; Bauer & Wing 2010 on this point). Moreover, they do not consider the elasticities of costs with changes in supply and demand (Southwick & Southwick 1992; Kevan & Phillips 2001). This important point is often overlooked by valuations done by ecologists: declines in crop production are often met by consequent increases in demand and prices.

Despite those limitations, Drucker (2004), using the above model, estimated the global annual service value of pollinators to be 65-70 billion US$. Newer (2005) estimates are approximately 225 billion US$, globally (Gallai et al. 2009). These figures reflect the simple value of the produced commodity, and not the total value of the traded commodity. A further salient point is a comparison between the value of crops that are pollinator-dependent and those grown independently of pollination: pollinated crops were valued at 761 €/ton (approx. 1050 $US in 2009), whereas pollinator-independent crops realized only 151 €/ton (approx. 208 $US) (Gallai et al. 2009). This finding is complemented by Ashworth et al. (2009), who reported that in Mexican agriculture pollinator-dependent crops generated both greater crop volume and crop revenue per unit area than pollinator-independent crops.

Southwick & Southwick (1992) and Drucker (2004) concluded that the value of insect pollination services are very large and many times the direct value of honey bee products. Furthermore, the value of commercially-managed insect pollination services is small compared to those provided by nature (Drucker 2004). Allsopp et al. (2008) conducted a study that determined the relative contribution of managed and wild pollinators to deciduous fruit production in South Africa and concluded that the contributions of both wild and managed pollinators had been grossly undervalued. In that study, managed pollinator services contributed 28-122 million US$, for which producers paid only 1.8 million US$. By contrast, wild pollinators' services were worth 49-310 million US$, for which producers paid nothing. This important study demonstrates well the extent to which wild pollinators have been overlooked and the bias towards managed pollinators. New research by Breeze et al. (2011) complements the conclusions of Allsopp et al. (2008), these authors reporting that in the UK, wild pollinators were providing about 66% of the agricultural crop pollination, and managed pollinators, only 34%. Yet the yields of UK crops pollinated by insects rose by approximately 54% since 1984, indicating the capacity of wild pollinators to sustain this increase. The evidence presented in this section indicates that there is a strong economic rationale for enhancing the role of insect pollination in global food production, especially that of unmanaged pollinating species (Kremen et al. 2007; Aizen et al. 2008; Klein et al. 2008; Garibaldi et al. 2011).

Scope and Rationale of the Present Study

The scope of the IPI-POA is necessarily broad and extends across many crops, species and regions. However the production, marketing, and trade of insect-pollinated crops tend to be commodity-specific, despite the ecological commonness of their pollination. We had the option of examining a broad range of commodities, their pollination and trade policy, or selecting a single commodity that benefits from insect pollination. For reasons of simplicity, coupled with the availability of extensive scientific and economic literature, we selected a single commodity, coffee. Should the scheme proposed for coffee, even with its being several species with different pollination needs for fruit/seed set, be successful, it could be adapted for application to other insect-pollinated crops.

Commercial coffee production is based mainly on two species; *Coffea arabica* and *C. canephora* (Ngo et al. 2011). The species *canephora* is pollinator dependent for fruit set. The species *arabica* is largely wind pollinated, but fruit set increases markedly with insect pollination (Roubik 2002a; Klein et al. 2003a,b). The production of coffee extends from Central and South America to the Caribbean, West Africa, Eastern Africa, and various regions of Asia, where it figures prominently in the economy of these nations. The global value of the coffee trade is large, approximately 80 billion US$/y (International Coffee Organization 2011). International trade in coffee is co-ordinated through the International Coffee Organization, which has strong interests in the manner in which coffee is grown and traded among its member nations (International Coffee Organization 2011). There is a well-developed literature on the economics of coffee production in the Americas and Asia, and a growing literature on the ecology of commercial coffee production (reviewed in Ngo et al. 2011). There is also a growing body of research on the relationships among coffee pollination and biodiversity production in wild adjacent regions (e.g. Ricketts 2004; Klein et al. 2008; Klein 2009; Jha & Dick 2010) and the biodiversity of different types of coffee ecosystems.

Assumptions of Approach and Related Considerations

It is assumed that insect pollination is a direct adjunct to production in both *C. arabica* and *C. canephora* by raising both the quantity and the quality of the crop, so enhancing trade. It is also assumed that these benefits may cover the costs of management of all pollinators, including their habitats. A further assumption is that a positive feedback can develop between the insect pollinator-enhanced commodity and its trade. Given that our paper considers coffee production as part of a regional ecosystem, it is further assumed that valuable externalities exist between pollination of coffee by insects and adjacent biodiversity maintenance. These assumptions are addressed later in the paper.

Other considerations for developing revised policy and certification involve the economics of pollination services and coffee production, and include, in no apparent order:

1. determining the value of contributions of both managed and wild pollinating species to coffee production in a given region in the manner of Allsopp et al. (2008) for deciduous fruit crops in South Africa. Related to this consideration is knowing what is(are) the cost(s) of replacing insect pollinators for different coffee crops. Should wild pollinating species be important pollinators of the crop, it is necessary to know the costs of maintaining their habitat and its forgone economic opportunity.

2. from a biodiversity conservation perspective, it is important to know what is(are) the contribution(s) of the pollinating species to adjacent productivity of wild plant species and the food webs they comprise.

3. given that a greater involvement of insect pollinators has an economic cost, it is important to know if subsidies (incentives, or credits) are available to producers for having wild lands next to coffee plantations to offset costs.

4. implementation must consider the type of plantation ownership (i.e. private or corporate), the level of technology used (i.e. organic operation or synthetic inputs), the size of plantation, and distance to native forest lands.

Consideration of the socio-economics of the coffee trade is critical in any proposed venture. Kitti et al. (2006) indicated that

"... policy measures such as trade-related standards, premiums, tax reliefs, or government institutions are necessary for adoption of biodiversity-friendly growing practices".

Similarly, Allsopp et al. (2008) commented that

"...monetary valuation dominates natural resource conservation management decisions and policy-making".

Those quotations are important because they address the heart of the issue, that ecosystem-based objectives require the vehicles of economics and policy to be realized successfully, especially in the face of unstable prices and other risk factors (Benítez et al. 2006). Thus it is necessary to relate any proposed venture to consumer demand and market prices for the product and whether or not any trade standards are in effect, either by the coffee industry or in regulations of importing nations. It is also important to consider whether or not a particular nation factors in the costs of insect-pollinated sustainable coffee production into a broader biodiversity conservation program. Perhaps the most important economic criterion determining the acceptance of greater insect-pollinated coffee production is if it is consistently profitable for the independent farmer. Thus, as Kitti et al. (2006) and Allsopp et al. (2008) indicate, an agro-ecological venture soon becomes an economic venture.

Evaluation of the Assumptions and Considerations

Assessing the efficacy of insect pollination of coffee species

Wilmer & Stone (1989) were among the first to report the pollination of *C. canephora* by solitary bees in Papa New Guinea and its role in fruit set, leading them to outline measures to increase this type of pollination in plantations. Klein et al. (2003a) reported a great increase in the fruit set of Indonesian *C. canephora* when flowers were exposed to open pollination from insects and/or wind and pollen from different plants, the increase being approximately from 15% to 70%. In a different study, Klein et al. (2003c) reported that fruit set increased directly with the number of bee species visiting the flowers, and with the number of individual bees across all species. All those authors reported that the degree of fruit set declined the further away *C. canephora* coffee trees were from adjacent wild forest. In keeping with this result, the number of social bee species visiting coffee flowers also declined with increasing distance from adjacent forest. These results and those of Klein et al. (2008) attest to the role of wild bees in effecting pollination for *C. canephora*, especially those for which forest is the natural habitat.

The flowers of *C. arabica* are largely self-pollinating, but fruit set benefits from insect cross-pollination. Klein et al. (2003a) reported an increase (12.3%) in fruit set in this coffee species when insect visitation was allowed, a smaller increase than that seen in *C. canephora*. The percentage fruit set in *C. arabica* increased directly with the number (or diversity) of bee species, but not with the number of individual bees. The number of social bee species visiting the *C. arabica* flowers declined significantly with increasing distance from neighbouring forest (Klein et al. 2003b). The results of these studies indicate that highly successful fruit set rests on a rich wild bee community dependent on wild forest habitat situated less than a half kilometer from coffee plantations.

Ricketts et al. (2004) conducted similar pollination-productivity experiments on *C. arabica* in Costa Rica. That study reported a 20% increase in coffee fruit yield based on pollination by wild bees living with in 1km of forested land. A further benefit of bee pollination was that the overall quality of the fruit was increased by reducing the prevalence of distorted fruit by 27%. Distorted fruits must be removed from the harvested crop, adding extra expense to the producer. Ricketts (2004) also found that bee visitation rates and amounts of pollen transferred were higher when coffee plants were located less that 100m from forest fragments, and concluded that forest fragments, by providing habitat for a great diversity of wild pollinators, enhanced the reliability of coffee flower pollination by reducing dependency on managed honey bees. Vergara & Badano (2009) measured fruit production in Mexican coffee plantations that had low and high-impact management systems. They reported that the pollinating insect species richness and diversity of the low-impact plantations was greater than that of the high-impact lands. Fruit yield was also positively related to the diversity and species richness of the pollinators, indicating that the pollination service value was amenable to direct human management.

Roubik (2002a) presented evidence of global increase in coffee (*C. arabica*) yields and attributed them to the presence of introduced Africanised honey bees. He reported that bee pollination generated heavier fruits. Where national declines in yields were noted, Roubik (2002a) attributed the decline to a lack of pollination service linked to intensive agricultural land management practices that eliminated pollinator habitats. Roubik (2002b) reported that in Panama, Africanised honey bees were the principal pollinators of different cultivars of *C. arabica*, and that coffee flowers near forests attracted more bees, both native and Africanised, than flowers distant from forest vegetation.

The flower visitation rates and efficiency of pollination of coffee flowers may vary among insect pollinating species and regions. This issue has been reviewed by Free (1993) and more recently by Ngo et al. (2011). In Central and South American countries (Brazil, Costa Rica, Ecuador, Jamaica, Mexico, Panama) the honey bee is seen to be the most important pollinator species for *C. arabica*. This is consistent with the origins of both honey bees and *C. arabica* in Africa. Amaral (1972, cited in Free 1993) reported that the larger-bodied bees (*A. mellifera* and *Melipona quadrifasciata*) were more efficient pollinators of *C. arabica* flowers than smaller bee species. For *C. canephora* grown in Papua New Guinea, a leafcutter bee *Creightonella frontalis* was reported as the main pollinator (Willmer & Stone 1989). Klein et al. (2008) reported on the importance of a range of pollinating insects visiting *C. canephora*, their relationships to each other, coffee flowers, and their local environment. These same authors indicated that pollination success may reflect more the diversity of pollinators' niches and their interactions in the coffee-pollinator community than upon species richness. A similar increase in crop yield with functional diversity of pollinators has been found for *Cucurbita moschata* (Pinkus-Rendon et al. 2005; Hoehn et al. 2008) and strawberries (*Fragaria* sp.) (Chagnon et al. 1993). However, in highly-impacted areas of Ecuador with few forest remnants, Veddeler et al. (2008) reported that coffee crop yield increased directly with the density of wild social bees visiting coffee shrubs, and not the number of flowers per plant.

A consistent picture emerges from all those independent studies conducted on several continents. The pollination ecology of coffee is very generalized (Klein et al. 2008) and pollination by diverse insect pollinators of both *C. arabica* and *C. canephora* can offset potential pollinator deficits and increase fruit yield and quality. The role of nearby forest fragments that support a diversity of wild pollinating insects is a direct adjunct to production of both coffee species, and should be a component of modern policy to manage coffee on a sustainable basis.

Scale of operations and feasibility

In Mexico the majority of coffee plantations are small, less than 5 ha (Moguel & Toledo 1999), often in association with forest fragments that provide a canopy of shade. By contrast, sun coffee monocultures are much larger, and are grown in the absence of shade-conferring native trees and shrubs. Given the results of Klein et al. (2003a,b,c) and Ricketts (2004) that insect-pollination is most effective when coffee plants are less 500m from forest, and the study of Vergara & Badano (2009) on rustic plantations, it would appear that most plantations (at least in Mexico and Costa Rica) are conducive to active management for insect pollinators, especially wild species. Larger areas of sun coffee monocultures might require use of managed hive bees interspersed through plantations to enhance pollination. Small plantations adjacent to forests would be inclined to use fewer synthetic inputs than large sun coffee monocultures, a feature that would favour greater involvement of pollinating insects in production.

Assessing benefits of insect pollination of coffee to adjacent biodiversity

Although the benefits of wild pollinators to coffee pollination have been established, the contribution of coffee pollinators to adjacent biodiversity has been little-researched by comparison. *Coffea arabica* has a short period of synchronous blooming each year, but *C. canephora* may bloom synchronously but over several months (Klein et al. 2008), thus coffee plantations do not provide a continuous source of nectar to bees. However, shade coffee plantations in southern Mexico contained native bee communities that transported the pollen of native forest trees far across the landscape, linking forest fragments with shade coffee groves, and making a vital contribution to gene flow among native forest trees (Jha & Dick 2010). Karanja et al. (2010) studied insect pollinator visitation in organic and conventional coffee plantations and surrounding unmanaged areas of Kenya. Pollinating insects on coffee flowers also visited flowers of 42 other wild plant species that provided alternative floral resources. That study also found highly significant ($P < 0.0001$) statistical relationships between the richness of plants and pollinating bees in both the organic and the conventionally-managed plantations. Karanja et al. (2010) raised a further important management issue. Maintenance of insect populations that pollinate coffee flowers is vital when coffee plants are not in bloom, and this role is provided by other diverse flowering plant species adjacent to coffee plantations. Thus there is a pollination reciprocity that is valuable to biodiversity production and agricultural production that should form part of any policy considerations, and be the basis of future research in coffee growing regions. The greatest biodiversity is reported to reside in rustic and shade-coffee-certified plantations (Mas & Dietsch 2004; Philpott et al. 2007): large sun coffee monocultures likely contain least native biodiversity. Thus the opportunity for significant ecological reciprocity is possible only in the rustic and shade coffee systems in proximity to native forest areas.

What constitutes optimal adjacent forest cover for coffee and its pollinating species?

There are four types of Mexican coffee plantation systems that comprise trees (Moguel & Toledo 1999; Hernández-Martínez et al. 2009). These are shaded monoculture, commercial polyculture, traditional polyculture, and rustic, the amount and size of native trees increasing with progression to the rustic category. Shade coffee has been associated mainly with the traditional polyculture in this classification scheme, and is grown mainly by small-scale community producers. Moguel & Toledo (1999) identified this forest category as important habitat for both native plants (trees and epiphytes) and animals (all Chordates and Arthropods). The degree of shade cover is an important determinant of coffee yield: there is a positive relationship between 23-38% shade, but production may decline when cover exceeds 50% (Soto-Pinto et al. 2000). This relates to the studies of Klein et al. (2003b, c) who showed that for both *C. arabica* and *C. canephora*, the numbers of solitary bee species increased with light intensity of the plantation (i.e. habitat openness), and correlated with this, an increase in fruit set with increasing light intensity for *C. canephora* (Klein et al. 2003c).

Recent studies have investigated the biodiversity attributes of different Mexican shade coffee plantations. Gordon et al. (2007) analyzed the avian and mammalian richness of coffee plantations, and López-Gómez et al. (2008) measured the tree species richness and vegetation physical structure on coffee farms. However, neither of these studies reports on the composition of the pollinator communities of these areas. Similarly, the pollinator community structure of coffee plantations other than traditional shade systems has yet to be measured. This is an important next step, especially if shade coffee plantations are to be identified as an important agroecosystem for sustainable certified coffee production involving greater insect pollination. Important research from Jha & Vandermeer (2010) indicates how the bee species richness and abundance can be correlated with the tree vegetation cover in Mexican shade coffee plantations. Their research findings applied to solitary bees, cavity and wood-nesting bees and ground nesting bees. Most importantly, these authors found that the composition of the pollinating bee community was related directly to local farm land management practices (Jha & Vandermeer 2010). Specifically, these practices provide niches for diverse species of the pollinating community and also provide a continuity of nectar resources across seasons. Veddeler et al. (2008) concluded that the deliberate retention of old trees as bee nesting sites, and the presence of flowering plants that provide nectar year-round might promote a high abundance of bees to pollinate coffee flowers, but Priess et al. (2007) remarked that these features of land management have still to develop at a wider scale. The species composition and structure of shade coffee habitats can vary, and the distinction between rustic forest and traditional shade polyculture may be blurred. Mas and Dietsch (2004) called for a greater rigour in the definition of what is termed shade coffee landscape as it pertains to biodiversity conservation and trade certification of the coffee produced within it. This definition ought also to include an assessment of the animal pollinator community (Ngo et al. 2011).

Economic assessment of pollination services in coffee plantation mosaic habitats

Ricketts et al. (2004) calculated the economic value of insect pollination services provided by Costa Rican forest patches to coffee production. They reported that the value of the pollination service was comparable to potential alternative land uses, but far exceeded conservation incentives available to farmers. These authors' calculations did not include the value of other forest environmental services (e.g. water retention or carbon sinks), and concluded that the forest habitat portion of the coffee landscape yielded a "win-win" situation for both forest biodiversity and coffee culture. Gordon et al. (2007) could not correlate profitability with biodiversity measures in Mexican coffee plantations, but reported that the most diverse plantations were among the most profitable, regardless of prevailing coffee prices.

Olschewski et al. (2006) were critical of the economic assessment of Ricketts et al. (2004), arguing that valuations based on market prices and fruit yields, alone, were inadequate. They reported from their Ecuadorian and Indonesian studies that the absolute value of pollination services was similar in the two regions, and that the pollination service value was influenced directly by the amount of intact forest within 400 m of the coffee trees. In both Sulawesi and Ecuador, coffee revenues decreased markedly with increasing distance to intact forest. However, these authors reported that the value of alternative land uses was greater per unit area than the pollination service value of the forest. Priess et al. (2007) calculated the pollination service value of forested land in Sulawesi to be 46€/ha (approx. 52 US$), based on 2001 prices and data. That does not create a strong incentive to invest more in insect pollinator forest habitat, especially during times of depressed coffee prices, unless the other service values of such lands (especially the increased stability of crop pollinating bee communities (Klein 2009), nutrient capture, soil and water rentention, carbon sinks, and contributions to regional and/or continental biodiversity) are considered as additional adjuncts to production. There are, moreover, additional local human uses of tropical forest (e.g. fuel, construction wood, plant fibre, foods, medicines) that generate considerable value beyond the environmental services. Priess et al. (2007), on the basis of simulated land use changes in Sulawesi, indicated that preservation of patches of natural forest in the coffee agriculture mosaic could benefit both local economy and ecology, noting that reduced yields of coffee from reduced pollination of the flowers leads to greater conversion of forested land to alternate crop production. Olschewski et al. (2006) concluded that a higher realized price paid by a public desiring the continuation of coffee production in shade-grown systems would create the incentive to maintain this practice. Kitti et al. (2006) conducted an economic analysis of shade and sun coffee production in Costa Rica in which prices and protection fee (forest subsidy) were factored. They also found that higher prices for shade coffee and very high forest subsidy fees (as high as 358 US $/ha in an optimal scenario) would be necessary for shade coffee production to thrive versus sun-grown coffee. From an ecological perspective, shade-grown coffee is preferable to sun-grown monocultures because most natural biodiversity occurs on shade-protected plantations (see Philpott et al. 2007 and 2008 for Chiapas, Mexico, and Sulawesi, respectively). However, as Kitti et al. (2006) indicate, applying ecological considerations must be reconciled with economic realites.

Other economic analyses of shade coffee production yield similar conclusions. Perfecto et al. (2005) contended that high consumer prices for shade coffee that has low market penetration (10.5% of all Latin American exports) might hurt consumption, and so regional production. These authors argued that subsidies for shade coffee, apart from higher consumer prices, would be needed to sustain shade coffee production. Moreover, based on their analyses, high price premiums for shade coffee should go directly to growers to facilitate forest conservation. Benítez et al. (2006) envisaged subsidies from certification programs as the only way to maintain environmentally-friendly coffee production in the face of sun monoculture and other forms of land use. This is especially in view of the lower per unit area production of the shade coffee system compared to the sun coffee monoculture (Perfecto et al. 2005).

The conclusions generated from global surveys of crop production also have direct implications for coffee production and policy makers. A transition to intensely managed crop land may have serious negative consequences for long-term crop production, and especially for pollinator-dependent crops (Aizen & Harder 2009). Relationships between the growth in yield and the stability of crop production of pollinator-dependent crops have been analyzed in detail for intensely-cultivated agricultural regions (Aizen et al. 2008; Garibaldi et al. 2011). Aizen et al. (2008) presented evidence for an impact of pollinator shortage on crop yield at the global scale. Garibaldi et al. (2011) endorsed the management of pollinators and their habitats for their service to fruit set because of their positive effect on the yield and production stability of pollinator-dependent crops. Those latter authors endorsed further managed use of pollinators for certain crops, suggesting that such a practice could improve yields and reduce the need for more land cultivation.

Ricketts et al. (2004) indicated that the significance of the "shade" in shade coffee systems was really the proximity to forests that provide habitat for wild pollinators. Thus discussion of the role of these forest "mosaics" or "landscapes" is as much about the pollinators' habitats as it is about other aspects of native plant and animal biodiversity, especially given the observed reciprocity between shade coffee production and pollen dispersal in forests. Linking the diverse values and ecological service roles of forests to coffee production should permeate thinking about certification processes, subsidies and other forms of incentive payments: it is about the entire regional ecosystem, and not just coffee or native forest production. The functional connections between agriculture and ecology were detailed in Kevan et al. (1997), in which the term "natural mutualism" was used to define the natural services that supported agriculture. Insect pollination of crops (including coffee species) is such a natural mutualism whose importance forms the basis of the conclusions of Garibaldi et al. (2011).

Certification and insect pollination of coffee

There are three main types of certified coffee (fair trade, organic, and shade coffee), and the criteria for each have been outlined in Philpott et al. (2007). The three certifications do not use insect pollination as an explicit criterion. However, it may be implied in organic and shade-grown varieties. From a policy perspective, it is important that consideration of pollination by managed and wild pollinating insects be an explicit part of the certification process, and that consumers be aware of how that type of coffee has been generated. This could be applied not only to coffee grown in rustic and traditional shade systems, but also to the sun monoculture, shaded monoculture and commercial polyculture production systems identified by

Moguel & Toledo (1999). In these more intensive systems, pollination involving more Africanized honey bees (Roubik 2002b), as opposed to wild bees is a possible management option.

The explicit mention of insect-mediated pollination in coffee production in certification is not likely to detract from the public's attitude to that type of coffee: it could enhance it, given the generally favourable esteem in which pollinators are held. Francis & Francis (2006) indicated that the public demand for higher quality coffee was growing. This would favour environmentally-friendly coffee production over that from sun monocultures. A strong public demand for a certified product, coupled with awareness of what it entailed, could be an important factor in driving policy to protect wild pollination services, especially as it would affect prices and, possibly, regional-national conservation incentives. This is the positive feedback between trade and local forest shade coffee production to which we alluded earlier in the paper. Certification may have positive socio-economic benefits to coffee growers, especially when low market prices prevail (Pefecto et al. 2005). Philpott & Dietsch (2003) identified social justice as a concept to be linked with certification, quite apart form conservation issues, and advocated combining the three certifications into a single, rigorous certification, together with participation of farmers to enhance the conservation and socio-economic aspects of coffee growing. The inclusion of pollination by unmanaged and managed insect communities in the certification process is consistent with these objectives.

Roles of governments and non-governmental organizations in promoting pollination services via policy

Concerns about sustainable coffee yields, producers, trade in coffee, and regional social welfare rest on the initial formation of coffee fruits via pollination, which is why a consideration of pollinators must pervade policy and decision-making. Coffee production's enormous annual traded value, combined with its contribution to national economies and human livelihoods, has the power to influence production policy through multi-lateral government organizations, individual governments and coffee trade organizations. Treweek et al. (2006) identified the major challenge for policy makers as the recognition of biodiversity as the basis of sustainable agriculture, and to ensure that all agriculture and its trade is conducted respectful of biodiversity conservation. To achieve this, Treweek et al. (2006) emphasized identifying opportunities to incorporate biodiversity into policy making. We contend that insect pollination services to coffee production and other food crops represent such an opportunity that could be taken by agencies representing sustainable agriculture, coffee production and trade, plant-pollination relations, biodiversity conservation, and community development. Work on how to implement this concept has already been initiated by Kremen et al. (2007) who developed a model showing how insect pollination systems may be influenced by market forces and their interactions with policies determining agricultural land management. This conceptual model could be integrated readily with the predictive models of Jha & Vandermeer (2010) to implement pollinator habitat management at the plantation level. We contend that these authors' approaches should be applied to different coffee producing regions when revising policy.

The International Coffee Organization already conducts much research into aspects of coffee production that benefits the industry, such as improving the genetics of coffee varieties for increased yield and disease resistance (International Coffee Organization 2011). This organization, apprised of the role of insect pollinators, could develop policy to promote insect pollination services in all coffee production. This could be combined with extension work on how to implement best management practices for pollinator conservation in different coffee growing systems. These need to be developed for coffee grown in South and Central America, Africa and Asia because of different growing landscapes and prevailing management in these regions (Philpott et al. 2008).

There is a large coincidence between the global regions of greatest natural biodiversity and those of coffee production. The coffee growing system has to be linked, conceptually, with its broader natural ecosystem, and the criteria for sustainable coffee production should be incorporated into broader conservation schemes in the same regions (Mas & Dietsch 2004). This is important because floral resources from wild plants are needed to sustain the wild pollinator community that provides coffee pollination services during periods of non-flowering by coffee. Given that financial support will be required to realize and maintain this goal, policy to achieve this could be developed in partnership with farmers and governments of producing nations, and especially those global agencies whose principal aim is preserving native biodiversity. This approach is best combined with a revised rigorous certification scheme that reflects support for both ecosystem and socio-economic benefits, as envisioned by Mas & Dietsch (2004) and Philpott et al. (2007). Complementary policy could be developed by importing nations, given that much of the tropical coffee production is exported. How this revision of certification criteria might affect coffee production in different countries is not known and is beyond the scope of this paper. However, "raising the bar" for certification would create competition among producing nations for the higher-end segment of the coffee market, and especially for *C. arabica* that supplies most of the coffee for the gourmet niche market. This would be beneficial in generating more revenue for those producers of that variety of coffee, and benefits to the environment might follow. Ultimately, it is the quality of the retailed coffee that determines consumer preference and willingness to pay, and not so much the ecological conditions that prevailed on plantations. It is assumed that the quality of post-harvest processing of coffee right up to the retail level would be commensurate with the care given to the cultivation of coffee in certified plantations.

The following examples indicate how governments of coffee–importing nations have already implemented policy and regulations dealing with the manner in which imported food items and natural resources are produced. The

European Parliament placed a legal ban on the import of wild seal pelts and other seal products from Canada in 2011 because of the perceived way the animal pelts were procured (European Commission 2011a). This applies to all member nations of the Union, although individual nations (e.g. The Netherlands) had applied their own, national, ban in earlier years. The European Parliament has also developed a regulatory framework applied to imports of crops that are grown from genetically-modified seeds, fearing what might be the consequences of such genes entering human foods and agriculture (European Commission 2011b). In these two cases, the concern is about the manner in which the products are procured for import. The same consideration can be applied for imported coffee, in which certification reflects particular aspects of production and sale (e.g. under shade, without synthetic chemical input or fairly-traded). Producing coffee under field conditions that promote pollination by unmanaged and managed insect communities is also procurement, and, as such, is open to consideration by the European Parliament and any of its member nations.

A large amount of the coffee produced in Mexico is marketed in Canada and the USA. The North American Free Trade Agreement (NAFTA) among Mexico, Canada, and the USA contains an agreement (under the Commission on Environmental Cooperation) on environmental co-operation to ensure that goods traded among the Parties conform to environmentally-acceptable trade standards. It is within the mandate of the Commission on Environmental Cooperation to develop regulation on the pollination requirements of traded coffee under the terms of the NAFTA. Thus any imported coffee from Mexico could be regulated consistent with plantations having management plans that ensure that appropriate habitat is present to maintain populations of unmanaged and managed pollinators. Again, deliberate incorporation of pollinator provisions into a revised, rigorous, certification process is the effective way to do this.

Both Europe and North America are realizing the consequences of potential pollinator deficits to their own food security (National Research Council of the National Academies 2006; Biesmeijer et al. 2011). On November 16, 2011, European Union lawmakers passed with a huge majority a resolution to protect European bees because of the problems pollinators and food security face at the regional level (World Environment News 2011). Thus, it is plausible that the European Parliament might support policy development to protect coffee pollinators at the international level. Such action is warranted and is completely consistent with the Precautionary Principle. It would also represent an example of a laudable policy initiative by importing nations assisting, simultaneously, the sustainable production of a commodity in producing nations and the biodiversity of their natural environment. While the above discussion relates to Europe and North America, coffee is produced and imported over a greater range of other countries that are free to devise their own policies on how imported coffee is generated. However, any progressive policy developments made by the European Parliament and the NAFTA could emerge for consideration at a regional level (as among the major producers of South and Central America), and at the global level by the World Trade Organization.

Conclusions and Considerations for Policy Makers

1. Managed and unmanaged insect pollinators, by their abundance and diversity, enhance fruit production and quality in commercial coffee production. This role could offset potential pollinator deficits, directly promote production and trade, and has relevance to all countries in which *C. arabica* and/or *C. canephora* is/are produced.

2. The insect pollination service provided for coffee can extend into adjacent forest regions and promote natural biodiversity maintenance, especially in rustic and shade grown coffee.

3. Maintenance of a community of pollinators that service coffee flowers requires that alternate floral resources are available when coffee shrubs are not blooming. This requires management of adjacent forest mosaic lands to provide pollinator nesting sites and a blooming continuum of other flowering species. This management objective is more readily attainable in rustic and shade coffee plantations than in large sun coffee monocultures.

4. Insect pollination services should be a mandatory component of all coffee certification processes. A single, rigorous, certification that combines the criteria for existing coffee certifications would enhance the ecological and socio-economic aspects of coffee growing.

5. The costs of conserving forest areas adjacent to plantations may need to be offset by higher prices for certified coffee varieties and coupled with incentives and subsidies for biodiversity conservation.

6. The value of the international trade in coffee and the revenues generated for nations and communities enable trade to influence policy and regulation at many levels, from plantations to importing nations. The insect pollination service to coffee is a critical "entry point" into the agricultural and environmental policy process and should be explored.

7. Consumers who seek higher quality coffee and who realize the role of pollinator services can assist the policy process through choice of purchase.

8. Policy development can be undertaken by coffee producing and importing nations and agencies representing the coffee trade. Importing nations could require certified standards of coffee production that favour forest conservation, its contained insect pollinator community, and other biodiversity. This is an opportunity for the NAFTA, the European Parliament, the World Trade Organization, and other agencies engaged in supporting large-scale biodiversity conservation.

Acknowledgements

Funding for this study came from NSERC-CANPOLIN for which this is Publication Number 37. Funds for the presentation of this material at the 10th International Pollination Symposium at Cholula, Mexico, were made available from the ICPBR, IUBS, and CONABIO. We are grateful for the highly-constructive comments of the two anonymous reviewers.

References

Aizen M, Harder LD (2009) The global stock of domesticated honey bees is growing slower than agricultural demand for pollination. Current Biology 19: 915-918.

Aizen MA, Garibaldi LA, Cunningham SA, Klein A-M (2008) Long-term global trends in crop yield and production reveal no current pollination shortage but increasing pollinator dependency. Current Biology 18: 1572-1575.

Aizen MA, Garibaldi LA, Cunningham SA, Klein A-M (2009) How much does agriculture depend on pollinators? Lessons from long-term trends in crop production. Annals of Botany 103: 1579-1588.

Allen-Wardell G, Bernhardt P, Bitner R, Burquez A, Buchmann S, Cane J, Cox PA, Dalton V, Feinsinger P, Ingram M, Inouye D, Jones CE, Kennedy K, Kevan PG, Koopowitz H, Medellin R, Nabhan GP (1998) The potential consequences of pollinator declines on the conservation of biodiversity and stability of food crop yields. Conservation Biology 12: 8-17.

Allsopp MH, de Lange WJ, Veldtman R (2008) Valuing insect pollination services with cost of replacement. PLoS ONE September 10: 3(9): e3128.

Ashworth L, Quesada M, Casas A, Aguilar R, Oyama K (2009) Pollinator-dependent food production in Mexico. Biological Conservation 142: 1050-1057.

Bauer DM, Wing IS (2010) Economic consequences of pollinator declines: a synthesis. Agricultural and Resource Economics Review 39/3: 368-383.

Benítez PC, Kuosmanen T, Olschewski R, van Kooten GC (2006) Conservation payments under risk: a stochastic dominance approach. American Journal of Agricultural Economics 88: 1-15.

Biesmiejer JC, Roberts SPM, Reemer M, Ohlermuller R, Edwards M, Peeters T, Schaffers AP, Potts SG, Kleukers R, Thomas CD, Settele J, Kunin WE (2006) Parallel declines in pollinators and insect-pollinated plants in Britain and the Netherlands. Science 313: 351-354.

Biesmeijer JC, Sorensen PB, Carvalheiro LG (2011) How pollination ecology research can help answer important questions. Journal of Pollination Ecology 4: 68-73.

Breeze TD, Bailey AP, Balcombe KG, Potts SG (2011) Pollination services in the UK: How important are honey bees? Agriculture, Ecosystems and Environment 142: 137-143.

Brittain C, Potts SG (2011) The potential impacts of insecticides on the life-history traits of bees and the consequences for pollination. Basic and Applied Ecology 12: 321-331.

Buchmann S, Nabhan G (1996) The forgotten pollinators. Island Press, Washington, D.C.

Byrne A, Fitzpatrick U (2009) Bee conservation policy at the global, regional and national levels. Apidologie 40: 194-210.

Cameron SA, Lozier JD, Strange JP, Koch JB, Cordes N, Solter LF, Griswold TL (2011) Patterns of widespread decline in North American bumble bees. Proceedings of the National Academy of Sciences 108: 662-667.

Chagnon M, Gingras J, de Oliveira D (1993) Complementary aspects of strawberry pollination by honey and indigenous bees (Hymenoptera). Journal of Economic Entomology 86: 416-420.

Drucker AG (2004) Economic value of bee pollination services: Implications for farm management and policy. Workshop on Solitary Bees: Conservation, Rearing and Management for Conservation, Beberibe, Ceará, Brazil.

Eardley C, Roth D, Clarke J, Buchmann S, Gemmill B (2006) Pollinators and pollination: a resource book for policy and practice. African Pollinator Initiative, African Research Council, Plant Protection Research Institute, Pretoria, South Africa.

European Commission (2011a) Trade in seal products: Regulation (EC) No 1007/2009 of the European Parliament and of the Council on trade in seal products. URL:http://ec.europa.eu/environment/biodiversity/animal_welfare/se (accessed December 1, 2011).

European Commission (2011b) Genetically modified food and feed. URL: http://ec.europa.eu/food/food/.biotechnology/index_en.htm (accessed December 1, 2011).

Francis M, Francis N (Producers) (2006) Black Gold: A film about coffee and trade. Fulcrum productions, London, UK.

Free JB (1993) Rubiaceae. Insect Pollination of Crops (2nd Edition). Academic Press, London, UK, pp 475-478.

Gallai N, Salles J-M, Settele J, Vaissière BE (2009) Economic valuation of the vulnerability of world agriculture confronted with pollinator decline. Ecological Economics 68: 810-821.

Garibaldi LS, Aizen MA, A-M Klein, Cunningham SA, Harder LD (2011) Global growth and stability of agricultural yield decrease with pollinator dependence. Proceedings of the National Academy of Science 108:5909-5914.

Ghazoul J (2005a) Buzziness as usual? Questioning the global pollination crisis. Trends in Ecology and Evolution 20: 367-373.

Ghazoul J (2005b) Response to Steffan-Dewenter et al.: Questioning the global pollination crisis. Trends in Ecology and Evolution 20: 652-653.

Gordon C, Manson R, Sundberg J, Cruz-Angón A (2007) Biodiversity, profitability, and vegetation structure in a Mexican coffee agroecosystem. Agriculture, Ecosystems and Environment 118:256-266.

Hernández-Martínez G, Manson RH, Hernández AC (2009) Quantitative classification of coffee agroecosystems spanning a range of production intensities in central Veracruz, Mexico. Agriculture, Ecosystems and Environment 134: 89-98.

Hoehn P, Tscharntke T, Tylianakis JM, Steffan-Dewenter I (2008) Functional group diversity of bee pollinators increases crop yield. Proceedings of the Royal Society B 275: 2283-2291.

International Pollinator Initiative (1999) The São Paulo declaration on pollinators. Report on the recommendations of the workshop on the conservation and sustainable use of pollinators in agriculture with emphasis on bees. Brazilian Ministry of the Environment (MMA), Brazilia.

International Coffee Organization (2011) URL : http://www.ico.org/ (accessed August 22, 2011).

International Risk Governance Council (IRGC) (2009) Concept note: Risk governance of pollination services. IRGC, Geneva.

Jha S, Dick CW (2010) Native bees mediate long-distance pollen dispersal in a shade coffee landscape mosaic. Proceedings of the National Academy of Sciences 107: 13760-13764.

Jha S, Vandermeer JH (2010) Impacts of coffee agroforestry management on tropical bee communities. Biological Conservation 143: 1423-1431.

Karanja RHN, Njoroge GN, Gikungu MW, Newton LE (2010) Bee interactions with wild flora around organic and conventional coffee farms in Kiambu district, central Kenya. Journal of Pollination Ecology 2: 7-12.

Kevan PG (1999) Pollinators as bioindicators of the state of the environment, activity and diversity. Agriculture, Ecosystems and Environment 74: 373-393.

Kevan PG (2001) Pollination: Plinth, pedestal, and pillar for terrestrial productivity. The why, how, and where of pollination protection, conservation, and promotion. In: Stubbs CS, Drummond FA (eds) Bees and crop pollination – Crisis, crossroads, conservation. Entomological Society of America, Lanham, MD, pp 7-68.

Kevan PG, Phillips TP (2001) The economic impacts of pollinator declines: an approach to assessing the consequences. Conservation Ecology 5(1): 8. [online] URL: http://www.consecol.org/vol5/iss1/art8/ (accessed December 1, 2011)

Kevan PG, Thomas VG, Belaoussoff S (1997) AgrECOlture: Defining the ecology in agriculture. Journal of Sustainable Agriculture 19: 109-129.

Kitti M, Heikkilä J, Huhtala A (2006) "Fair" policies for the coffee trade – protecting people or biodiversity? MTT Discussion Papers 7, MTT Economic Research, Luutnantintie, Helsinki.

Klein A-M (2009) Nearby rain forest promotes coffee pollination by increasing spatio-temporal stability in bee species. Forest Ecology and Management 258: 1838-1845.

Klein A-M, Steffan-Dewenter I, Tscharntke T (2003a) Bee pollination and fruit set of *Coffea arabica* and *C. canephora* (Rubiaceae). American Journal of Botany 90: 153-157.

Klein A-M, Steffan-Dewenter I, Tscharntke T (2003b) Fruit set of highland coffee increases with the diversity of pollinating bees. Proceedings of the Royal Society, London B 270: 955-961.

Klein A-M, Steffan-Dewenter I, Tscharntke T (2003c) Pollination of *Coffea canephora* in relation to local and regional agroforestry management. Journal of Applied Ecology 40: 837-845.

Klein A-M, Vaissière BE, Cane JH, Steffan-Dewenter I, Cunningham SA, Kremen C, Tscharntke T (2007) Importance of pollinators in changing landscapes for world crops. Proceedings of the Royal Society B 274: 303-313.

Klein A-M, Cunningham SA, Bos M, Steffan-Dewenter I (2008) Advances in pollination ecology from tropical plantation crops. Ecology 89: 935-943.

Kremen C, Williams NM, Aizen MA, Gemmill-Herren B, LeBuhn G, Minckley R, Packer L, Potts SG, Roulston T, Steffan-Dewenter I, et al. (2007) Pollination and other ecosystem services produced by mobile organisms: a conceptual framework for the effects of land-use change. Ecology Letters 10: 299-314.

López-Gómez AM, Williams-Linera G, Manson RH (2008) Tree species diversity and vegetation structure in shade coffee farms in Veracruz, Mexico. Agriculture, Ecosystems and Environment 124: 160-172.

Mas AH, Dietsch TV (2004) Linking shade coffee certification to biodiversity conservation: butterflies and birds in Chiapas, Mexico. Ecological Applications 14: 642-654.

Mayer C, Adler L, Armbruster WS, Dafni A, Eardley C, Huang S, Kevan PG, Ollerton J, Packer L, Ssymank A, Stout JC, Potts SG (2011) Pollination ecology in the 21st Century: Key questions for future research. Journal of Pollination Ecology 3: 8-23.

Moguel P, Toledo VM (1999) Biodiversity conservation in traditional coffee systems of Mexico. Conservation Biology 13: 11-21.

National Research Council of the National Academies (2006) Status of pollinators in North America. National Academic Press, Washington D.C.

Ngo HT, Mojica AC, Packer L (2011) Coffee plant-pollinator interactions: a review. Canadian Journal of Zoology 88: 647-660.

Ollerton J, Winfree R, Tarrant S (2011) How many flowering plants are pollinated by animals? Oikos 120: 321-326.

Olschewski R, Tscharntke T, Benítez PC, Schwarze S, Klein A-M (2006) Economic evaluation of pollinator services comparing coffee landscapes in Ecuador and Indonesia. Ecology and Society 11(1): article 7. URL: http://www.ecologyandsociety.org/vol11/iss1/art7/ (accessed December 1, 2011).

Perfecto I, Vandermeer J, Mas A, Soto-Pinto L (2005) Biodiversity, yield, and shade coffee certification. Ecological Economics 54: 435-446.

Philpott SM, Dietsch T (2003) Coffee and conservation: a global context and the value of farmer involvement. Conservation Biology 17: 1844-1846.

Philpott SM, Bichier P, Rice RA, Greenberg R (2007) Field-testing ecological and economic benefits of coffee certification programs. Conservation Biology 21: 975-985.

Philpott SM, Bichier P, Rice R, Greenberg R (2008) Biodiversity conservation, yield, and alternative products in coffee agroecosystems in Sumatra, Indonesia. Biodiversity Conservation 17:1805-1820.

Pinkus-Rendon MA, Parra-Tabla V, Melendez-Ramírez V (2005) Floral resource use and interactions between *Apis mellifera* and native bees in cucurbit crops in Yucatán, México. Canadian Entomologist 137: 441-449.

Potts SG, Biesmeijer JC, Kremen C, Neumann P, Schweiger O, Kunin WE (2010) Global pollinator declines: trends, impacts and drivers. Trends in Ecology and Evolution 25: 345-353.

Priess JA, Mimler M, Klein A-M, Schwarze S, Tscharntke T, Steffan-Dewenter I (2007) Linking deforestation scenarios to pollination services and economic returns in coffee agroforestry systems. Ecological Applications 17: 407-417.

Richards, AJ (2001) Does low biodiversity resulting from modern agricultural practices affect crop pollination and yield? Annals of Botany 88: 165-172.

Ricketts TH (2004) Tropical forest fragments enhance pollinator activity in nearby coffee crops. Conservation Biology 18: 1262-1271.

Ricketts TH, Daily GC, Ehrlich PR, Michener CD (2004) Economic value of tropical forest to coffee production. Proceedings of the National Academy of Sciences 101: 12579-12582.

Roubik DW (2002a) The value of bees to the coffee harvest. Nature 417: 708.

Roubik DW (2002b) Feral African bees augment neotropical coffee yield. In: Kevan PG, Imperatriz-Fonseca V (eds) Pollinating bees: the conservation link between agriculture and nature. Ministry of Environment, Brasilia, Brazil, pp 255-266.

Sisk TD, Singh G, Tam J, Chan KMA, Klain S, Mach M, Martone R (2011) Barriers and incentives to engagement in public policy and science-based advocacy. Bulletin of the Ecological Society of America 92: 276-280.

Soto-Pinto L, Perfecto I, Castillo-Hernandez J, Cabellero-Nieto J (2000) Shade effect on coffee production at the northern Tzeltal zone of the state of Chiapas, Mexico. Agriculture, Ecosystems and Environment 80: 61-69.

Southwick EE, Southwick Jr L (1992) Estimating the economic value of honey bees (Hymenoptera; Apidae) as agricultural pollinators in the United States. Journal of Economic Entomology 85: 621-633.

Steffan-Dewenter I, Potts SG, Packer L (2005) Pollinator diversity and crop pollination services are at risk. Trends in Ecology and Evolution 20: 651-652.

Tang J, Wice J, Thomas VG, Kevan PG (2007) Assessment of Canadian federal and provincial legislation to conserve native and managed pollinators. International Journal of Biodiversity Science and Management 3: 46-55.

Treweek JR, Brown C, Bubb P (2006) Assessing biodiversity impacts of trade: areview of challenges in the agriculture sector. Impact Assessment and Project Appraisal 24: 299-309.

Veddeler D, Olschewski R, Tscharntke T, Klein A-M (2008) The contribution of non-managed social bees to coffee production: new economic insights based on farm-scale yield data. Agroforestry Systems 73: 109-114.

Vergara CH, Badano EI (2009) Pollinator diversity increases fruit production in Mexican coffee plantations: The importance of rustic management systems. Agriculture, Ecosystems and Environment 129: 117-123.

Williams IH (2003) The Convention on Biological Diversity adopts the International Pollinator Initiative. Bee World 84: 27-81.

Willmer PG, Stone GN (1989) Incidence of entomophilous pollination of lowland coffee (*Coffea canephora*); the role of leaf cutter bees in Papua New Guinea. Entomologia, Experimentalis et Applicata 50: 113-124.

World Environment News (2011) EU lawmakers call for action to protect bees. URL: http://planetark.org/environews/item/63898 (accessed December 1, 2011).

11

Potential pollinators and robbers: a study of the floral visitors of *Heliconia angusta* (Heliconiaceae) and their behaviour

Katharina Stein* and Isabell Hensen

Martin-Luther-University Halle-Wittenberg, Institute for Biology/Geobotany and Botanical Garden, Am Kirchtor 1, 06108 Halle/Saale, Germany

Abstract—Floral syndromes are traditionally thought to be associated with particular pollinator groups. Ornithophilous flowers tend to have traits that facilitate bird pollination such as having long, narrow, tubular corollas, often vivid coloration and diluted, sucrose-rich nectar. However, recent studies have shown that flowers attract a broader spectrum of visitors than might be expected. Furthermore, the classification of floral visitors as 'robbers' or 'pollinators' often is not as simple as it seems, as pollinators can at times act as robbers and vice versa. We studied the species composition, behaviour and ecology of floral visitors, including potential pollinators and robbers, of *Heliconia angusta* (Heliconiaceae), an endemic understory herb of the Atlantic Rainforest of Brazil. In addition, the impact of the plant inflorescence attractiveness and of weather and light conditions on visitor abundance and frequency was investigated.

Flower visitors were found to be scarce with a total of only 151 visits being observed during 120 h of field observations. A stingless bee species (*Trigona* sp.) appeared to be the most abundant visitor to the ornithophilous flowers of *H. angusta,* along with four different species of hummingbirds and two species of butterflies. We consider *Trigona* sp. rather as pollen robber, but which still has the potential to be a secondary pollinator, whereas the hummingbirds were the principle legitimate visitors. Most flower visitors were recorded between 9.00 am and 1.00 pm with a higher number visiting under semi-shaded conditions than in full shade. Hummingbird numbers increased with flower abundance while the other visitor group numbers were not affected.

Keywords: hummingbirds, stingless bees, Trigona, pollen robbers, nectar, inflorescence attractiveness

Introduction

Tropical flowering plants depend overwhelmingly on animals as vectors of pollen transfer. The vast majority of pollinators are represented by insects, such as bees and butterflies, as well as by some vertebrate groups including mainly birds and bats (Proctor et al. 1996). Despite the great variety of plant-pollinator systems, it is possible to associate floral traits with particular pollinator groups as a series of pollination syndromes (e.g. Campbell et al. 1996; Galetto 1998). As an example, ornithophilous flowers tend to have traits that facilitate bird pollination (Smith et al. 1996), such as long, narrow, tubular corollas and often vivid coloration (Willmott & Burquez 1996). Neotropical *Heliconia* species (Heliconiaceae) are associated with "bird-flower" pollination syndromes in respect to their long, white, tubular flowers and intensive red coloured bracts, attributes which may otherwise infer that they have no other types of visitors (Stiles 1975).

However, the pollination syndrome concept has recently been criticized with researchers finding that flowers often attract a broader spectrum of visitors than that predicted from their respective syndromes (Fenster et al. 2004; Dias da Cruz 2006; Ollerton et al. 2009; Schmid et al. 2011) and that many plants have more than one type of pollinator (Waser et al. 1996; Waser & Ollerton 2006). Ollerton et al. (2003) reported that generalist insects often visit more specialized plants, a situation that was also found for bird-adapted mistletoe flowers being visited by bees (Robertson et al. 2005). In addition to representing secondary or alternative pollinators for more specialized plants (Canela & Sazima 2005; Schmid et al. 2011) other studies have shown that more generalist insects may frequently be regarded as pollen and nectar robbers, in particular to flowers of many long-tubed, nectar rich, hummingbird pollinated species (McDade & Kinsman 1980; Renner 1983). The interaction of plants with robbers and pollinators may be influenced by factors such as the morphology of the visitors, the availability and quality of rewards, and spatio-temporal variation in the abundance of flowers, robbers and pollinators (Horvitz & Schemske 1990; Thompson & Pellmyr 1992). This complexity and interdependence of factors can lead to the assumption, that the classification of floral visitors as either 'robbers' or 'pollinators' may be simplistic (Arizmendi et al. 1996), as some robbers can act as pollinators (e.g. Graves 1982, Slaa et al. 2006) and pollinators sometimes can act as robbers (e.g. Willmer & Corbet 1981; Roubik 1989)

The principle reward for most flower visitors is nectar (Proctor et al. 1996), and the particular nectar sugar

*Corresponding author; email: steinkatharina@hotmail.com,

composition has often been related to the pollinators of a plant. Flowers pollinated by hummingbirds, butterflies or long-tongued bees often secrete sucrose-rich nectar (Baker & Baker 1990; Perret et al. 2001; Krömer et al. 2008). In addition to the floral reward, in terms of number of flowers displayed or open, flower attractiveness may have a positive influence on the abundance and frequency of the flower visitors (Bosch & Waser 2001; Steven et al. 2003; Harder et al. 2004). In addition, weather and light conditions influence the flower visitors′ behaviour. Studies by Renner (1983), Stone and Jenkins (2008), Vicens and Bosch (2000) and Döll et al. (2007) revealed impacts of temperature, solar radiation and time of day on the number of floral visitors and their frequency.

This study focuses on *H. angusta* Vell. (Heliconiaceae), an understory herb endemic to the Atlantic Rainforests of Brazil. Wild populations are classified as vulnerable by the World Conservation Union, mostly because of the conversion of their diminishing habitats into agricultural land. Studies have been carried out concerning the morphology and anatomy of the species (Simão & Scatena 2001), or the role of the species as a nectar source for hummingbirds from an ornithological point of view (Sazima et al. 1995; De Castro & Araujo 2004). However, knowledge on its ecology is still lacking. This study focuses on the pollination biology of *H. angusta*. Thus we address the following questions:

1. What are the floral visitors of *H. angusta* and what is their behaviour in terms of being potential pollinators or flower robbers?

2. Does the floral attractiveness (number of bracts, flowers) have an impact on the number of visitors or their frequency?

3. Is the flower visitation rate influenced by the time of day, weather and/or light conditions?

4. What is the sugar composition of the nectar produced and is it related to the preferences of the observed flower visitors?

MATERIAL AND METHODS

Species and study region

The family Heliconiaceae comprises a single genus, *Heliconia* L., with 250-300 species distributed mainly throughout neotropical areas from northern Mexico to southern Brazil (Dahlgren et al. 1985; Kress 1990). In Brazil, there are approximately 40 species covering two primary areas of distribution: the Amazon basin and the Atlantic coastal rainforest (Kress 1990).

H. angusta is an ornithophilous, perennial, clonally growing, understory herb endemic to the Atlantic Rainforest in south-eastern Brazil. Like other *Heliconia* spp., it is a common component of the understory of Neotropical forests and has a patchy distribution. *H. angusta* displays a "steady state" flowering strategy (Gentry 1974b; Stiles 1975) producing flowers from April to October, making it a crucial nectar resource for hummingbirds (De Castro & Araujo 2004). The single-terminal inflorescence consists of five to eight red coloured bracts and about five hermaphroditic flowers per cincinnus. The approximately 4 cm long tubular corolla of the flowers is white, while the ovary and pedicel are orange in colour. Flower anthesis usually lasts for only one day (Stein, pers. obs.); flowers open in the early morning and are abscised in the evening, or at the latest, the next morning. Over a period of 2 – 4 months (between June and October), each ramet produces one or two new flowers roughly every 2nd day. In the Atlantic Rainforest, *H. angusta* is found mainly in shaded, moist places inside the forest, sometimes at the forest edge close to rivers (Simão & Scatena 2001; Stein, pers. obs.).

The study was conducted in the Atlantic Rainforest ("Mata Atlântica") of the state of Rio de Janeiro, Brazil in the private reserve "Reserva Ecológica de Guapiaçu" (REGUA – 22°25'53"S, 42°45'20"W) in the municipality of Cachoeiras de Macacu. The 5500 ha reserve is located on the south-facing slopes of the Serra dos Órgãos Mountain range, about 100 km from the city of Rio de Janeiro. The mean annual temperature for this region is about 23 °C with a mean annual rainfall of about 2560 mm. There is a hot and rainy season from October to March and a cooler and drier season from April to September (Kurtz & de Araújo 2000). The vegetation can be classified as "evergreen dense ombrophilous forest" (Veloso et al. 1991), which is typical for the lower and medium elevations of the coastal mountain range (Morellato & Haddad 2000; Oliveira-Filho & Fontes 2000).

Data collection

Data collection was carried out from July to September 2009 and from June to August 2010. Flower visitor observations were carried out at three different 2 x 2m plots at elevations of 50 m to 200 m a.s.l. The three plots were chosen to have replication. The distance between observation plots ranged between 800 m (minimum) and 4 km (maximum) as a result of the patchy distribution of the plants. The mean number of open flowers per plot and observation day was 2.08 (standard deviation SD: 1.56), with a minimum of 1 and a maximum of 6 open flowers. Mean density of ramets per observation plot and day was 2.58 (SD: 1.44). Each ramet was considered an individual due to the fact that an assignment of ramets to a certain individual was not possible in the field. Flower visitors were observed on a total of twelve days (6 d in 2009, 6 d in 2010, 2 d per plot and year), each continuously from 7:00 am to 5:00 pm, resulting in a total observation period of 120 hours. Care was taken to ensure that weather conditions on all observation days were comparable (sunny to cloudy). In order to test for any potential impact on the number/frequency of flower visitors, the data recorded on each observation day included the number of flowering conspecifics within the 2 x 2m observation plot, the total number of displayed and open flowers respectively and the total number of bracts per plot.

Young flowers tend to be hidden among the bracts and, shortly before anthesis, the flowers (still closed) become erect and visible. Open flowers show a lower lip that curls downwards and provides access to the corolla interior with the reproductive organs and the nectar reward. In addition to the flowers, the bracts of the inflorescences might play an important role in attracting flower visitors due to their bright reddish colour (assumed to be very attractive to hummingbirds, e.g. Waser 2006), which makes them highly visible. We counted flowers and bracts separately because young flowers remain hidden in the bracts and potentially do not contribute to the attractiveness of an inflorescence, whereas the bracts appear to be very "showy". Open flowers were counted separately because they offer floral rewards such as nectar and pollen. The number of displayed flowers includes the number of open flowers, thus being the total number of visible flowers to visitors. Flower visitors entering an open flower or touching the anthers/stigma respectively were counted as legitimate flower visitors (potential pollinators). Only one "nectar thieve" species (see Inouye 1980) was observed during the whole observation period and not included in this study: a butterfly species that stole nectar eight times without touching the flowers' reproductive organs. Flower visitors that only collected pollen without touching reproductive parts of the flower were counted as "pollen robbers" and included in this study. Each time a flower visitor entered the observation plot, it was counted, with any visitor leaving the plot and entering it again being counted as a "new" visitor. Marking individuals was not possible, thus some visits may have been repeat visits by the same individual. Flower visitors were counted in observation units of 20 min each. Three observation units per hour were conducted, resulting in 30 units per observation day of 10 hours (7:00 am-5:00 pm) and in 360 observation units in total for all twelve observation days. In addition, for each unit, weather (sunny, bright, cloudy, overcast) and light conditions (half shade, shade) were recorded as well as the guild of visitor (Trochilidae, Hymenoptera, Lepidoptera) and the number of visited flowers in a row per plot. Hummingbird species were identified using a fieldguide and confirmed by ornithologists from the University of Rio de Janeiro. Insect visitors were caught by sweep netting. Whereas the butterfly species remained unidentified, the genus of the stingless bee species could be identified with the help of a binocular microscope and an identification key for neotropical insects.

Nectar sugar composition

Twenty nectar samples of *H. angusta* were collected in the late afternoon from flowers bagged before anthesis using micro-pipettes (Drummond "Microcaps" 10 μl). The nectar was transferred onto small strips of filter paper, which were then dried and stored in silica gel (Schwerdtfeger 1996). The laboratory analysis was carried out using a "Merck Hitachi-HPLC" with RI-Detector. The solvent used was 30 % water and 70 % acetonitril (ACN); the column was a Macherey-Nagel Nucleodur 100-5 NH2, 250 x 4.6 mm. S/H ratio was calculated as [sucrose] / [fructose + glucose].

Statistical analyses

In order to test the homogeneity of the recorded number of total visits and number of visits of each visitor group for all days and sites we performed a Two-Way-ANOVA.

The impact of plant inflorescence attractiveness (number of open and displayed flowers, number of bracts) on the number of visitors per group and visitation frequency (number of visitors / flower / hour) per plot was analysed by way of a Best Subsets Regression. Best Subsets Regression is a technique for selecting variables in a multiple linear regression by systematically searching through the different combinations of the independent variables (number of flowers displayed, number of flowers open, number of bracts) and selecting the subsets of variables that best contribute to predicting the dependent variable (number of visitors; visitation frequency, respectively).

To statistically determine the potential influence of the time of day, light- and weather-conditions on the number and frequency of all visits, a Kruskall-Wallis-ANOVA on ranks with Student-Newman-Keuls multiple comparison procedure was used, as the data were not normally distributed. To test for differences between the number of visits of each visitor group in respect to the time of day, a One-Way-ANOVA was carried out. The levels of significance are as followed: $p < 0.05$ = *; $p < 0.01$ = ** and $p < 0.001$ = ***. All analyses were performed using SigmaStat 3.0.1, SPSS Inc. 2003 and SigmaPlot 2000, SPSS Inc. 2000.

RESULTS

Visitor species and their behaviour

The numbers of overall visits for each observation day and site were comparable (total visits vs. observation day, $F = 0.16$, $p = 0.965$; total visits vs. site, $F = 0.33$, $p = 0.67$). Furthermore, no influence of the observation day or site could be observed on the number of visits of each visitor group (Hymenoptera: day vs. visits $F = 0.04$, $p = 0.99$; site vs. visits $F = 0.28$, $p = 0.69$; Trochilidae: day vs. visits $F = 5.37$, $p = 0.32$; site vs. visits $F = 0.64$, $p = 0.57$; Lepidoptera: day vs. visits $F = 0.02$, $p = 0.88$; site vs. visits $F = 0.16$, $p = 0.76$).

TAB. 1. Total number of floral visits to the three 2 x 2 m observation plots of *H. angusta* based on 360 20 min observation periods in the Atlantic Rainforest of Brazil.

visitor species	number of visits
Meliponini (stingless bees) *Trigona* sp.	81
Trochilidae (hummingbirds) in total	60
Phaethornis ruber	*41*
Phaethornis squalidus	*10*
Aphantochroa cirrochloris	*8*
Thalurania glaucopis	*1*
Lepidoptera (butterflies), 2 species	10
Total number of visits	151

During the 360 observation units of 20 minutes each we recorded a total of 151 flower visits to the flowers of *H. angusta*, with a total of seven different visitor species being noted (Tab. 1). The most frequent visitor to *H. angusta* was *Trigona* sp., a small black stingless bee (Hymenoptera, Apidae, Apinae, Meliponini), which accounted for 54 % of all visits and collected only pollen. Four different species of nectar feeding hummingbirds accounted for 40 % of all visits, of which two species of traplining hermits (*Phaethornis ruber* and *Phaethornis squalidus*) visited *H. angusta* most frequently

(Tab. 1). All hummingbird species made legitimate visits to the flowers. Pollen was deposited on the forehead of the birds and they touched the stigma each time when visiting a flower. Therefore the observed hummingbird species can be considered as potential pollinators. *Trigona* sp. was observed collecting pollen 81 times in contrast to the hummingbirds that only were observed 60 times visiting the flowers during the whole observation period. Out of all visits of *Trigona* sp. only 16.05 % were legitimate visits, meaning that in 13 visits out of a total of 81 visits the stingless bees touched the stigmas and thus might have pollinated the flowers of *H. angusta*. 83.95 % of all visits can be considered as illegitimate, because the stingless bees were only collecting pollen without touching the stigmas and thus were robbing the flowers. Two unidentified species of butterfly were rare flower visitors, together accounting for only 6 % of all visits. When visiting the flowers they touched the anthers and pollen got deposited to their body.

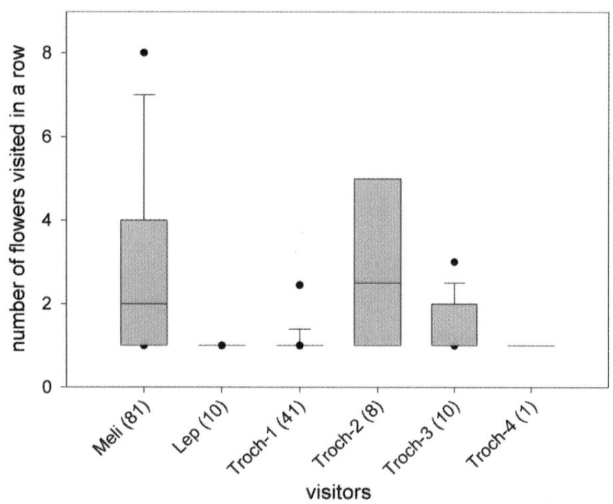

FIG. 1. Box-plots showing the number of *H. angusta* flowers visited in a row by each visitor species per 20min observation period for a total of 120 h of observation (figures in parentheses indicate number of visits; the black line is the median; the upper/lower lines of the box are the 1st and 3rd quartiles; the vertical lines are the 5th and 95th percentiles and the dots represent outliers within the 5th and 95th percentiles). Meli = stingless bees (Meliponini), single species: *Trigona sp.*; Lep = butterflies (Lepidoptera), two unidentified species; Troch = hummingbirds, Troch-1: *Phaethornis ruber*, Troch-2: *Aphantochroa cirrochloris*, Troch-3: *Phaethornis squalidus*, Troch-4: *Thalurania glaucopis*

Trigona sp. visited the flowers per plot up to 12 times in a row (mean: two flowers, Fig. 1). The Sombre Hummingbird (*Aphantochroa cirrochloris*) visited an average of three flowers in a row, with *P. ruber* and *P. squalidus* visiting one and two, respectively. The Violet-capped Woodnymph (*Thalurania glaucopis*) was recorded only once visiting one flower (Fig. 1).

Influence of plant inflorescence´ attractiveness on visitor abundance

The mean number of open flowers per plot was 2.08 (SD: 1.56); mean number of flowers displayed per plot 21.16 (SD: 13.45), and mean number of bracts per plot and day 17.00 (SD: 9.74). Inflorescence attractiveness (parameters: number of open and displayed flowers, number of bracts) had an effect on the number of floral visitors for hummingbirds (Tab. 2), but not for *Trigona* sp. or butterflies. Hummingbird visits per plot significantly increased with more flowers displayed ($p < 0.001$, $R^2 = 0.68$, Tab. 2). The same positive interrelation was found when the two parameters of flowers displayed and number of bracts were taken together ($p < 0.01$, $R^2 = 0.71$), and when all three parameters (plus number of open flowers) were included in the analysis ($p < 0.05$, $R^2 = 0.71$, Tab. 2). There was no influence of inflorescence attractiveness on visitor frequency in any of the three visitor groups ($p \geq 0.19$, $R^2 \leq 0.20$).

TAB. 2. Best Subsets Regression of the influence of the plant inflorescence attractiveness on the number of visits. Each visitor group was tested separately (R^2 = coefficient of determination, p = level of significance; $p < 0.05$ = *; $p < 0.01$ = **; $p < 0.001$ = ***)

	P number of flowers displayed	P number of bracts	P number of open flowers	R^2
Meliponini	-	-	0.13	0.21
	-	0.53	0.12	0.25
	0.86	0.63	0.18	0.25
Lepidoptera	-	-	0.45	0.06
	0.97	-	0.55	0.06
	0.99	0.99	0.58	0.06
Trochilidae	<0.001 ***	-	-	0.68
	0.01 **	0.4	-	0.71
	0.03 *	0.43	0.83	0.71

Influence of time of day, weather and light conditions on number and frequency of visits

Most total visits were recorded in the morning between 9:00 am and 11:00 am (morning vs. early morning: 7:00 am to 9:00 am, $q = 4.06$, $p < 0.05$, Fig. 2) and between 11:00 am and 1:00 pm (Fig. 2). The total number of visits decreased throughout the course of the day and was lowest in the late afternoon between 3:00 pm and 5:00 pm (morning vs. late afternoon, $q = 7.51$ $p < 0.05$, Fig. 2). Stingless bees (Meliponini) responded strongest to the time of day with a clear peak around noon (Fig. 2, Tab. 3). Significantly more bees than hummingbirds ($q = 3.27$, $p < 0.05$) and butterflies ($q = 6.18$, $p < 0.001$) were observed between 9:00 am and 11:00 am. The number of visits of hummingbirds and bees within the time intervals of 11:00 am to 1:00 pm and 1:00 pm to 3:00 pm did not differ significantly from each other. No differences in the number of visits of all visitor groups could be found for the time intervals 7:00 am - 9:00 am and 3:00 pm - 5:00 pm (Tab. 3).

Light conditions had a significant impact on visitor frequency, but not on visitor number. The number of visits per flower increased under half-shade conditions in comparison to full shade at the observation plots (half-shade vs. shade, $q = 4.28$, $p < 0.05$). There was no significant influence of weather conditions on number and frequency of visits.

Nectar sugar composition

The one-day flowers of *H. angusta* produced 44.4 µl (SD: 18.4) of nectar on average per day (maximum 80 µl). The nectar was sucrose-rich with an average of 10 % sucrose (SD: 2.1), but the hexose level was rather low with 3 % fructose (SD: 0.8) and 3 % glucose (SD: 0.9). The total nectar concentration accounted for 16 % (SD: 3.5) and the sugar ratio (sucrose/hexose) was 1.63 (SD: 0.33).

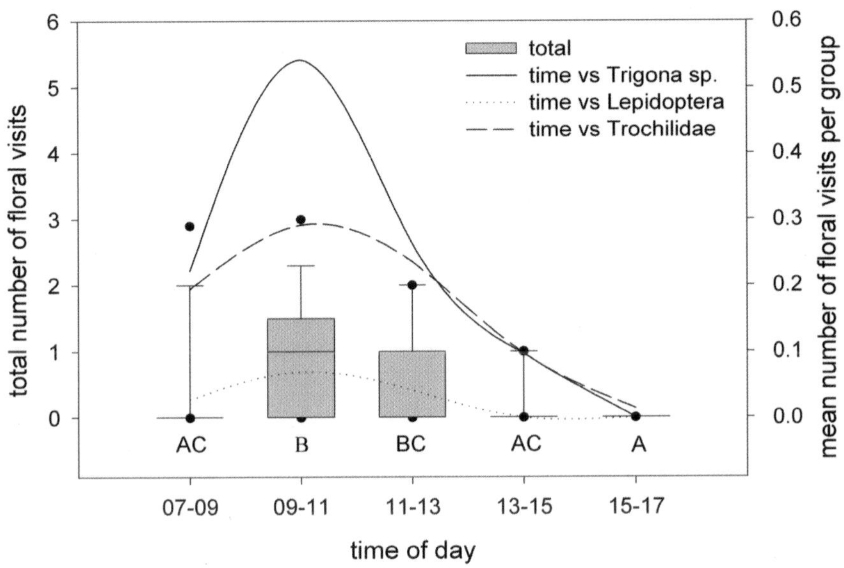

FIG. 2. Boxplots (corresponding to left y-axis) showing the total number of floral visits per flower per 2 x 2m observation plot to *H. angusta* at different times of day (number of 20 min observation units was 72 for each time span; dots, outliers within 5th and 95th percentiles). Box-plots with equal letters do not differ significantly from each other (ANOVA testing). Curves (corresponding to right y-axis) show the mean number of floral visitors per group. *Trigona* = stingless bees (Meliponini, one species); Lepidoptera = butterflies; Trochilidae = hummingbirds.

TAB. 3. Mean number of visits (± SD) of each visitor group during the course of day (different time intervals of 2 hours). ns = not significant.

time period (hrs)	7 – 9	9 – 11	11 – 13	13 – 15	15 – 17
visitor group:					
Hymenoptera	0.22 ± 0.75	0.54 ± 0.93	0.26 ± 0.63	0.1 ± 0.30	-
Trochilidae	0.19 ± 0.49	0.29 ± 0.57	0.24 ± 0.52	0.1 ± 0.34	0.01 ± 0.12
Lepidoptera	0.03 ± 0.45	0.07 ± 0.26	0.04 ± 0.20	-	-
F*	2.84	9.57	4.5	3.3	1.0
p*	0.06 ns	< 0.001	< 0.05	< 0.05	0.37 ns

* One-Way-ANOVA

DISCUSSION

The investigated plant species was found to be visited by two dominant flower visitor species (a stingless bee *Trigona* sp. and a hermit hummingbird *P. ruber*), which is in accordance with the results of other studies in tropical and temperate regions. De Castro and De Oliveira (2002) also reported visitation by one or two dominant flower-visitor species for several species of Rubiaceae while Gao et al. (2004) revealed different flower visitors for *Curcumorpha longiflora* (Zingiberaceae), Mitchell et al. (2004) for *Mimulus ringens* (Phrymaceae) and Tomimatsu & Ohara (2003) for *Trillium camschatcense* (Trilliaceae). However, care has to be taken when interpreting the most common visitor as the most important pollinator (Fenster et al. 2004). Schmid et al. (2011) reported that although bees most frequently visited the ornithophilous flowers of *Aechmea nudicaulis* (Bromeliaceae), a rare visiting hummingbird had the highest relative pollination effectiveness and was significantly more effective than all the bees combined. Only experimental tests can reveal the pollination effectiveness of each visitor species (Freitas & Paxton 1998; Botes et al. 2009). As this study only deals with the flower visitor species of *H. angusta* and their behaviour, our data do not allow for conclusions to be drawn on the effectiveness of the visitors as pollinators. Further reports on many tropical long-tubed, nectar rich, hummingbird-pollinated species suggest that a more complicated relationship frequently exists involving floral parasitism, e.g. by pollen and nectar collecting insects (Gentry 1974a, Janzen 1975, McDade & Kinsman 1980). Many species of stingless bees, in particular of the genus *Trigona* are known to rob flowers (Almeda 1977; Roubik 1982; Renner 1983) by collecting nectar and pollen. A study of McDade and Kinsman (1980) revealed that species of the genus *Trigona* took pollen from the ornithophilous *Aphelandra golfodulcensis* (Acanthaceae) flowers in Costa Rica without contributing to pollination. Roubik (1982) postulated that robbing *Trigona* sp. may present a reproductive hazard to many species of flowering plants throughout the tropics (see also Hubbel & Johnson 1978; Michener 1979). On the other hand stingless bees are considered important pollinators of the native flora in tropical and subtropical parts of the world, and they have been found to contribute to the effective pollination of 18 crops and many wild plants (Heard 1999; Slaa et al. 2006). Especially in the Neotropics stingless bees were the only bees for a long time (before the introduction of European honey bees) and often make up half of all flower visitors (Biesmeijer JC, 2011, pers. comm.). The review of Slaa et al. (2006) on the importance of stingless bees in applied pollination specifies ten different species of *Trigona* alone, which effectively pollinated different crops. Clearly the objective of bees (and all other flower visitors) is not to pollinate but to get reward, but in the process they might pollinate (Biesmeijer JC, 2011, pers. comm.). In our study, *Trigona* sp. observed at flowers of *H. angusta* only collected pollen, and the typical robbing behaviour of piercing of the corolla to steal nectar was not observed. *Trigona* sp. was observed at times touching the stigmas of *H. angusta* flowers when collecting pollen. Although these legitimate visits only account for 16.05 % of all visits of the bees, they might have resulted in a transfer of pollen to the stigma and the pollination of flowers on occasion. However, 83.95 % of all bee visits were illegitimate, when *Trigona* sp. only collected pollen without touching the stigma. Thus, *Trigona* sp. may rather be considered a robber to *H. angusta* due to the frequent removal of pollen that otherwise would have been transferred to stigmas by more-effective pollinators. Nevertheless species of the genus *Trigona* still have the potential to be secondary pollinators, even if pollination may happen occasionally and rare during pollen collection. In contrast, all visits of hummingbirds were legitimate. Pollen was deposited to the forehead of the birds and they touched the stigmas each time they visited a flower. Thus the observed hummingbird species are considered as principle potential pollinators of *H. angusta*.

Although the stingless bees and the main hummingbird visitors visited several flowers in a row, the mean visitation rate of all visitor groups was very low at 0.7 visits per flower per hour. Similar results were found by Bruna et al. (2004) for *Heliconia acuminata* (Heliconiaceae) from Amazonia and Singer and Sazima (2000) also recorded between zero and four visits per day to the flowers of *Stenorrhynchos lanceolatus* (Orchidaceae), a hummingbird pollinated orchid from the Atlantic rainforest.

Our study revealed a significant correlation between plant inflorescence attractiveness and the number of hummingbird visits per plot. Hummingbird visits increased with the number of flowers displayed and the influence of this parameter on the number of hummingbird visits was statistically the strongest. The number of bracts or that of open flowers had no individual influence on the number of hummingbird visits; however, when they were assessed together with the number of displayed flowers, a significantly higher number of hummingbird visits were recorded. The contrast of the red bracts with the long white flowers (ornithophilous flowering syndrome, see Vogel 1996; Waser 2006) seems to be very attractive to hummingbirds, as this colour display is highly visible in the tropical forest vegetation environment. A positive correlation between number of flower visitors and increasing attractiveness of the plant (e.g. number of flowers, inflorescence display size) was also reported by Brody and Mitchell (1997) for the hummingbird-pollinated plant *Ipomopsis aggregata* (Polemoniaceae), by Rodriguez-Robles et al. (1992) for the ornithophilous orchid *Comparettia falcate* (Orchidaceae), and by Bosch and Waser (2001) for *Aconitum columbianum* (Ranunculaceae).

The time of day had a significant influence on the number of visitors of each floral visitor group, with most visitors appearing between 9:00 am and 1:00 pm (Fig. 2, Tab. 3). Similar results were reported by Renner (1983) for *Trigona* spec. at Melastomataceae and by Stone and Jenkins (2008) for a solanaceous shrub in Costa Rica with a concentration of pollinator visits in the late morning, while Singer and Sazima (2001) observed most visitors appearing between 10:00 am and 2:00 pm for three orchid species from south-eastern Brazil. In addition, better light conditions promoted higher visitor frequencies and significantly more visitors per flower per hour were observed under half shade conditions than under full shade. Our results are in accordance with Vicens and Bosch (2000), who found bee activity to be positively related to temperature and light conditions, and with Singer and Sazima (2001), who recorded more visitors during sunnier weather conditions. A feasible explanation for these results might be the activity dependence of ectothermic insects on external heat. Furthermore the body colour and body size influence a bee´s thermoregulation (Biesmeijer et al. 1999). The observed species of *Trigona* has a thorax width of > 2 mm, thus it can be considered as rather large in comparison to other stingless bee species. Larger insects gain and lose heat more slowly but attain higher temperature excesses than smaller insects (Pereboom & Biesmeijer 2003). This might explain the clear activity peak around noon, when temperatures are highest. More visits of these bees were recorded under half-shade conditions than under full shade, which might be due to a better avoidance of temperature excesses under half-shade conditions. Higher temperatures, co-correlated with time of day, and sunlight also increased the visitation rates of the Florida scrub, belonging to the family of Lamiaceae, *Dicerandra frutescens* (Deyrup & Menges 1997) and for the understory herb *Justicia rusbyi* (Acanthaceae) in eastern Bolivia (Döll et al. 2007).

The number of visits by hummingbirds showed a clear peak between 9:00 am and 11:00 am and then decreased during the day. A similar pattern of foraging activity was reported by Garrison and Gass (1999) for the traplining hermit hummingbird *Phaethornis longirostris* that uses highly productive flowers such as *Heliconia pogonantha* (Heliconiaceae) in Costa Rica. The birds visited the flowers less in early morning and late afternoon than in mid-morning with a peak at 9:00 am. Because nectar production rates of many flowers used by traplining hummingbirds (including *H. pogonantha*) are high in the early morning and declines rapidly during the day (Stiles 1975; Stiles & Freeman 1993) the hummingbirds decrease their feeding frequency during the day to maximize net energy intake (Gass & Garrison 1999) rather than attempting to maintain constant net intake (Hainsworth et al. 1981; Stiles & Wolf 1979). The low visitation rate during the early morning hours may reflect satiation due to an abundance of food (Gill 1988). This early morning surplus of food could affect visitation rates in two ways: birds would be full after visiting few flowers and would not have to visit many others to meet their energy requirements. After the morning surplus is depleted, foraging effort should increase because birds must visit more flowers to take the same amounts of nectar. As nectar production rates continue to decrease and standing crop is diminished, hermits should decrease feeding efforts to save energy (Garrison & Gass 1999). This foraging pattern was also the case in our study with two traplining hermits being the most abundant hummingbird visitors to *H. angusta.* Tiebout (1992) and Stiles (1995) also reported that total observed foraging activity was lower in the afternoon than in the morning for several species of hermit and nonhermit hummingbirds in Costa Rica. These tropical species also feed primarily on flowers that have a decreasing nectar production rate over the day (Feinsinger 1976). This suggests that unlike in temperate systems, where foraging effort is bimodal with peaks of visitation rates in the morning and afternoon (Gass & Montgomerie 1981), tropical hummingbirds decrease foraging efforts after mid-morning to conserve energy because nectar production and standing crop are quite low in the afternoon (Stiles 1975, Stiles & Wolf 1979).

The nectar of *H. angusta* is rich in sucrose, whereas hexoses (glucose and fructose) are rare. The total sugar concentration is 16 %. Nectar sugar composition is claimed to be a useful predictor of pollinators, with sucrose-dominant nectar being indicative of butterflies, hummingbirds and large bees (Stiles & Freeman 1993; Baker et al. 1998). Stingless bees are usually included in the sucrose-preferring group (Schwerdtfeger 1996, Biesmeijer et al. 1999a,b). Our data are in accordance with these findings and all observed floral visitor groups fit well with the nectar sugar characteristics. The sucrose-hexose ratio of 1.63 is typical of hummingbird and large bee flowers (Baker & Baker 1990). Similar results were found by Perret et al. (2001) for ornithophilous neotropical Gesneriaceae and by Schmid et al. (2011) for the ornithophilous bromeliad *Aechmea nudicaulis* (Bromeliaceae). According to Roubik et al. (1995), stingless bee foragers tend to use all nectar concentrations, even as low as 5 % or as high as 67 %, although they prefer higher sugar concentrations of about ≤ 45 % (e.g. Roubik & Buchmann 1984; Fidalgo & Kleinert 2010). They are true generalists, collecting nectar and pollen from a vast array of plants (Ramalho et al. 1990; Biesmeijer et al. 2005). Pollen is used for larval development and nectar as the energy source. Since we did not observe *Trigona* bees collecting nectar, because they could not enter the narrow corolla tube nor did they pierce the corolla, we consider them visiting the flowers only to collect pollen. Thus the nectar characteristics fit to stingless bee foragers, but the nectar reward seems not to be the reason why *Trigona* bees visit the flowers of *H. angusta.*

We conclude that the specialized flowers of *H. angusta* not only attract hummingbirds, but as well generalistic insects, that have a certain potential to be pollinators and that the classification of floral visitors as either 'robbers' or 'pollinators' may be simplistic as some robbers can at times be considered potential pollinators.

The study confirms the importance of direct field observations to investigate the visitor spectrum of a particular plant species and to determine potential pollinators and antagonistic floral visitors. Further studies should focus on the pollination effectiveness of each visitor group separately in terms of e.g., fruit set, seed set, pollen deposition rate and the role of *Trigona* sp. in particular as pollen robber and/or secondary alternative pollinator of the flowers of *H. angusta.*

ACKNOWLEDGEMENTS

We would like to thank Nickolas and Raquel Locke from the NGO "Regua" for logistical support, Martin Hoepke and Jens Rabenstein for help in the field as well as Anja Wodak from the Institute of Pharmacy for analysing the nectar samples. Special thanks to Koos Biesmeijer for rich information on stingless bees and Josef Settele for valuable hints. We also thank three anonymous reviewers for their valuable comments on the manuscript. KS is funded by Studienstiftung des Deutschen Volkes.

REFERENCES

Almeda F (1977) Systematics of the neotropical genus *Centradenia* (Melastomataceae). Journal of the Arnold Arboretum 58:73-108.

Arizmendi MC, Domínguez CA, Dirzo R (1996) The role of an avian nectar robber and of hummingbird pollinators in the reproduction of two plant species. Functional Ecology 10:119-127.

Baker HG, Baker I, Hodges SA (1998) Sugar composition of nectar and fruits consumed by birds and bats in the tropics and subtropics. Biotropica 30:559-586.

Baker HG, Baker I (1990) The predictive value of nectar chemistry to the recognition of pollinator types. Israel Journal of Botany 39:157-166..

Bawa KS (1990) Plant-pollinator interactions in tropical rain forests. Annual Review of Ecology, Evolution, and Systematics 21:399-422.

Bawa KS, Perry DR, Bullock SH, Coville RE, Grayum MH (1985) Reproductive biology of tropical lowland rain forest trees. II. Pollination mechanisms. American Journal of Botany 72:346-356.

Biesmeijer JC, Slaa EJ, Siqueira de Castro M, Viana BF, Kleinert A, Imperatriz-Fonseca VL (2005) Connectance of Brazilian social bee – food plant networks is influenced by habitat, but not by latitude, altitude or network size. Biota Neotropica 5:1-9.

Biesmeijer JC, Richter JAP, Smeets MAJP, Sommeijer MJ (1999a) Niche differentiation in nectar-collecting stingless bees: the influence of morphology, floral choice and interference competition. Ecological Entomology 24:380-388.

Biesmeijer JC, Smeets MAJP, Richter JAP, , Sommeijer MJ (1999b) Nectar foraging by stingless bees in Costa Rica: botanical and climatological influences on sugar concentration of nectar collected by *Melipona.* Apidologie 30:43-55.

Bosch M, Waser NM (2001) Experimental manipulation of plant density and its effect on pollination and reproduction of two confamilial montane herbs. Oecologia 126:76-83.

Botes C, Johnson SD, Cowling RM (2009) The birds and the bees: using selective exclusion to identify effective pollinators of African tree *Aloes.* International Journal of Plant Sciences 170: 151-156.

Brody AK, Mitchell RJ (1997) Effects of experimental manipulation of inflorescence size on pollination and pre-dispersal seed predation in the hummingbird-pollinated plant *Ipomopsis aggregata*. Oecologia 110:86-93.

Bruna EM, Kress WJ, Marques F, da Silva OF (2004) *Heliconia acuminata* reproductive success is independent of local floral density. Acta Amazonica 34:467-471.

Campbell DR, Waser NM, Price MV (1996) Mechanisms of hummingbird-mediated selection for flower width in *Ipomopsis aggregata*. Ecology 77:1463-1472.

Canela MBF, Sazima M (2005) The pollination of *Bromelia antiacantha* (Bromeliaceae) in southeastern Brazil: ornithophilous *versus* melittophilous features. Plant Biology 7:411-416.

Dahlgren RMT, Clifford HT, Yeo PF (1985) The families of the monocotyledons. Springer-Verlag, Berlin.

De Castro CC, De Oliveira PE (2002) Pollination biology of distylous Rubiaceae in the Atlantic rain forest, SE Brazil. Plant Biology 4:640-646.

De Castro CC, Araujo AC (2004) Distyly and sequential pollinators of *Psychotria nuda* (Rubiaceae) in the Atlantic rain forest, Brazil. Plant Systematics and Evolution 244:131-139.

Deyrup M, Menges E (1997) Pollination ecology of the rare scrub mint *Dicerandra frutescens* (Lamiaceae). Florida Scientist 60:143-157.

Dias da Cruz D, Mello MAR, Van Sluys M (2006) Phenology and floral visitors of two sympatric *Heliconia* species in the Brazilian Atlantic forest. Flora 201:519-527.

Döll S, Hensen I, Schmidt-Lebuhn AN, Kessler M (2007) Pollination ecology of *Justicia rusbyi* (Acanthaceae), a common understory plant in a tropical mountain forest in eastern Bolivia. Plant Species Biology 22:211-216.

Feinsinger P (1976) Organization of a tropical guild of nectarivorous birds. Ecological monographs 46:257-291.

Fenster CB, Armbruster WS, Wilson P, Dudash MR, Thomson JD (2004) Pollination syndromes and floral specialization. Annual Review of Ecology, Evolution, and Systematics 35:375-403.

Fidalgo A, Kleinert A (2010) Floral preferences and climate influence in nectar and pollen foraging by *Melipona rufiventris* Lepeletier (Hymenoptera: Meliponini) in Ubatuba, Sao Paulo State, Brazil. Neotropical Entomology 39:879-884.

Freitas BM, Paxton RJ (1998) A comparison of two pollinators: the introduced honey bee *Apis mellifera* and an indigenous bee *Centris tarsata* on cashew *Anacardium occidentale* in its native range of NE Brazil. Journal of Applied Ecology 35:109-121.

Galetto L (1998) Estructura floral y composición química del néctar en tres especies de Rubiaceae. Kurtziana 26:83-98.

Gao JY, Zhang L, Deng XB, Ren PY, Kong JJ, Li QJ (2004) The floral biology of *Curcumorpha longiflora* (Zingiberaceae): a ginger with two day flowers. American Journal of Botany 91:289-293.

Garrison JSE, Gass CL (1999) Response of a traplining hummingbird to changes in nectar availability. Behavioural Ecology 6:714-725.

Gass CL, Garrison JSE (1999) Energy regulation by traplining hummingbirds. Functional Ecology 13:483-492.

Gass CL, Montgomerie RD (1981) Hummingbird foraging behaviour: decision-making and energy regulation. In: Karnil AC, Sargent TD (eds.) Foraging behaviour: ecological, ethological, and psychological approaches. Garland STPM Press, New York, pp 157-194.

Gentry AH (1974a) Coevolutionary patterns in Central American Bignoniaceae. Annals of the Missouri Botanical Garden 61:728-759.

Gentry AH (1974b) Flowering phenology and diversity in tropical Bignoniaceae. Biotropica 6:54-68.

Gill FB (1988) Trapline foraging by hermit hummingbirds: competition for an undefended, renewable resource. Ecology 69:1933-1942.

Graves GR (1982) Pollination of *Tristerix* mistletoe (Loranthaceae) by *Diglossa* (Aves, Thraupidae). Biotropica 14:316-317.

Gross C (2005) Pollination efficiency and pollinator effectiveness. In: Dafni A, Kevan P, Husband B (eds). Practical Pollination Biology. Ontario Enviroquest, Cambridge, pp 354–363 .

Hainsworth FR, Tardiff M, Wolf LL (1981) Proportional control for daily energy regulation in hummingbirds. Physiological Zoology 54:452-462.

Harder LD, Jordan CY, Gross WE, Routley MB (2004) Beyond floricentrism: The pollination function of inflorescences. Plant Species Biology 19:137-148.

Horvitz CC, Schemske DW (1990) Spatio temporal variation in insect mutualism of a neotropical herb. Ecology 71:1085-1097.

Hubbell SP, Johnson LK (1978) Comparative foraging behaviour of six stingless bee species exploiting a standardized resource. Ecology 59:1123-1136.

Inouye DW (1980) The terminology of floral larceny. Ecology 61:1251-1253.

Janzen DH (1975) Ecology of plants in the tropics. Edward Arnold, London.

Kress WJ (1990) The diversity and distribution of *Heliconia* (Heliconiaceae) in Brazil. Acta Botanica Brasilica 4:159-167..

Krömer T, Kessler M, Lohaus G, Schmidt-Lebuhn AN (2008) Nectar sugar composition and concentration in relation to pollination syndromes in Bromeliaceae. Plant Biology 10:502-511.

Kurtz BC, de Araújo DSD (2000) Composição florística e estrutura do componente arbóreo de um trecho de Mata Atlântica na Estaoção Ecológica Estadual do Paraíso, Cachoeiras de Macacu, Rio de Janeiro, Brasil. Rodriguésia 51:69-112.

McDade LA, Kinsman S (1980) The impact of floral parasitism in two neotropical hummingbird-pollinated plant species. Evolution 34:944-958.

Michener CD (1979) Biogeography of the bees. Annals of the Missouri Botanical Garden 66:277-347.

Mitchell RJ, Karron JD, Holmquist KG, Bell JM (2004) The influence of *Mimulus ringens* floral display size on pollinator visitation patterns. Functional Ecology 18:116-124.

Morellato LPC, Haddad CFP (2000) Introduction: The Brazilian Atlantic Forest. Biotropica 32:786-792.

Oliveira-Filho AT, Fontes MAL (2000) Patterns of floristic differentiation among Atlantic Forests in southeastern Brazil and influence of climate. Biotropica 32:793-810.

Ollerton J, Alarcón R, Waser NM, Price MV, Watts S, Cranmer L, Hingston A, Peter CI, Rottenberry J (2009) A global test of the pollination syndrome hypothesis. Annals of Botany 103:1471-1480.

Ollerton J, Johnson SD, Cranmer L, Kellie S (2003) The pollination ecology of an assemblage of grassland Asclepiads in South Africa. Annals of Botany 92:807-834.

Pereboom JJM, Biesmeijer JC (2003) Thermal constraints for stingless bee foragers: the importance of body size and coloration. Oecologia 137:42-50.

Perret M, Chautems A, Spichiger R, Peixoto M, Savolainen V (2001) Nectar sugar composition in relation to pollination syndromes in Sinningieae (Gesneriaceae). Annals of Botany 87:267-273.

Proctor M, Yeo P, Lack A (1996) The natural history of pollination. Timber Press, Portland.

Ramalho M, Kleinert-Giovannini A, Imperatriz-Fonseca VL (1990) Important bee plants for stingless bees (Melipona and Trigonini) and Africanised honeybees (Apis mellifera) in neotropical habitats: a review. Apidologie 21:469-488.

Renner S (1983) The widespread occurrence of anther destruction by *Trigona* bees in Melastomataceae. Biotropica 15:251-256.

Robertson AW, Ladley JI, Kelly D (2005) Effectiveness of short-tongued bees as pollinators of apparently ornithophilous New Zealand mistletoes. Austral Ecology 30:298-309.

Rodriguez-Robles JA, Melendez EJ, Ackermann JD (1992) Effects of display size, flowering phenology, and nectar availability on effective visitation frequency in *Comparettia falcata* (Orchidaceae). American Journal of Botany 79:1009-1017.

Roubik DW, Yanega D, Aluja SM, Buchmann SL, Inouye DW (1995) On optimal nectar foraging by some tropical bees (Hymenoptera: Apidae). Apidologie 26:197-211.

Roubik DW (1989) Ecology and Natural History of Tropical Bees. Cambridge University Press, Cambridge.

Roubik DW, Buchmann SL (1984) Nectar selection by *Melipona* and *Apis mellifera* (Hymenoptera: Apidae) and the ecology of nectar intake by bee colonies in a tropical forest. Oecologia 61:1-10.

Roubik DW (1982) The ecological impact of nectar-robbing bees and pollinating hummingbirds on a tropical shrub. Ecology 63:354-360.

Sazima I, Buzato S, Sazima M (1995) The Saw-billed Hermit *Ramphodon naevius* and its flowers in southeastern Brazil. Journal of Ornithology 136:195-206.

Schmid S, Schmid VS, Zillikens A, Harter-Marques B., Steiner J (2011) Bimodal pollination system of the bromeliad *Aechmea nudicaulis* involving hummingbirds and bees. Plant Biology 13:41-50.

Schwerdtfeger, M (1996). Die Nektarzusammensetzung der Asteridae und ihre Beziehung zu Blütenökologie und Systematik. Dissertationes Botanicae 264..

Singer RB, Sazima M (2001) Flower morphology and pollination mechanism in three sympatric Goodyerinae Orchids from Southeastern Brazil. Annals of Botany 88:989-997.

Singer RB, Sazima M (2000) The pollination of *Stenorrhynchos lanceolatus* (Aublet) L.C. Rich. (Orchidaceae: Spiranthinae) by hummingbirds in southeastern Brazil. Plant Systematics and Evolution 223:221-227.

Simão DG, Scatena VL (2001) Morphology and anatomy in *Heliconia angusta* Vell. and *H. velloziana* L. Emygd. (Zingiberales: Heliconiaceae) from the Atlantic forest of southeastern Brazil. Revista Brasileira de Botânica 24:415-424.

Slaa EJ, Sánchez Chaves LA, Malagodi-Braga KS, Hofstede FE (2006) Stingless bees in applied pollination: practice and perspectives. Apidologie 37:293-315.

Smith CE, Stevens JT, Temeles EJ, Ewald PW, Hebert RJ, Bonkovsky RL (1996) Effect of floral orifice width and shape on hummingbird-flower interactions. Oecologia 106:482-492.

Steven JC, Rooney TP, Boyle OD, Waller DM (2003) Density-dependent pollinator visitation and self-incompatibility in upper Great Lakes populations of *Trillium grandiflorum*. Journal of the Torrey Botanical Society 130:23-29.

Stiles FG, Freeman CE (1993) Patterns in floral nectar characteristics of some bird-visited plant species from Costa Rica. Biotropica 25:191-205.

Stiles FG, Wolf LL (1979) Ecology and evolution of lek mating behaviour in the long-tailed hermit hummingbird. Ornithological Monographs 27:1-78.

Stiles FG (1975) Ecology, flowering phenology, and hummingbird pollination of some Costa Rican *Heliconia* Species. Ecology 56:285-301.

Stone JL, Jenkins EG (2008) Pollinator abundance and pollen limitation of a solanaceous shrub at premontane and lower montane sites. Biotropica 40:55-61.

Thompson JN, Pellmyr O (1992) Mutualism with pollinating seed parasites amid co-pollinators: constraints on specialization. Ecology 73:1780-1791.

Tiebout HM (1992) Comparative energetics of divergent foraging modes: a doubly labelled water experiment on hummingbird competition. Animal Behaviour 44:895-906.

Tomimatsu H, Ohara M (2003) Floral visitors of *Trillium camschatcense* (Trilliaceae) in fragmented forests. Plant Species Biology 18:123-127.

Veloso HP, Rangel-Filho AL, Lima JCA (1991) Classificação da vegetação brasileira, adaptada à um sistema universal. IBGE/CDDI, Departamento de Documentação e Biblioteca, Rio de Janeiro.

Vicens N, Bosch J (2000) Weather-dependent pollinator activity in an apple orchard, with special reference to *Osmia cornuta* and *Apis mellifera* (Hymenoptera: Megachilidae and Apidae). Environmental Entomology 29:413-420.

Vogel S (1996) Christian Konrad Sprengel's theory of the flower: The cradle of floral ecology. In: Lloyd DG, Barrett SCH (eds). Floral ecology, studies on floral evolution in animal-pollinated plants. Chapman and Hall, New York, pp 44-62.

Waser NM (2006) Specialization and Generalization in Plant-Pollinator Interactions: A Historical Perspective. In: Waser NM, Ollerton J (eds). Plant-Pollinator Interactions: from Specialization to Generalization. The University of Chicago Press, Chicago, pp 3-17.

Waser NM, Ollerton J (2006) Plant-pollinator interactions: from specialization to generalization. The University of Chicago Press, Chicago.

Waser NM, Chittka L, Price MV, Williams NM, Ollerton J (1996) Generalization in pollination systems, and why it matters. Ecology 77:1043-1056.

Waser NM, Price MV (1990) Pollination efficiency and effectiveness of bumble bees and hummingbirds visiting *Delphinium nelsonii*. Collectanea Botanica 19:9-20.

Willmer PG, Corbet SA (1981) Temporal and microclimatic partitioning of the floral resources of *Justicia aurea* amongst a concourse of pollen vectors and nectar robbers. Oecologia 51:67-78.

Willmott AP, Búrquez A (1996) The pollination of *Merremia palmeri* (Convolvulaceae): can hawk moths be trusted? American Journal of Botany 83:1050-1056.

12

HONEY BEES ENSURE THE POLLINATION OF *PARKIA BIGLOBOSA* IN ABSENCE OF BATS

Kristin Marie Lassen[1], Moussa Ouédraogo[2], Yoko Luise Dupont[3], Erik Dahl Kjær[1], Lene Rostgaard Nielsen[1]

[1]*Department of Geosciences and Natural Resource Management, University of Copenhagen, Rolighedsvej 23, 1958 Frederiksberg C, Denmark*
[2]*Centre National de Semences Forestières, Route de Kaya, 01 BP 2682 Ouagadougou, Burkina Faso*
[3]*Institute of Bioscience, Aarhus University, Vejlsøvej 25, 8600 Silkeborg, Denmark*

Abstract—One of the socio-economically important West African parkland trees, *Parkia biglobosa* yields pods consumed by people and animals. Being animal-pollinated, it is unknown how climate change will affect the species if the guild of pollinators changes. Here, we compare the pollinator guilds and the pod production at two climatically different study sites: the first site, Tiba was drier with lower tree density and fewer fruit bats, whereas the second site, Pinyiri was more humid, with higher tree density and more fruit bats. We carried out a pollinator exclusion trial with bags separating the flower-visitors and made observations of the flower-visitors. Furthermore, we calculated pollination distances based on paternity analysis of the seeds produced within the trial. The numbers of immature fruits were similar for the sites, but the drier site, Tiba experienced more abortion and decreased pod set. At both sites, exclusion of bats led to marginally reduced pod set, while exclusion of bats and honey bees led to significantly reduced pod set. We found a small effect of stingless bees and solitary bees but only at the more humid site, Pinyiri. Tiba experienced a higher level of self-pollination, fewer pollen donors per tree, and longer median distances of pollen flow, compared with Pinyiri. Implications of the results for management of the species and its pollinators in the face of climate change are discussed.

Keywords: Agroforestry parklands; Bats; Honey bees; Pollination ecology; Solitary bees; Stingless bees

INTRODUCTION

Animal pollination is a requirement for successful fruit set for more than 90% of the flowering species in the tropics (Ollerton et al. 2011), but both the species richness (Patiny et al. 2009) and composition of pollinator guilds (Franzén & Öckinger 2012) may be influenced by climate change through several mechanisms. The local presence of animal pollinators is affected by many environmental factors including temperature and annual precipitation (Fryxell 1957), which impact the quantity and quality of food and nesting places (Kearns et al. 1998). Furthermore, mismatching of periods of flowering and pollinator activity may occur (Parmesan 2007; Petanidou et al. 2014). Climatic changes are therefore expected to have consequences for the interactions between plants and pollinators (Abrol 2012; Settele et al. 2014), although different species of pollinators can be differentially affected. For instance, bats, honey bees (*Apis mellifera* ssp.) and some species of sunbirds are known to migrate when resources become scarce (Hepburn & Radloff 1998; Cheke et al. 2001; Monadjem et al. 2010), whereas e.g. stingless bees do not migrate (Roubik 2006). The yield of animal-pollinated plants may thus depend on whether important pollinating species can be substituted, e.g. honey bees for bats or stingless bees for honey bees.

*Corresponding author: kmla@ign.ku.dk

In the present study we investigated if denied access of different types of animal pollinators led to reduced fruit set of an important fruit tree, the West African Sudano-Sahelian parkland tree *Parkia biglobosa* (Jacq.) R. Br. ex G. Don (English: African locust bean, French: néré). Specifically, we assessed the role of honey bees, because abundance of these pollinators can be increased by beekeeping and *P. biglobosa* is an important honey bee forage tree in the savannah (Dukku 2010).

Burkina Faso, where the present study took place, has a strong north-south increasing rainfall gradient, which divides the country into three climatic zones with different vegetation. From 1950 until the mid-1980s rainfall declined steeply but recovered in the 1990s and stabilised at a level 15% lower than that of 1920-1969. Since 1975, air temperatures have increased about 0.6°C through most of Burkina Faso (Funk et al. 2012), and this trend is projected to persist or even increase depending on the future concentration of greenhouse gasses in the atmosphere (Niang et al. 2014). Climate models are very uncertain regarding predictions for the future annual rainfall (Ibrahim et al. 2014). In connection with the declining rainfall, the composition of vegetation has changed in recent decades with xeric species replacing mesic species (Maranz 2009; Gonzalez et al. 2012). Although not documented, these changes may also impact pollination interactions of mesic species, including the legume *P. biglobosa* (Boffa 1999). Presently, the northern limit of the distribution of *P.*

biglobosa is in the Sudano-Sahelian climatic zone, and increased drought is therefore likely to move the border of the distribution further south (Ouedraogo et al. 2012).

Parkia biglobosa is animal-pollinated. The flowers yield nectar and pollen during night, and are pollinated by bats (Harris & Baker 1959; Hopkins 1983) in the period from dusk till dawn, after which several species of bees takes over in the early morning (Ouédraogo 1995; Lassen et al. 2012). Six species of frugivorous bats (Pteropodidae) are known from Burkina Faso (Kangoyé et al. 2012) and three of them are known to visit the flowers of *P. biglobosa* (Hopkins 1983). However, in the drier part of the distribution area of *P. biglobosa* in Burkina Faso, bats are rare, and different species of bees are expected to be the main pollinators (Ouédraogo 1995). It is likely that the widespread beekeeping activities in Burkina Faso (Schweitzer et al. 2013) act as a pollination buffer against the negative effects of low abundance of bats at the drier areas, especially if the bees are given access to water in order to limit their seasonal migration (B. Svensson 2014, pers. comm.). Generally, honey bees are believed to have smaller flight ranges than bats (Linhart 1995). Thus, pollination by honey bees may decrease distances of pollen transportation and hence limit gene flow across a fragmented landscape, relative to bat-pollination. Geitonogamous pollination is possible and any shift in pollinators that results in more selfing may reduce pod production because *P. biglobosa* is mainly outcrossing (Ouédraogo 1995; Sina 2006) and assumed to possess a partial self-incompatibility system (Sina 2006).

On the above background, the overall objective of the present study was to evaluate if pollinators can be substituted by others, when assessed by pod set in two different climatic zones in Burkina Faso. One study site, near the northern distribution limit of *P. biglobosa*, was characterised by a dry climate while the other site was characterised by higher precipitation and located within the core of the natural distribution area of *P. biglobosa*. More specifically, we studied: 1) differences between the sites in flower-visitors and pod yield, 2) the effect of excluding different pollinator types on yield and pollination distances, and 3) pod yield in relation to distances to bee hives. Based on the results, we have discussed the potential management of pollinators of *P. biglobosa* in a drier environment.

MATERIALS AND METHODS

Plant species

Parkia biglobosa is a medium sized leguminous tree widespread in the savannah woodlands in a belt from Senegal to Uganda (Hopkins & White 1984). It is limited by 500-700 mm rainfall but prospers with up to 1,200 mm annual precipitation (Haq 2008). In Burkina Faso, flowering takes place in the dry season between January and May. The bright red, tiny flowers are closely packed in capitula (with around 2,000 flowers), borne on long peduncles. Close to the peduncle are about 85 staminodial flowers, and below are around 250 sterile flowers that produce nectar, which accumulates in a nectar ring (Hopkins 1983). Each capitulum flowers only for one night (Hopkins 1983; Ouédraogo 1995), and is here treated as a single unit.

The hermaphroditic flower consists of a bract, calyx, corolla, ten stamens, and an ovary of one carpel ending in a style with a cup-shaped stigma. *Parkia biglobosa* is generally protandrous and pollen is shed in the evening in polyads of around 32 pollen grains (Nombré, 2003; pers. obs.). The styles elongate during the night of flowering, but in some flowers the styles do not elongate and these flowers are thus functionally male. A pilot trial with controlled cross-pollination showed that the stigmas were receptive and polyads fertile from at least 21:30 to 07:30 (Lassen 2016). An ovary contained 16-29 ovules (mean = 23, SE = 0.63, N = 15). At our study sites, pods matured in around two months, producing mostly a few pods per capitulum, and rarely up to 20 pods per capitulum.

Study sites and sampled trees

A pollinator exclusion trial of *P. biglobosa* was carried out at two sites in Burkina Faso in 2012. A relatively dry northern site was located at two neighbouring villages, Tiba and Gam (12°42'26.26"N, 1°18'2.04"W) in the Sudano-Sahelian climatic zone with a mean annual precipitation (1981-2010) of 600-700 mm (Sanfo 2012), which henceforth will be called Tiba for convenience. Another more humid southern site was located at the village Pinyiri (syn. Kacheli) (11°14'34.89"N, 1°8'1.73"W) in the Sudanian climatic zone with a mean annual precipitation (1981-2010) of 900-1,000 mm (Sanfo 2012). The distance between the two sites is approximately 160 km. In 2011, the annual rainfall (preceding the fruiting season of *P. biglobosa* in 2012) was 566 mm for the dry site (nearest weather station was at Guilongou 9 km from Tiba) whereas the more humid site obtained 927 mm (measured at Pô, 8 km from Pinyiri) (Météo 2015). Mean annual temperatures (2012) were 29.3°C and 28.4°C for Tiba and Pinyiri, respectively (Météo 2015).

The sites were 360 ha (Tiba) and 220 ha (Pinyiri), respectively. Tree characteristics were similar between the two sites (diameter, height, crown area and age), but tree density was higher at the more humid site, Pinyiri (1.23 trees/ha relative to 0.26 trees/ha). Further, only around half of the trees flowered at the drier site, Tiba, giving a density of 0.14 flowering trees/ha. GPS coordinates of trees were recorded, leaves were sampled for DNA extraction, and flowering periods were recorded. A total of 93 and 271 trees were sampled at Tiba and Pinyiri, respectively. These included mother-trees, which were used for experimental pollination studies.

Pollinator observations

At both study sites, flower-visitors were observed in different periods between dusk and dawn, during five nights at Tiba (total of 11 hours) and six nights at Pinyiri (total of 16 hours). Capitula were observed from the ground using a Yukon Ranger 5x42 digital night vision monocular (model 28041VK) with a mobile recorder and a compact camera (Canon PowerShot S95). Insect observations were supplemented by video recordings, which we paused regularly (every ten seconds) and counted the number of insects on these frames. Based on these counts we estimated mean visitation number per capitulum per site during early morning. Samples of insects were preserved in alcohol for later identification. Visiting bats and birds were registered, but it was not possible to identify the species. At both study sites, we registered the GPS coordinates of colonies of the most abundant flower-visitors, namely, bats and honey bees (both hived and feral honey bees).

Pollinator exclusion trial

Flower-visitors were separated according to their size by bags with different mesh size. Four treatments were applied with means of 18 and 20 replicates of each treatment on each tree, respectively for Tiba (12 trees) and Pinyiri (16 trees). All treatments were distributed randomly in the tree crown but applied only to the most apical capitulum (bud) on the compound inflorescences (consisting of many capitula), since these have the greatest chance of developing into hermaphroditic capitula (Lassen 2016). The treatments consisted of an open treatment, T0, where capitula were accessible to all visitors, two semi-open treatments, T1, where capitula were protected by a bag of chicken wire with ca. 10 mm × 10 mm holes, excluding bats and birds, and T2, where the capitula were protected by bags of mosquito netting with ca. 1.8 mm × 2.0 mm holes, only accessible to small bees and ants. Furthermore, a closed treatment, T3, excluded all potential visitors by bags of cheesecloth with ca. 200 μm × 350 μm holes. The bags in treatments T2 and T3 were kept in shape by an inner band of chicken wire, in order to keep the fabric from touching the flowers and the flower-visitors from reaching the flowers from the outside of the bags.

Trees were tended on a daily basis during the experiment, and abscised capitula were collected and examined for sex (hermaphroditic or functionally male) and signs of pods, pests, and diseases. Based on earlier results (Lassen 2016), we considered a capitulum as functionally male if the lengths of the anthers in the dissected flowers were more than 5 mm longer than the lengths of the corresponding stigmas.

The bags were removed around one week after flowering, and the number of developing pods was counted. At maturity we harvested the complete infructescenses from the trees. The pods were weighted and length measured excluding the pedicel. Two pods per receptacle (one if single) were randomly chosen and the number of healthy seeds, aborted seeds, and seeds consumed by birds (creating shallow holes on one side of the pods) were assessed. A seed was registered as aborted when the weight was below 0.05 g and/or the shape was flat.

DNA extraction and genotyping

We extracted genomic DNA from dried leaf samples of mother-trees (i.e. providers of pods for the study, $N = 12$ and 16 for Tiba and Pinyiri, respectively), potential pollen donors ($N = 81$ and 255 respectively), and from a sub-sample of seeds from the randomly chosen pods from the exclusion trial ($N = 61$ and 498 from Tiba and Pinyiri, respectively). Only one seed per pod was selected as we had previously found that all seeds per pod were sired by a single pollen donor (Lassen et al. 2014). The raw de-shelled seeds were utilised without prior germination. For the extraction, we used the DNeasy 96 Plant Kit (QIAGEN, Hombrechtikon, Switzerland) following the manufacturer's protocol.

The extracted DNA samples were genotyped using ten microsatellite markers developed for *P. biglobosa* (Lassen et al. 2014), and divided into three primer mixes (Mix 1: PbL03, PbL04, PbL15. Mix 2: PbL09, PbL11, PbL21. Mix 3: PbL02, PbL05, PbL12, PbL22). PCR reactions were done with Qiagen Multiplex PCR Kit in 10 μl reactions including 1 μl of extracted DNA. PCR conditions, fragment separation and genotype scoring followed Lassen et al. (2014).

Paternity analysis

Paternity analysis was performed on seeds from the exclusion trial treatments with Cervus v3.0.3 (Kalinowski et al. 2007), where we assigned the most likely father-trees to offspring given the known genotype of the mother-tree. Only individuals with a minimum of eight scored loci were included in the analysis. Selfing was allowed in the model. The critical values for assignment with 95% confidence were estimated by 100,000 simulations assuming a genotyping error-rate of 0.01 and sampling of 75% of the potential pollen donors. Only trees assigned with confidence level of 95%, positive trio (mother-father-offspring) LOD scores, and a maximum of two mismatching loci were accepted as fathers.

For each assigned paternity, the pollen transfer distance was calculated as the distance between the known mother-tree and the assigned father-tree based on their GPS coordinates.

For the adult trees at the two study sites, we tested deviations from expected genotype frequencies under Hardy-Weinberg equilibrium using Genepop v4.3 (Rousset 2008). The level of genetic differentiation between the two sites was estimated using AMOVA (Analysis of Molecular Variance) (Excoffier et al. 1992) as implemented in GenAlEx v6.501 (Peakall & Smouse 2006; Peakall & Smouse 2012) including test of significance based on 999 permutations.

Statistical analyses

For numerical variables (Y_{ijk}), the average value was calculated per tree and treatment excluding non-flowering trees. The differences between treatments, trees or sites were tested based on a general linear model as implemented in the GLM procedure in the SAS software v.9.4 (SAS Institute 2011) corresponding to the model:

$$Y_{ijk} = \text{Treatment}_i + \text{Site}_j + \text{Tree}_k(\text{Site}_j) + \mathcal{E}_{ijk}$$

, where Y_{ijk} is the response variable, treatment i = (T0, T1, T2, T3), site j = (1, 2) and tree k = (1 - 28). Treatment$_i$ and site$_j$ were considered fixed effects while Tree$_k$(Site$_j$) was considered a random effect with residual $\mathcal{E}_{ijk}$ assumed independent and $N(0,\sigma_e^2)$. The model assumptions were assessed and accepted by visual inspection of the residuals.

For T0 data (open-pollinated flowers) and T1 data (vertebrate excluded flowers), the relationship between fecundity variables (immature and mature fruits) and number of bee colonies within a certain range of each tree were analysed. We tested this for number of colonies within 9 different distances from the studied trees; d = 200, 300, 400, 500, 600, 700, 800, 900, and 1,000 m. The significance levels were sequential Bonferroni corrected with table-wide N = 18 (9 distance classes at each of the two sites).

Results

Pollinator observations

The first animals visiting capitula with accessible pollen were nocturnal animals, such as bats and moths. During the first hour after sunrise we observed mainly honey bees, different species of smaller bees, and birds.

At Tiba, we observed two small colonies of bats of about ten individuals each. At Pinyiri we registered two small colonies of around five and fifteen individuals and a large colony of about 500 individuals. However, bats from roosting sites outside the study sites may also visit the trees during the night. At both sites, we observed bats visiting *P. biglobosa* capitula in the period from ca. 20:00 until ca. 06:00. At Tiba, bats were observed visiting *P. biglobosa* in three out of five nights and at Pinyiri in four nights out of six. Around 45 min of video recordings at Tiba captured no bats visiting the capitula, and only 5 bats were recorded under the crowns of *P. biglobosa*. In comparison, about one hour video recordings at Pinyiri showed three bats at the capitula and 53 bats under the crowns. The bats moved quickly and silently, and visitation rates of bats were difficult to assess. Nevertheless, we observed more bats visiting capitula of *P. biglobosa* at the more humid site, Pinyiri, and they also stayed longer in the trees, compared to Tiba.

Honey bees were common visitors of *P. biglobosa*. At the study sites, the honey bees belong to the species *Apis mellifera jemenitica* Ruttner (Apidae, tribe Apini) (Al-Ghamdi et al. 2013). We registered 69 colonies of honey bees at Tiba and 65 colonies at Pinyiri. However, at the drier site, Tiba, the area was larger and the population of flowering *P. biglobosa* trees was smaller compared with the more humid site, Pinyiri. A few honey bee colonies were feral (without hive) while the majority were managed by beekeepers. The honey bees arrived at *P. biglobosa* around dawn. The number of visiting honey bees peaked around 15 minutes after the first honey bee had arrived (with up to 24 honey bees simultaneously on one side of a capitulum), and after another 15 minutes only very few honey bees remained. Despite the differences between sites, the mean visitation numbers were very similar with 1.9 honey bees per capitulum (SD = 1.64, N = 137 counts) at Tiba compared with 1.6 honey bees per capitulum (SD = 1.66, N = 101 counts) at Pinyiri during the early morning.

We observed two species of stingless bees (Apidae, tribe Meliponini) at Pinyiri; *Hypotrigona* sp. 1 (large type) and *Hypotrigona* sp. 2 (small type), while at Tiba, only *Hypotrigona* sp. 1 was found. Additionally, one species of solitary bees, *Compsomelissa borneri* Alfken (Apidae, tribe Allodapini) was observed at both study sites. The small bees generally arrived after the peak of honey bees but stayed longer and peaked in numbers around one hour after the honey bees had left. Up to 15 small bees were observed collecting pollen simultaneously on one side of a capitulum. The mean visitation number was 1.1 small bees per capitulum (SD = 0.55, N = 14 counts) at Tiba and 6.5 small bees per capitulum (SD = 3.16, N = 85 counts) at Pinyiri. However, at both study sites we observed large differences among trees in numbers of small bees visiting the capitula.

Several species of moths were observed. During the already mentioned video recordings, we counted a total of 17 moths visiting capitula at Tiba, whereas only 1 moth was observed at a capitulum at Pinyiri.

Other occasional visitors observed were: eight species of beetles, four species of flies, three species of ants, one dragonfly, one true bug, two spiders, in addition to a few sunbirds.

Effect of pollinator exclusion on pod and seed production

Three of the 12 flowering mother-trees at the drier site, Tiba, did not produce any pods, whereas all mother-trees at the more humid site, Pinyiri, produced pods. Predation by birds affected 6% of the pods within the trial at Pinyiri compared to 79% at Tiba, and one of the nine pod-producing trees at Tiba had all its pods eaten by birds. An additional 14% of the harvested pods at Tiba were flat and with aborted seeds and no pulp.

The average numbers of immature and mature pods per hermaphroditic capitulum (incl. capitula without pods) varied significantly among treatments and trees (Tab. 1). For both sites, the number of pods per capitulum in treatments T0 and T1 were statistically similar. For Pinyiri, we found a trend towards fewer pods per capitulum with smaller mesh size in bags for T2 and T3, whereas for Tiba, treatment T2 and T3 resulted in the same low number (Tab. 1). The number of mature pods per hermaphroditic capitulum was significantly lower at Tiba compared to Pinyiri, in particular for T0, which resulted in around five times more pods being harvested at Pinyiri compared to Tiba (Tab. 1).

Pod size at Tiba was not analysed because of the heavy predation. At Pinyiri, pod weight and pod length ranged from 0.4 g to 24.5 g and from 1.8 cm to 35.4 cm, respectively, while the number of healthy seeds per pod varied from 0-27. The pods from T2 were heavier and longer compared with T0 and T1 (Tab. 2), but T2 also had fewer pods per capitulum as mentioned above. The numbers of healthy seeds per pod followed the trend but were not statistically different (Tab. 2).

Paternity analysis

The paternity analysis assigned most of the genotyped seeds (each representing one pod) to a potential father-tree

TABLE 1. Average number of immature and mature pods per hermaphroditic capitulum (incl. capitula without pods) for *Parkia biglobosa*, and total number of pods shown per treatment and site, including *F*-tests and their significance levels.

Treatment	No. of immature pods/ hermaphroditic capitulum[a]		No. of mature pods/ hermaphroditic capitulum[a]		Total no. of mature pods	
	Tiba	Pinyiri	Tiba	Pinyiri	Tiba	Pinyiri
T0, open	2.52 (0.37) a	3.31 (0.22) a	1.16 (0.26) a	2.88 (0.21) a	97	579
T1, chicken wire	1.84 (0.40) a	2.73 (0.22) a	1.16 (0.28) a	2.32 (0.21) a	107	441
T2, mosquito net	0.02 (0.35) b	0.39 (0.22) b	0.01 (0.25) b	0.28 (0.21) b	1	56
T3, closed	0.02 (0.35) b	0.03 (0.22) c	0.01 (0.25) b	0.01 (0.21) c	1	1
All treatments	1.11 (0.16)	1.61 (0.13)	0.61 (0.14)	1.37 (0.11)	206	1077
$F_{Treatment}$	$F = 57.0$ (***)		$F = 40.8$ (***)			
$F_{Tree(Site)}$	$F = 1.7$ (*)		$F = 1.7$ (*)			
F_{Site}	$F = 3.8$ (ns)		$F = 10.6$ (**)			

[a]Values are least squares (LS) means with the standard error (SE) of the LS estimate in brackets. Values within a column followed by different letters are significantly different at $P < 0.05$. Significance level: *** = $P < 0.001$, ** = $P < 0.01$, * = $P < 0.05$, and ns = $P > 0.05$.

TABLE 2. Pod weight and length, and the number of healthy seeds per pod for *Parkia biglobosa* shown per treatment at Pinyiri, including *F*-tests and their significance levels (excluding treatment T3 due to a single pod).

Treatment	Pod weight, g	Pod length, cm	Healthy seeds/pod
T0, open	10.5 (0.53) a	19.0 (0.63) a	12.9 (0.64) a
T1, chicken wire	11.9 (0.53) b	20.3 (0.63) a	13.6 (0.64) a
T2, mosquito net	13.5 (0.68) c	23.0 (0.82) b	14.9 (0.83) a
T3, closed	8.0 (2.42) -	19.7 (2.90) -	3.2 (2.92) -
Treatments T0,T1,T2	12.0 (0.34)	20.8 (0.42)	13.8 (0.41)
$F_{Treatment}$	$F = 6.4$ (**)	$F = 7.3$ (**)	$F = 1.8$ (ns)
F_{Tree}	$F = 4.3$ (***)	$F = 4.5$ (***)	$F = 2.3$ (*)

Values are least squares (LS) means with the standard error (SE) of the LS estimate in brackets. Values within a column followed by different letters are significantly different at $P < 0.05$. Significance level: *** = $P < 0.001$, ** = $P < 0.01$, * = $P < 0.05$, and ns = $P > 0.05$.

TABLE 3. Percentage of assigned pods (with number of assigned pods in brackets) for *Parkia biglobosa*, percentage of self-pollination (of all genotyped pods, including the unassigned), mean number of pollen donors per tree (incl. self-pollination), and the average distance to pollen donors (without selfing), shown per treatment and site. Treatments with single pods are excluded. *F*-tests and their significance levels are shown for number of pollen donors and average distance to pollen donors.

Treatment	Assigned pods, % (No. pods)		Self-pollination, %		No. of pollen donors per tree[a]		Average distance to pollen donors[a], m	
	Tiba	Pin[b]	Tiba	Pin[b]	Tiba	Pin[b]	Tiba	Pin[b]
T0, open	70 (23)	84 (195)	21	2	1.8 (0.37) a	7.0 (0.60) a	155 (70.5) a	238 (19.1) a
T1, chic[c]	96 (25)	93 (207)	23	4	2.2 (0.33) a	7.5 (0.60) a	275 (64.1) a	183 (18.9) a
T2, mos[d]	- (1)	98 (41)	-	19	- (na) -	2.0 (0.77) b	- (na) -	173 (43.8) a
T3, clos[e]	- (1)	- (1)	-	-	- (na) -	- (na) -	- (na) -	- (na) -
All treatments	82 (50)	89 (444)	23	4	0.5 (0.69)	5.5 (0.36)	212 (55.8)	198 (18.7)
$F_{Treatment}$					$F = 19.2$ (***)		$F = 2.0$ (ns)#	
$F_{Tree(Site)}$					$F = 2.9$ (**)		$F = 3.3$ (***)	
F_{Site}					$F = 19.0$ (***)		$F = 0.1$ (ns)	

[a]Values are least squares (LS) means with the standard error (SE) of the LS estimate in brackets. [b]Pin means Pinyiri. [c]T1, 'chic' is chicken wire. [d]T2, 'mos' is mosquito net. [e]T3, 'clos' is closed. Values within a column followed by different letters are significantly different at $P < 0.05$. Significance level: *** = $P < 0.001$, ** = $P < 0.01$, * = $P < 0.05$, and ns = $P > 0.05$. # Treatment T0 versus T1 is significantly different ($P < 0.04$) at Pinyiri, when excluding treatment T2.

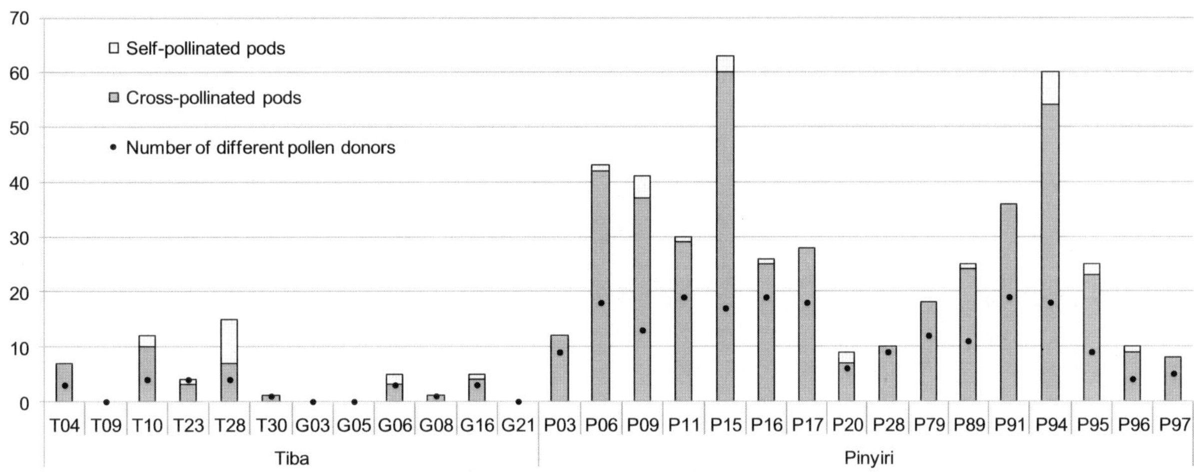

FIGURE 1. Number of genotyped self- and cross-pollinated pods across treatments in *Parkia biglobosa* shown per tree and site. The points show the total number of dissimilar pollen donors (one father-tree counts only once) across treatments per mother-tree.

incl. self-pollination (Tab. 3). Some flat pods contained only aborted seeds, but these did also yield DNA suitable for the paternity analysis. A father-tree was assigned to 82% of genotyped seeds (50 pods) for Tiba and 89% (444 pods) for Pinyiri (Tab. 3). The majority of unassigned pods came from the open (T0) treatment (10 pods from Tiba and 38 pods from Pinyiri). At the drier site, Tiba, all unassigned pods had negative trio LOD scores (i.e. pollen most likely came from outside the study site), whereas it was 76% at the more humid site, Pinyiri.

Self-pollination in T0 and T1 was significantly higher (X^2 = 28.3; P < 0.001) at Tiba (21-23%) compared to Pinyiri (2-4%). At Pinyiri, 19% of the pods from T2 were self-pollinated (Tab. 3). Self-pollinated pods were found on five trees at Tiba and on ten trees at Pinyiri (Fig. 1).

The number of different pollen donors (incl. self-pollination) per studied mother-tree in Tiba was only 1-4 (based on 50 assigned pods from 8 mother-trees) compared to 4-19 (based on 444 assigned pods from 16 mother-trees) at Pinyiri (Tab. 3, Fig. 1). At Tiba, 55% (= 11 capitula) of the capitula with two studied pods revealed more than one pollen donor, while the corresponding fraction was 74% (= 164 capitula) in Pinyiri. The number of pollen donors per tree differed significantly among treatments, with treatments T0 and T1 resulting in a higher number of pollen donors compared to T2 where the honey bees were excluded (Tab 3).

Genetic diversity

The microsatellite genotyping of the adult trees revealed genotypic distributions corresponding to Hardy-Weinberg proportions at both sites, which suggests a history of random mating. Only limited genetic differentiation between the two study sites was revealed by the AMOVA (Φ_{st} = 0.011, P < 0.001).

Spatial pollen dispersal at the two study sites

The distances between the studied mother-trees and their nearest flowering *P. biglobosa* neighbour ranged from 16 m to 263 m (mean 98 m, SD = 65 m) at Tiba and from 10 m to 68 m (mean 32 m, SD = 18 m) at Pinyiri. Still, the average distance of realised cross-pollination per tree across treatments did not differ significantly between the sites (Tab. 3). A specific comparison of treatments T0 against T1 at Pinyiri showed significant longer mean cross-pollen transport within the open treatment T0 compared to T1 at this site (P < 0.04).

The median distances of pollen transport (excluding selfing) including all treatments were 143 m and 115 m for Tiba and Pinyiri, respectively. About half of the pollen donors were within a distance of 200 m from the mother-trees, while approximately 20% of the donors were more than 300 m away (Fig. 2). Long distance pollen movement (> 500 m) were estimated as 6% for treatment T0 and 5% for T1 at Tiba, whereas at Pinyiri long distance pollination occurred four times more frequently for T0 (8%) compared to T1 (2%). However, many of the non-assigned pollen donors (mostly found in T0) are likely to be trees outside the sampled area and therefore reflect additional long distance pollen movements.

The DNA documented pollination events at the two sites are presented in Fig. 3. At Tiba, only a small fraction of trees were identified to serve as pollen donors and self-pollination was relatively frequent (Fig. 3 A). At Pinyiri, the pollination events involved many trees and often included movement of pollen across the main road that divides the village (Fig. 3 B).

Distance to honey bee colonies

The average distance from a mother-tree to the closest honey bee colony was 402 m (SD = 323 m) for Tiba and 108 m (SD = 59 m) for Pinyiri. Within a radius of 1,000 m from a given mother-tree, Tiba had a mean of 4 colonies while Pinyiri had a mean of 35 colonies. However, average number of honey bee colonies per flowering *P. biglobosa* trees was higher at Tiba compared to Pinyiri (1.3 versus 0.2), due to the low density of flowering *P. biglobosa* trees (Fig. 3). We found no correlation between the number of honey bee colonies within a given radius and the pod

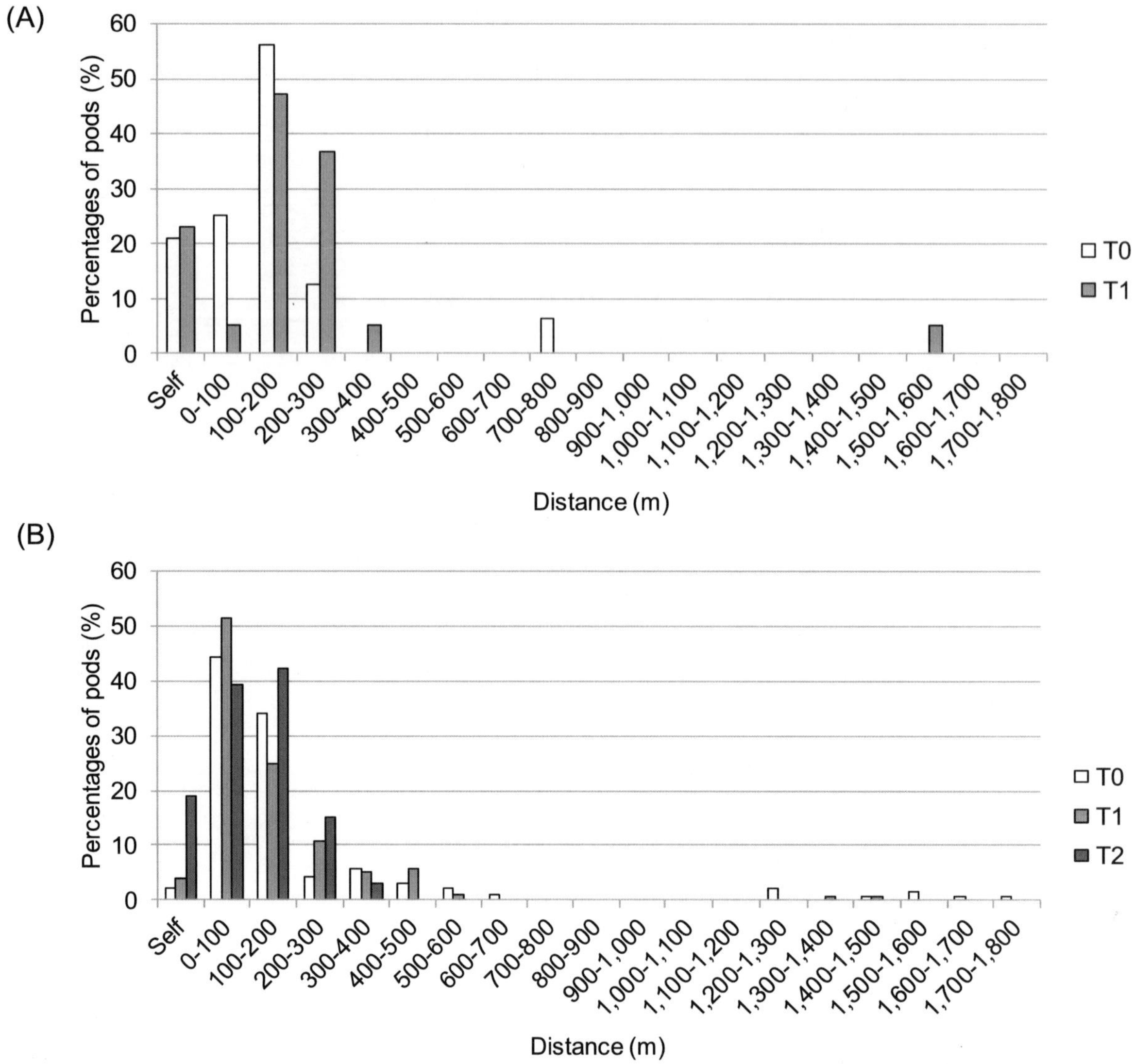

FIGURE 2. Percentages of self-pollination and distances of cross-pollen transport from donor tree to mother-tree in *Parkia biglobosa* divided on treatments and study sites (A) Tiba (drier site); (B), Pinyiri (more humid site).

production for neither the open T0-treatment nor the chicken wire T1-treatment at any of the two sites.

DISCUSSION

In the current study, pollinator exclusion trials showed that pod and seed set were only slightly reduced when excluding vertebrates, while pod and seed set were highly reduced when also excluding larger insects such as honey bees from visiting the flowers. Genetic analysis of seeds revealed a higher level of selfing, fewer pollen donors per tree, and longer median pollen transport distances at Tiba, compared to the population at Pinyiri.

Differences in flower-visitors and pod yield between the sites

We found the same functional groups of flower-visitors at the two study sites. However, bats were more frequent visitors of *Parkia biglobosa* at the more humid site, Pinyiri, compared to the drier site, Tiba, while honey bee visitation rates were similar despite differences in tree densities between sites. Visitation rates by smaller bees varied among individual trees.

The number of mature pods per hermaphroditic capitulum was significantly lower at Tiba, but the difference in number of immature pods per hermaphroditic capitulum was less noticeable and higher abortion rate rather than pollen limitation thus seemed to be the major factor behind the lower pod production per capitulum at the drier site. However, the higher level of self-pollination and fewer pollen donors per tree at Tiba compared to Pinyiri, suggest that poor conditions for cross-pollination played a role at Tiba, and might have contributed to the high abortion rate and therefore poor fruit set. Similar findings of low tree density increasing self-pollination rates have been observed in other studies (Franceschinelli & Bawa 2000; Eckert et al. 2010).

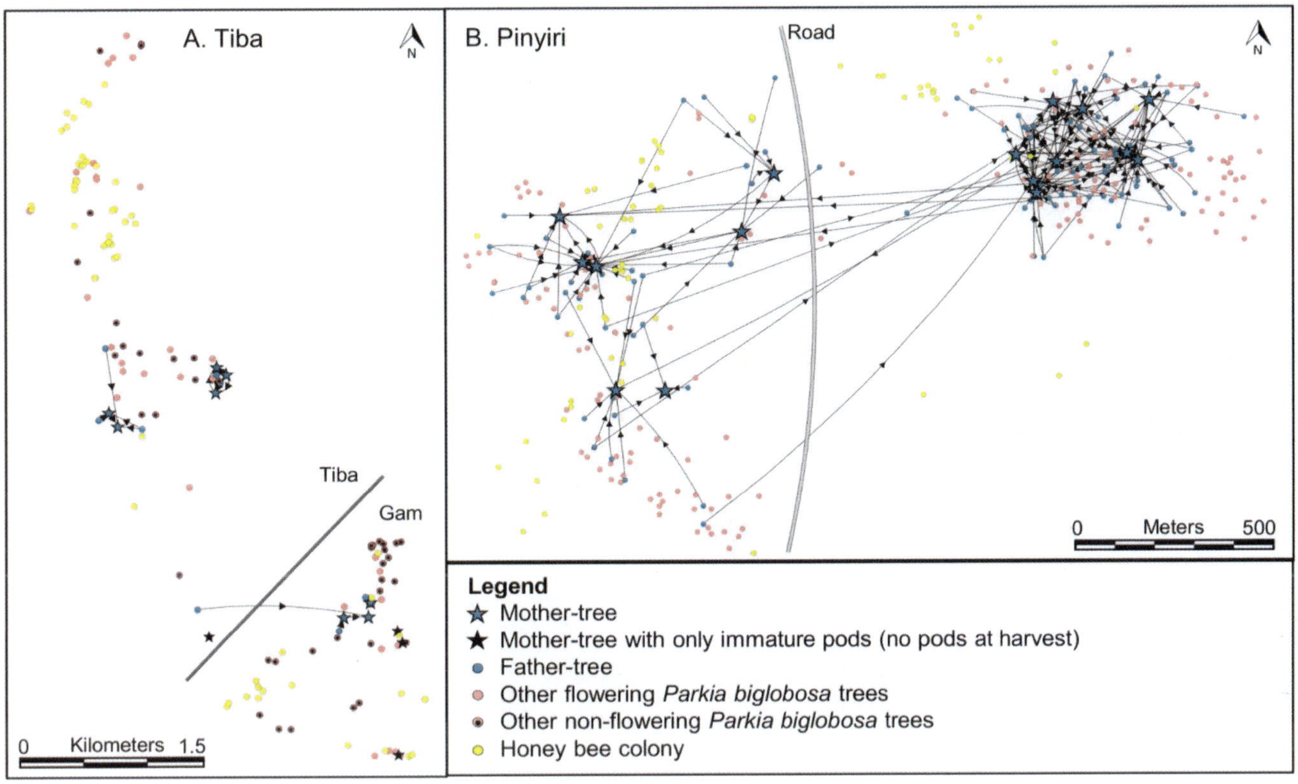

FIGURE 3. Illustration of the cross-pollination events for *Parkia biglobosa* across treatments (A, Tiba (drier site, also showing Gam); B, Pinyiri (more humid site)). The arrows show the direction of the pollen transport. One arrow may represent more than one pod. Events of self-pollination are not shown.

Additionally, differences in climate at the two sites may have contributed to fewer flowering *P. biglobosa* trees and higher abortion rate of pods at the drier site, Tiba, which is located close to the northern border of the natural distribution area of the species (Ouédraogo 1995). Here, the annual precipitation in 2011 leading up to the fruiting season of *P. biglobosa* in 2012 was below average (only ca. 570 mm). Furthermore, many *P. biglobosa* trees found north of the 700 mm isohyet are in bad condition due to water stress (Timmer et al. 1996). Thus, the growth conditions at Tiba were less favourable for the species compared with those at Pinyiri. The paternity analysis revealed that half of the underdeveloped pods with large, flat seeds and no pulp at Tiba was cross-pollinated, which indicate lack of resources as cause of aborted seeds (Stephenson 1981; Lee 1988). In comparison, we found no flat pods at Pinyiri.

The trend towards fewer but larger pods with decreasing mesh size (T0<T1<T2) observed at Pinyiri suggests the presence of competition for maternal resources for the development of the pods and seeds in line with results from other studies (Lloyd 1980; Weinbaum et al. 2001; Minchin et al. 2010).

Effects of excluding certain types of pollinators

The result that the closed treatment (T3) inhibited fruit set confirms the need of animal pollinators for *P. biglobosa* as previously found in a similar study in The Gambia (Lassen et al. 2012). However, a major finding of the present study was that excluding bats and birds had no or little effect on the yield of *Parkia biglobosa,* whereas excluding larger insects such as honey bees reduced the yield significantly and resulted at the drier site, Tiba, in as few pods as the closed treatment.

As the open treatment T0 allowed all pollinators to visit the capitula, while the chicken wire in treatment T1 excluded larger visitors (bats and birds), any difference between the two treatments will reveal the effect of bats and birds as pollinators. We suspect that mainly bats were responsible for the few differences between T0 and T1. Sunbirds were mostly interested in the nectar ring and therefore considered to be insignificant pollinators (Hopkins 1983), and during field work we only observed birds visiting *P. biglobosa* on very few mornings. In general the differences between T0 and T1 were small at Tiba, and although the mean cross-pollen transport distance was much higher for T1 than T0 the difference was statistically non-significant. However, it was not possible to assign fathers for 30% of the pods from T0 while this was the case for only 4% of the pods from T1, thus indicating that the fathers of T0 pods more often were found outside the study site compared with the insect-pollinated pods (T1). At Pinyiri, the cross-pollen transport occurred over significantly longer distances for the open treatment T0 compared with treatment T1. This fits well with the observations of considerably more bats at Pinyiri than at Tiba. At Pinyiri, the percentage of pods from T0 and T1 that had been sired by a father outside the study site was smaller than at Tiba. However, this may be due to the difference in densities of *P. biglobosa* at the two sites.

The differences between treatments T1 and T2 are assumed mainly to be due to the absence of honey bees in T2. Pod and seed set measures were significantly larger for T1 than for T2 at both sites suggesting that honey bees were important for proper pollination of *P. biglobosa* flowers. This result is also in line with a similar trial performed in The Gambia, where a comparable T2 treatment yielded more aborted seeds than a comparable T1 treatment (Lassen et al. 2012).

Our observations at both sites showed that the smaller bees were responsible for the main pollination in treatment T2. At the drier site, Tiba, it is likely that the smaller bees carried out relatively more self-pollination due to the lower density of flowering *P. biglobosa*. The distances of pollen transport for T2 were similar to the findings from the tropical tree species yellow mombin (*Spondias mombin* L.) and fig (*Ficus* spp.), where different small insect pollinators regularly transported pollen 100-200 m among flowering trees (Nason & Hamrick 1997).

Influence by pollinator type on the number of pollen donors has also been observed for Japanese chestnut (*Castanea crenata* Siebold & Zucc.), where analysis of individual pollen grains from different pollinators revealed that the average effective number of pollen donors was around three times higher for bumblebees than for small bees (Hasegawa et al. 2015).

Self-pollination rates and pollination distances

In a controlled pollination trial with *P. biglobosa* in Burkina Faso, Ouédraogo (1995) found around 5% fruit set after self-pollination compared with around 37% fruit set after cross-pollination. Sina (2006) estimated about 5% self-pollination in open-pollinated *P. biglobosa* based on enzyme electrophoresis at two sites close to Ouagadougou, Burkina Faso, which is similar to our findings of 2% for the open treatment at Pinyiri. It is thus likely that *P. biglobosa* possesses a self-incompatibility system (SI), which is also supported by a controlled pollination trial with *P. biglobosa* in Burkina Faso (Lassen et al., *in prep.*). Asian species of *Parkia* (*P. speciosa* and *P. timoriana*) have also been reported as self-incompatible (Bumrungsri et al. 2008). The relatively high percentages (21% and 23%) of self-pollination in both the open (T0) treatment and the chicken wire (T1) treatment at Tiba, however, imply that selfing is important and possible under certain conditions. The difference in selfing between sites cannot simply be explained by genetic differentiation as the two populations were very similar. Nor can it be explained by the limited number of bats at Tiba, since treatment T1 (exclusion of bats) also had significantly more self-pollination at Tiba compared with Pinyiri. Hence, the differences in percentages of self-pollination are most likely due to limited mating opportunities allowing some selfing to occur.

In animal-pollinated populations of neo-tropical tree species, pollen dispersal occurs over longer distances in low-density populations compared to high-density populations (Ward et al. 2005). We found a similar tendency for *P. biglobosa* for cross-pollination (medians were 143 m and 115 m for Tiba and Pinyiri, respectively). Hence, a lower density may impact the reproduction in two opposite directions; one is a higher percentage of self-pollination and the other is an increased distance of pollen flow due to lack of close neighbours (Dick et al. 2003; Ward et al. 2005). Nevertheless, the pollen dispersal distributions (Fig. 2) are very similar for the two study sites. In several cases the assigned father-trees were more distantly placed than the neighbouring potential pollen donors (Fig. 3), in contrast to the findings by Levin & Kerster (1974). However, other studies have also reported random pollen movements (Finkner 1954; Larsen 2010; Stingemore 2014). The patterns may be driven by pollinator movements (Thomson & Plowright 1980; Ward et al. 2005) but also by the presence of a SI system, which may blur the real pollen flow since a given self- or cross-pollination event due to genetic SI may fail to result in fertilisation and pod production. The reduced but continuous outcrossing despite increased distances among flowering trees is likely to counteract potential negative effects of inbreeding (Breed et al. 2015).

Increased selfing as a consequence of low mating possibilities despite SI has been observed for other species (e.g. Willi, 2009). Breakdown of an SI system is common in pioneer species and marginal populations where compatible mates are lacking or where pollinators are scarce (Baker 1955; Barrett 2002; Xiong et al. 2013). This will ensure seed set, although severe consequences may occur due to inbreeding depression (Charlesworth & Charlesworth 1987; Whisler & Snow 1992). A controlled pollination trial with *P. biglobosa* reported around half the number of healthy seeds for the self-pollinated pods compared with the cross-pollinated ones (Lassen et al., *in prep.*).

Distances to honey bee colonies and pod yield

In spite of the importance of honey bees in both The Gambia and Burkina Faso, we did not find a correlation between distances to honey bee colonies and yield, which perhaps was because the distance to any colony was much shorter than the normal flight range of the honey bees (Abrol 2012). Furthermore, we have neither inspected the strength of the colonies (i.e. estimated the numbers of individuals) or registered competing flower species at the two study sites, both of which may have interfered with a given correlation.

Management of pollinators of P. biglobosa in face of climate change

The present study found that honey bees are capable of securing the pollination of *Parkia biglobosa* in the absence of bats, at least with distances between flowering trees as found in this study. However, fruit bats are still important pollinators of *P. biglobosa* especially due to their longer flight range (Linhart 1995; Monadjem et al. 2010; Fahr et al. 2015), which may connect scattered populations. Still, honey bees are almost equally good as bats in pollinating *P. biglobosa* due to their large numbers and relatively long flight range.

Besides being hunted as bushmeat (Kamins et al. 2011), fruit bats are influenced by fluctuations in the climate (Welbergen et al. 2008) and the availability of food (pollen, nectar, fruits) (Kalko 1997; Richter & Cumming 2006). Furthermore, the different bat species have different requirements to roosting sites (caves, trees, bushes etc.) (Kangoyé et al. 2012). The acknowledgement of their value as creating gene flow both as pollinators (e.g. Acharya et al., 2015) and as seed dispersers (e.g. Djossa et al., 2008) is important and likewise is their protection (Fujita & Tuttle 1991; Kunz et al. 2011).

If the climate change proceeds as some models foresee and the number of bats available for pollination is reduced in West Africa, it is important that honey bees are present and maintained in good conditions, so that the honey bees can supplement the needed pollination services. However, climate change will also impact the populations of honey bees (Le Conte & Navajas 2008) and other pollinating insects, and it may be necessary to provide them with water in order to reduce migration and increase survival. In spite of this, pod yield decline may be unavoidable, since reduced rainfall may hamper the blooming of *P. biglobosa* and increase the pod abortion, while birds may predate on what little is produced. However, a reduced pod set will have adverse effects for the rural people, because both the sweet pulp of the pods and the protein-rich seeds are widely used (Hall et al. 1997). The pods mature at the end of the dry season, where food supplies may be low (Longhurst 1985) and many women gain income from fermenting the seeds into a popular condiment known as 'soumbala' (Kessler 1994; Gausset et al. 2005).

Maintaining genetic variation in the future is important due to the challenge of adaptation to new situations (climate, pollinators, etc.). Another reason is the potential for breeding the species. *Parkia biglobosa* is considered semi-domesticated due to selective logging of trees when clearing new farm land (Kreike 2013). Hall et al. (1997) list examples of countries where different ethnic groups practice sowing of *P. biglobosa* seeds. However, due to the low natural regeneration (Ræbild et al. 2012), it is possible that more farmers will have to begin planting this species (Nikiéma 1993), despite problems with land tenure (Etongo et al. 2015), and cultural beliefs (Hall et al. 1997; Buchmann et al. 2009). Seeds or seedlings from high yielding trees producing pods with sweet pulp and many seeds will probably be in demand (Hall et al. 1997).

Acknowledgements

This research was conducted together with Centre National de Semences Forestières (CNSF) and the authors wish to thank the staff at CNSF for their support. Many thanks to Alassane Ouédraogo, Madi Tiemtoré, Philbert Zoungrana, Kassoum Sawadogo, and Jeoany Sawadogo for field assistance and to the farmers at Tiba, Gam, and Pinyiri for permitting us to use their trees. We wish to thank Claus Rasmussen, Aarhus University, Department of Bioscience, Denmark, for identifying and measuring samples of bees. Furthermore, we acknowledge Lars Vilhelmsen, Cecilie Svenningsen, and Thomas Pape from the Research Section of BioSystematics, Natural History Museum of Denmark, University of Copenhagen for identifying samples of mirid bugs, ants, and flies, respectively and Jan Bezdek from Mendel University, Brno, Czech Republic for identifying samples of beetles. Thanks to Annalise Metz, Lene Hasmark Andersen and Albin Lobo from University of Copenhagen for measuring the pods and seeds. The present paper is part of a PhD study financed by the Consultative Research Committee for Development Research (FFU) under the Danish International Development Agency (DANIDA) (research project no. 10-106-LIFE).

References

Abrol DP (2012) Pollination biology. Biodiversity conservation and agricultural production. Springer, New York, USA

Acharya PR, Racey PA, Sotthibandhu S, Bumrungsri S (2015) Feeding behaviour of the dawn bat (*Eonycteris spelaea*) promotes cross pollination of economically important plants in Southeast Asia. Journal of Pollination Ecology 15:44-50

Al-Ghamdi AA, Nuru A, Khanbash MS, Smith DR (2013) Geographical distribution and population variation of *Apis mellifera jemenitica* Ruttner. Journal of Apicultural Research 52:124-133

Baker HG (1955) Self compatibility and establishment after long distance dispersal. Evolution 9:347-349. doi: 10.2307/2405656

Barrett SCH (2002) The evolution of plant sexual diversity. Nature Reviews Genetics 3:274-284. doi: 10.1038/nrg776

Boffa J-M (1999) Agroforestry parklands in sub-Saharan Africa. FAO Conservation Guide, vol. 34. FAO, Rome, Italy, p 230

Breed MF, Ottewell KM, Gardner MG, Marklund MHK, Dormontt EE, Lowe AJ (2015) Mating patterns and pollinator mobility are critical traits in forest fragmentation genetics. Heredity 115:108-114. doi: 10.1038/hdy.2013.48

Buchmann C, Prehsler S, Hartl A, Vogl C (2009) To plant or not to plant? Considering the cultural context of adoptive transplantation of Baobab (*Adansonia digitata* L.) and Tamarind (*Tamarindus indica* L.) in West Africa. In: Tielkes E (ed) Tropentag 2009 – Biophysical and Socio-Economic Frame Conditions for the Sustainable Management of Natural Resources. DITSL GmbH (German Institute for Agriculture in the Tropics and Subtropics), Hamburg, Germany, p 426

Bumrungsri S, Harbit A, Benzie C, Carmouche K, Sridith K, Racey P (2008) The pollination ecology of two species of *Parkia* (Mimosaceae) in southern Thailand. Journal of Tropical Ecology 24:467-475. doi: Doi 10.1017/S0266467408005191

Charlesworth D, Charlesworth B (1987) Inbreeding depression and its evolutionary consequences. Annual Review of Ecology and Systematics 18:237-268. doi: 10.1146/annurev.ecolsys.18.1.237

Cheke RA, Mann CF, Allen R (2001) Sunbirds. A guide to the sunbirds, flowerpeckers, spiderhunters, and sugarbirds of the world. Christopher Helm Publishers, London, UK

Dick CW, Etchelecu G, Austerlitz F (2003) Pollen dispersal of tropical trees (*Dinizia excelsa*: Fabaceae) by native insects and African honeybees in pristine and fragmented Amazonian rainforest. Molecular Ecology 12:753-764. doi: 10.1046/j.1365-294X.2003.01760.x

Djossa BA, Fahr J, Kalko EKV, Sinsin BA (2008) Fruit selection and effects of seed handling by flying foxes on germination rates of shea trees, a key resource in northern Benin, West Africa. Ecotropica 14:37–48

Dukku UH (2010) Plants for Bees. *Parkia biglobosa*. An important honeybee forage in the Savanna. Bee World 87:28-29

Eckert CG, Kalisz S, Geber MA, Sargent R, Elle E, Cheptou PO, Goodwillie C, Johnston MO, Kelly JK, Moeller DA, Porcher E, Ree RH, Vallejo-Marin M, Winn AA (2010) Plant mating systems in a changing world. Trends in Ecology & Evolution 25:35-43. doi: 10.1016/j.tree.2009.06.013

Etongo D, Djenontin INS, Kanninen M, Fobissie K (2015) Smallholders' tree planting activity in the Ziro province, southern

Burkina Faso: Impacts on livelihood and policy implications. Forests 6:2655-2677. doi: 10.3390/f6082655

Excoffier L, Smouse PE, Quattro JM (1992) Analysis of molecular variance inferred from metric distances among DNA haplotypes: Application to human mitochondrial DNA restriction data. Genetics 131:479-491

Fahr J, Abedi-Lartey M, Esch T, Machwitz M, Suu-Ire R, Wikelski M, Dechmann DKN (2015) Pronounced seasonal changes in the movement ecology of a highly gregarious central-place forager, the African straw-coloured fruit bat (Eidolon helvum). Plos One 10:e0138985. doi: 10.1371/journal.pone.0138985

Finkner MD (1954) Random activity of pollen vectors in isolated plots of upland cotton. Agronomy Journal 46:68-70

Franceschinelli EV, Bawa KS (2000) The effect of ecological factors on the mating system of a South American shrub species (*Helicteres brevispira*). Heredity 84:116-123. doi: 10.1046/j.1365-2540.2000.00636.x

Franzén M, Öckinger E (2012) Climate-driven changes in pollinator assemblages during the last 60 years in an Arctic mountain region in Northern Scandinavia. Journal of Insect Conservation 16:227-238. doi: 10.1007/s10841-011-9410-y

Fryxell PA (1957) Mode of reproduction of higher plants. Botanical Review 23:135-233. doi: 10.1007/bf02869758

Fujita MS, Tuttle MD (1991) Flying foxes (Chiroptera: Pteropodidae): Threatened animals of key ecological and economic importance. Conservation Biology 5:455-463. doi: 10.1111/j.1523-1739.1991.tb00352.x

Funk C, Rowland J, Adoum A, Eilerts G, White L (2012) A climate trend analysis of Burkina Faso U.S. Geological Survey Fact Sheet, vol. 2012-3084. U.S. Geological Survey, South Dakota, USA, p 4

Gausset Q, Yago-Ouattara EL, Belem B (2005) Gender and trees in Péni, South-Western Burkina Faso. Women's needs, strategies and challenges. Geografisk Tidsskrift (Danish Journal of Geography) 105:67-76

Gonzalez P, Tucker CJ, Sy H (2012) Tree density and species decline in the African Sahel attributable to climate. Journal of Arid Environments 78:55-64.

Hall JB, Tomlinson HF, Oni PI, Buchy M, Aebischer DP (1997) *Parkia biglobosa*: a monograph. School of Agricultural and Forest Sciences Publications Number 9, University of Wales, Bangor, UK

Haq N (2008) *Parkia biglobosa* African locust bean. In: Janick J, Paull RE (eds) The encyclopedia of fruits & nuts. CABI, Wallingford, UK, pp 395-398

Harris BJ, Baker HG (1959) Pollination of flowers by bats in Ghana. The Nigerian Field 24:151-159

Hasegawa Y, Suyama Y, Seiwa K (2015) Variation in pollen-donor composition among pollinators in an entomophilous tree species, *Castanea crenata*, revealed by single-pollen genotyping. Plos One 10:e0120393. doi: 10.1371/journal.pone.0120393

Hepburn HR, Radloff SE (1998) Honeybees of Africa. SpringerVerlag, Berlin, Germany

Hopkins HC (1983) The taxonomy, reproductive biology and economic potential of *Parkia* (Leguminosae: Mimosoideae) in Africa and Madagascar. Botanical Journal of the Linnean Society 87:135-167. doi: 10.1111/j.1095-8339.1983.tb00987.x

Hopkins HC, White F (1984) The ecology and chorology of *Parkia* in Africa. Bulletin du Jardin Botanique National de Belgique / Bulletin van de National Plantentuin van België 54:235-266. doi: 10.2307/3667874

Ibrahim B, Karambiri H, Polcher J, Yacouba H, Ribstein P (2014) Changes in rainfall regime over Burkina Faso under the climate change conditions simulated by 5 regional climate models. Climate Dynamics 42:1363-1381. doi: 10.1007/s00382-013-1837-2

Kalinowski ST, Taper ML, Marshall TC (2007) Revising how the computer program cervus accommodates genotyping error increases success in paternity assignment. Molecular Ecology 16:1099-1106. doi: 10.1111/j.1365-294X.2007.03089.x

Kalko EKV (1997) Diversity in tropical bats. In: Ulrich H (ed) Tropical biodiversity and systematics. Proceedings of the International Symposium on Biiodiveristy and Systematics in Tropical Ecosystems. Zoologisches Forschungsinstitut und Museum Alexander Koenig, Bonn, Germany, pp 13-43

Kamins AO, Restif O, Ntiamoa-Baidu Y, Suu-Ire R, Hayman DTS, Cunningham AA, Wood JLN, Rowcliffe JM (2011) Uncovering the fruit bat bushmeat commodity chain and the true extent of fruit bat hunting in Ghana, West Africa. Biological Conservation 144:3000-3008. doi: 10.1016/j.biocon.2011.09.003

Kangoyé NM, Oueda A, Thiombiano A, Guenda W (2012) Bats (Chiroptera) of Burkina Faso: preliminary list with fifteen first record species. International Journal of Biological and Chemical Sciences 6:6017-6030

Kearns CA, Inouye DW, Waser NM (1998) Endangered mutualisms: The conservation of plant-pollinator interactions. Annual Review of Ecology and Systematics 29:83-112. doi: 10.1146/annurev.ecolsys.29.1.83

Kessler JJ (1994) Agroforestry in Burkina Faso: a careful balance. International Ag-Sieve 6:4-5

Kreike E (2013) Environmental Infrastructure in African History. Examining the Myth of Natural Resource Management in Namibia. Cambridge University Press, New York, USA

Kunz TH, de Torrez EB, Bauer D, Lobova T, Fleming TH (2011) Ecosystem services provided by bats. In: Ostfeld RS, Schlesinger WH (eds) Year in Ecology and Conservation Biology, vol 1223, pp 1-38

Larsen AS (2010) Gene flow and genetic structure of wild fruit trees: DNA-marker based studies in *Adansonia digitata, Vitellaria paradoxa, Parkia biglobosa* and *Malus sylvestris*. PhD thesis University of Copenhagen, Denmark

Lassen KM (2016) Pollination strategies to increase productivity of the African fruit trees *Vitellaria paradoxa* subsp. *paradoxa* & *Parkia biglobosa*. PhD thesis, University of Copenhagen, Copenhagen, Denmark

Lassen KM, Kjær ED, Ouédraogo M, Nielsen LR (2014) Microsatellite primers for *Parkia biglobosa* (Fabaceae: Mimosoideae) reveal that a single plant sires all seeds per pod. Applications in Plant Sciences 2:1400024. doi: 10.3732/apps.1400024

Lassen KM, Ræbild A, Hansen H, Brødsgaard CJ, Eriksen EN (2012) Bats and bees are pollinating *Parkia biglobosa* in The Gambia. Agroforestry Systems 85:465-475. doi: 10.1007/s10457-011-9409-0

Le Conte Y, Navajas M (2008) Climate change: impact on honey bee populations and diseases. Revue Scientifique Et Technique-Office International Des Epizooties 27:499-510

Lee TD (1988) Patterns of fruit and seed production. In: Lovett Doust J, Lovett Doust L (eds) Plant reproductive ecology. Patterns and strategies. Oxford University Press, New York, USA, pp 179-202

Levin DA, Kerster HW (1974) Gene flow in seed plants. In: Dobzhansky T, Hecht MK, Steere WC (eds) Evolutionary Biology, vol 7. Plenum Press, New York, USA, pp 139-220

Linhart YB (1995) Restoration, revegetation, and the importance of genetic and evolutionary perspectives. In: Roundy BA, McArthur ED, Haley JS, Mann DK (eds) Proceedings: Wildland Shrub and Arid Land Restoration Symposium. Intermountain Research Station, Forest Service, U.S. Department of Agriculture, Ogden, Utah, USA, pp 271-287

Lloyd DG (1980) Sexual strategies in plants I. An hypothesis of serial adjustment of maternal investment during one reproductive session. New Phytologist 86:69-79. doi: 10.1111/j.1469-8137.1980.tb00780.x

Longhurst R (1985) Cropping systems and households food security: Evidence from three West African countries. Food and Nutrition 11:10-16

Maranz S (2009) Tree mortality in the African Sahel indicates an anthropogenic ecosystem displaced by climate change. Journal of Biogeography 36:1181-1193

Météo (2015) Pluviometrie mensuelle (mm) dans Guilongou (Ziniare) et Pô 2010-2012 et temperature mensuelle (°C) dans Ouagadougou Aéroport et Pô 2011-2012. Direction Générale de la Météorologie du Burkina, Ouagadougou, Burkina Faso

Minchin PEH, Snelgar WP, Blattmann P, Hall AJ (2010) Competition between fruit and vegetative growth in Hayward kiwifruit. New Zealand Journal of Crop and Horticultural Science 38:101-112. doi: 10.1080/01140671003781728

Monadjem A, Taylor PJ, Cotterill FPDW, Schoeman MC (2010) Bats of Southern and Central Africa: A biogeographic and taxonomic synthesis. Wits University Press, Johannesburg, South Africa

Nason JD, Hamrick JL (1997) Reproductive and genetic consequences of forest fragmentation: Two case studies of neotropical canopy trees. Journal of Heredity 88:264-276

Niang I, Ruppel OC, Abrdrabo MA, Essel A, Lennard C, Padgham J, Urquhart P (2014) Africa. In: Barros VR, Field CB, Dokken DJ, Mastrandrea MD, Mach KJ, Bilir TE, Chatterjee M, Ebi KL, Estrada YO, Genova RC, Girma B, Kissel ES, Levy AN, MacCracken S, Mastrandrea PR, White LL (eds) Climate Change 2014: Impacts, Adaptation, and Vulvnerability. Part B: Regional Aspects. Contribution of Working Group II to the Fifth Assessment Report of the Intergovernmental Panel on Climate Change. Cambridge University Press, Cambridge, UK and New York, NY, USA, pp 1199-1265 pp.

Nikiéma A (1993) Regeneration of *Parkia biglobosa* (Jacq) R. Br. ex G. Don. in an agroforestry system: a pilot study in Burkina Faso. MSc Forestry thesis, Department of Forestry, Wageningen Agricultural University, Wageningen, the Netherlands

Nombré I (2003) Etude des potentialités mellifères de deux zones du Burkina Faso, Garango (Province du Boulgou) et Nazinga (Province du Nahouri). PhD Thesis, Université de Ouagadougou, Ouagadougou, Burkina Faso

Ollerton J, Winfree R, Tarrant S (2011) How many flowering plants are pollinated by animals? Oikos 120:321-326

Ouédraogo AS (1995) *Parkia biglobosa* (Leguminosae) in West Africa; biosystematics and improvement. PhD thesis, Landbouwuniversiteit Wageningen (Wageningen Agricultural University), Wageningen, The Netherlands

Ouedraogo M, Raebild A, Nikiema A, Kjaer ED (2012) Evidence for important genetic differentiation between provenances of *Parkia biglobosa* from the Sudano-Sahelian zone of West Africa. Agroforestry Systems 85:489-503. doi: 10.1007/s10457-011-9463-7

Parmesan C (2007) Influences of species, latitudes and methodologies on estimates of phenological response to global warming. Global Change Biology 13:1860-1872. doi: 10.1111/j.1365-2486.2007.01404.x

Patiny S, Michez D, Kuhlmann M, Pauly A, Barbier Y (2009) Factors limiting the species richness of bees in Saharan Africa. Bulletin of Entomological Research 99:337-346. doi: 10.1017/s0007485308006433

Peakall R, Smouse PE (2006) GenAlEx 6: Genetic analysis in Excel. Population genetic software for teaching and research. Molecular Ecology Notes 6:288-295. doi: 10.1111/j.1471-8286.2005.01155.x

Peakall R, Smouse PE (2012) GenAlEx 6.5: Genetic analysis in Excel. Population genetic software for teaching and research - an update. Bioinformatics 28:2537-2539. doi: 10.1093/bioinformatics/bts460

Petanidou T, Kallimanis AS, Sgardelis SP, Mazaris AD, Pantis JD, Waser NM (2014) Variable flowering phenology and pollinator use in a community suggest future phenological mismatch. Acta Oecologica-International Journal of Ecology 59:104-111. doi: 10.1016/j.actao.2014.06.001

Richter HV, Cumming GS (2006) Food availability and annual migration of the straw-colored fruit bat (*Eidolon helvum*). Journal of Zoology 268:35-44. doi: 10.1111/j.1469-7998.2005.00020.x

Roubik DW (2006) Stingless bee nesting biology. Apidologie 37:124-143

Rousset F (2008) GENEPOP ' 007: a complete re-implementation of the GENEPOP software for Windows and Linux. Molecular Ecology Resources 8:103-106. doi: 10.1111/j.1471-8286.2007.01931.x

Ræbild A, Hansen UB, Kambou S (2012) Regeneration of *Vitellaria paradoxa* and *Parkia biglobosa* in a parkland in Southern Burkina Faso. Agroforestry Systems 85:443-453. doi: 10.1007/s10457-011-9397-0

Sanfo JB (2012) Apport de la Direction Générale de la Météorologie dans le processus d'information sur l'eau 13ème sommet de l'information sur l'eau, 11-13 April 2012, Ouagadougou, Burkina Faso

SAS Institute (2011) The SAS system for Windows. Release 9.4, Cary, North Carolina, USA

Schweitzer P, Nombré I, Aidoo K, Boussim IJ (2013) Plants used in traditional beekeeping in Burkina Faso. Open Journal of Ecology 3:354-358

Settele J, Scholes R, Betts R, Bunn S, Leadley P, Nepstad D, Overpeck JT, Taboada MA (2014) Terrestrial and inland water systems. In: Field CB, Barros VR, Dokken DJ, Mach KJ, Mastrandrea MD, Bilir TE, Chatterjee M, Ebi KL, Estrada YO, Genova RC, Girma B, Kissel ES, Levy AN, MacCracken S, Mastrandrea PR, White LL (eds) Climate Change 2014: Impacts, Adaptation, and Vulnerability. Part A: Global and Sectoral Aspects. Contribution of Working Group II to the Fifth Assessment Report of the Intergovernmental Panel on Climate Change. Cambridge University Press, Cambridge, UK and New York, USA, pp 271-359

Sina S (2006) Reproduction et diversité génétique chez *Parkia biglobosa* (Jacq.) G.Don. PhD thesis, Wageningen University, Wageningen, the Netherlands

Stephenson AG (1981) Flower and fruit abortion: Proximate causes and ultimate functions. Annual Review of Ecology and Systematics 12:253-279. doi: 10.1146/annurev.es.12.110181.001345

Stingemore JA (2014) Plant density effects on genetic variation and dispersal for two co-occuring *Persoonia* species. PhD thesis, The University of Western Australia, Australia

Thomson JD, Plowright RC (1980) Pollen carryover, nectar rewards, and pollinator behavior with special reference to *Diervilla lonicera*. Oecologia 46:68-74. doi: 10.1007/bf00346968

Timmer LA, Kessler JJ, Slingerland M (1996) Pruning of néré trees (*Parkia biglobosa* (Jacq) Benth) on the farmlands of Burkina Faso, West Africa. Agroforestry Systems 33:87-98. doi: 10.1007/bf00122891

Ward M, Dick CW, Gribel R, Lowe AJ (2005) To self, or not to self... a review of outcrossing and pollen-mediated gene flow in neotropical trees. Heredity 95:246-254

Weinbaum SA, DeJong TM, Maki J (2001) Reassessment of seed influence on return bloom and fruit growth in 'Bartlett' pear. Hortscience 36:295-297

Welbergen JA, Klose SM, Markus N, Eby P (2008) Climate change and the effects of temperature extremes on Australian flying-foxes. Proceedings of the Royal Society B-Biological Sciences 275:419-425. doi: 10.1098/rspb.2007.1385

Whisler SL, Snow AA (1992) Potential for the loss of self-incompatibility in pollen-limited populations of mayapple (*Podophyllum peltatum*). American Journal of Botany 79:1273-1278. doi: 10.2307/2445055

Willi Y (2009) Evolution towards self-compatibility when mates are limited. Journal of Evolutionary Biology 22:1967-1973. doi: 10.1111/j.1420-9101.2009.01806.x

Xiong YZ, Fang Q, Huang SQ (2013) Pollinator scarcity drives the shift to delayed selfing in Himalayan mayapple *Podophyllum hexandrum* (Berberidaceae). Aob Plants 5:8. doi: 10.1093/aobpla/plt037

13

The bumblebee *Bombus hortorum* is the main pollinating visitor to *Digitalis purpurea* (Common Foxglove) in a U.K. population

Arthur A. D. Broadbent[1] and Andrew F. G. Bourke[2*]

[1]*School of Environmental Sciences and* [2]*School of Biological Sciences, University of East Anglia, Norwich Research Park, Norwich, UK*

Abstract—Specialization in plant-pollinator systems represents an important issue for both the ecological understanding and conservation of these systems. We investigated the extent to which the bumblebee *Bombus hortorum* (Linnaeus) is the main potential pollinator of Common Foxglove, *Digitalis purpurea* L. Twenty *D. purpurea* patches were selected in North Yorkshire, U.K., ten each in woodland and garden or park habitat. All insects visiting *D. purpurea* within the patches were recorded over seventy 30-min bouts. The relative frequency of insect visitors to other flowering plant species within 15 m of each patch was also determined. *B. hortorum* and *B. pascuorum* were the two most frequent visitors to *D. purpurea*, accounting for 82 - 92% and 3 -17%, respectively, of all insect visits (n = 1682), depending on habitat. *B. hortorum* showed a significant preference for visiting *D. purpurea* relative to its frequency of visits to other available plant species. The relationship of *D. purpurea* with *B. hortorum*, which pollinates several plant species with long corollas, therefore represents a potential case of asymmetric specialization, albeit one that may vary spatially. Because *D. purpurea* reproduction appears dependent on insect pollination, *B. hortorum* and *B. pascuorum* may help underpin the viability of *D. purpurea* populations.

Keywords: Bumblebee, Foxglove, mutualism, pollination, specialization

Introduction

The degree to which plant-pollinator systems are specialized (reliant on one or a few plant or pollinator partners) is a major issue in the study of insect pollination (Waser and Ollerton 2006). Specialization informs our understanding of the ecology and evolution of these systems (Ollerton et al. 2006) and affects the provision of pollination services and the risk of extinction in both pollinators and plants in the face of population declines in their partners (Memmott et al. 2004; Waser and Ollerton 2006; Kleijn and Raemakers 2008). Investigating potential cases of specialization in plant-pollinator systems is particularly important when, unlike the situation in species such as insect-mimicking orchids (Proctor et al. 1996), floral morphology does not immediately suggest the presence of specialization.

Bumblebees (*Bombus* spp) are major insect pollinators of both wild flowers and flowering crops (Goulson 2010). In the U.K. *Bombus* fauna, *B. hortorum* (Linnaeus) is unusual in having the longest tongue (proboscis) among the remaining, widespread species (Prys-Jones and Corbet 1991; Goulson et al. 2011). Correspondingly, it is well known to favour flowering plant species with long corollas (Benton 2006). Long-tongued bumblebees represent a particularly important group in conservation terms, because wild flowers and flowering crops with long corollas may not receive effective pollination in their absence (Goulson 2010). It is therefore worrying that, among bumblebees as a whole, long-tongued species, including *B. hortorum*, appear disproportionately prone to population decline and range contraction in modern agricultural landscapes (Goulson 2010; Bommarco et al. 2012). These considerations render the investigation of plant-pollinator specializations involving long-tongued bumblebees of particular interest.

One flowering plant with long corollas is Common Foxglove, *Digitalis purpurea* L., a biennial with elongate (42 - 43 mm), bell-shaped flowers that have basal nectaries and hang from tall, vertical inflorescences, widespread in its native Eurasia including the U.K. (Percival and Morgan 1965; Best and Bierzychudek 1982; Hill et al. 2004). *D. purpurea* appears to lack vegetative reproduction (Hill et al. 2004) and bagging experiments show that it requires insect pollinators to set seed (Nazir et al. 2008), suggesting that it is obligately dependent on insect pollination for reproduction. Although *D. purpurea* is not itself scarce (Hill et al. 2004), it therefore represents a potentially useful model with which to investigate plant-pollinator specialization. Previous studies have noted an association between *B. hortorum* and *D. purpurea* in the U.K. (Brian 1951, 1957; Edwards and Jenner 2005; Benton 2006) and elsewhere in Europe (Grindeland et al. 2005). However, none of these studies quantified the relative visitation rate of *B. hortorum* to *D. purpurea*. For example, Grindeland et al. (2005) found *B. hortorum* to account for > 52% of all *Bombus* visits to natural stands of *D. purpurea* in Norway, but, because this study had a different focus, whether this result simply reflected relative local abundances of *Bombus* species was not investigated. We therefore sought to investigate potential

*Corresponding author; email: a.bourke@uea.ac.uk

specialization in the relationship between *B. hortorum* and *D. purpurea* by quantifying the visitation rate of *B. hortorum* to *D. purpurea* relative to the visitation rate of other *Bombus* to *D. purpurea* and of *B. hortorum* and other *Bombus* to other available plant species.

MATERIALS AND METHODS

Twenty flowering patches of *Digitalis purpurea* in North Yorkshire, U.K., were selected, ten in woodland and ten in domestic gardens or parks ('non-woodland'). Only patches with five or more flowering plants per patch were included, a plant being defined as one or more inflorescences arising from a single basal rosette of leaves. Patches were spatially mixed with respect to habitat type and all patches lay within a radius of 11.5 km of the position, 54°01'25" N, 01°09'12" W. To maximize spatial independence, each was at least 2 km from its nearest neighbour. Within each patch, five *D. purpurea* plants were randomly chosen and individually labelled with a small tag near ground level. The white-flowered morph of *D. purpurea* (Blamey and Grey-Wilson 1989) was present at an overall mean frequency of 15%, with the number of patches having white morphs being the same (3/10) in both woodland and non-woodland habitat. Neither patch size nor within-patch plant density (i.e. for both morphs considered together) differed significantly between the two habitats (size: means = 6.1 and 6.2 m^2, respectively; two-sample *t*-test with equal variances, t_{18} = 0.074, P = 0.942; density: medians = 2.0 and 2.7 plants m^{-2}, respectively; Mann-Whitney *U*-test, U = 47.0, $n_1 = n_2 = 10$, P = 0.82).

Plants within each patch were observed simultaneously for 2 - 4 separate bouts of 30 min each between 11:00 and 20:00 from 25 June to 13 July 2010 (mean total observation = 105 min per patch). These dates include the peak annual flowering time of *D. purpurea* in central U.K. (Percival and Morgan 1965). Bout dates were rotated across patches and habitats, but, because of the limited flowering period of *D. purpurea*, some bouts (14/70) for a given patch took place on the same day (no more than two per day). Overall, therefore, the *D. purpurea* sample consisted of 100 plants (20 patches × 5 focal plants per patch) observed over 35 h (20 patches × 3.5 30-min observation bouts per patch). Bouts were conducted by recording, from 1.5 - 2.0 m away, each time an insect fully entered any corolla on the focal plants. Only insects approximately > 10 mm long and > 5 mm at their widest point were recorded, as smaller insects were considered unlikely to receive deposits of pollen by visiting *D. purpurea* flowers. Bumblebees were identified to species level (but not to sex or caste) where possible, using Edwards and Jenner (2005). Other insects were identified at least to family level.

To determine the relative frequency of all insect pollinators within the vicinity of the patches, eight equally spaced, linear 15 m transects (control transects) were laid out in a radial pattern centred on each patch. Each transect was walked for 2 min and every insect (again, approximately > 10 mm long and > 5 mm wide) seen visiting a flower within 1 m either side of the transect was recorded. The entire 16 minute-set of control transects was walked twice, on separate days for each patch, between 25 June and 13 July 2010 (83% of control transects) or between 14 and 20 July 2010 (17% of control transects). The plant species on control transects most frequently visited by insects were, in woodland habitat, *Arctium* spp, *Cirsium* spp, *Impatiens glandulifera*, *Rhododendron* spp and *Rubus* spp, and, in non-woodland habitat, *Cirsium* spp, *Crataegus* spp, *Lavandula* spp, *Trifolium repens* and a variety of other cultivated plant species.

For analysis of the proportion of visits made by focal *Bombus* species to *D. purpurea* relative to total visits made by *Bombus* species to *D. purpurea* (proportionate data; Fig. 1), bouts were pooled within patches to create datasets of 10 patches for each of woodland and non-woodland habitat. For analysis of the numbers of visits made by *Bombus* species to *D. purpurea* relative to total visits made by *Bombus* species to other plant species within the patches (Fig. 2), numbers of visits by bee species to *D. purpurea* were divided by the mean number of flowers that single *Bombus* workers visit per plant of *D. purpurea*, which was calculated as 23.2 (Best and Bierzychudek 1982; Nazir et al. 2008). This correction was applied to minimize effects of pseudoreplication arising from individual insects visiting multiple flowers of single *D. purpurea* plants within bouts. It was not applied to proportionate data, which were unaffected by multiple visits by single insects (i.e., if two *Bombus* species made 80% and 20%, respectively, of all *Bombus* visits to *D. purpurea*, these proportions would be unchanged whether measured at the plant or the flower level, provided the two species did not differ systematically in the number of flowers per plant visited by single bees). Datasets were analysed non-parametrically whenever they proved to be significantly non-normal (Kolmogorov-Smirnov tests).

RESULTS

In both habitats, *Bombus hortorum* was by a large margin the most frequent visitor to *Digitalis purpurea* (Fig. 1). Overall, *B. hortorum* accounted for 82% and 92% of total visits to *D. purpurea* (n = 1682) in woodland habitat (Fig. 1a) and non-woodland habitat (Fig. 1b), respectively, whereas the next most frequent visitor in both habitats, *B. pascuorum*, accounted for 17% and 3% of total visits in each respective habitat. Together, therefore, *B. hortorum* and *B. pascuorum* accounted for 95 - 99% of total visits to *D. purpurea* across both habitats. Within each of *B. hortorum* and *B. pascuorum*, the proportions of visits to *D. purpurea* in woodland and non-woodland habitat were not significantly different (Mann-Whitney *U*-tests: *B. hortorum*, U = 54.5, $n_1 = n_2 = 10$, P = 0.73; *B. pascuorum*, U = 59.0, $n_1 = n_2 = 10$, P = 0.48; Figs. 1a, 1b). Hence remaining data analyses were conducted using data from both habitats pooled.

In contrast to other insects, including *B. pascuorum*, *B. hortorum* was recorded visiting *D. purpurea* at a significantly much higher frequency than expected from its abundance as a visitor to other plant species on the control transects (chi-squared test, χ^2 = 248.9, d.f. = 5, P < 0.001; Fig. 2). Therefore, *B. hortorum* was not the most abundant bumblebee at all plants in the vicinity of the patches but visited *D. purpurea* preferentially.

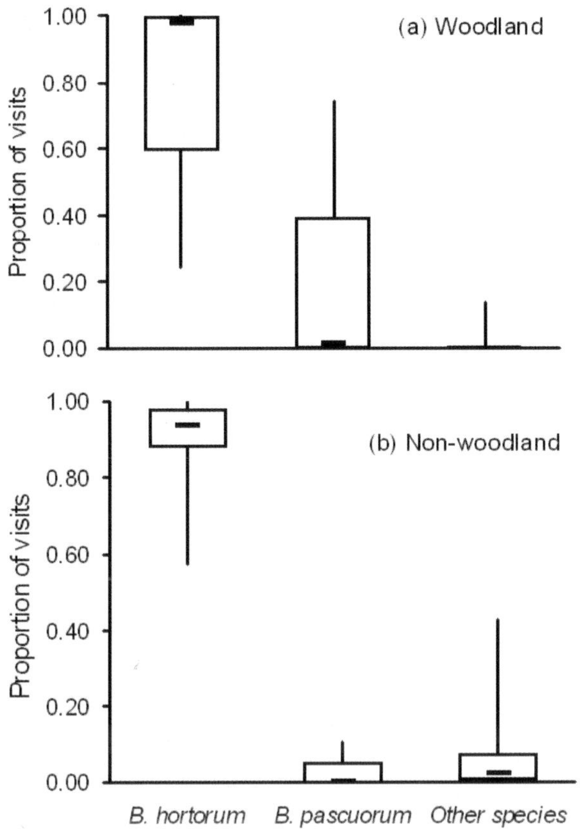

FIGURE. 1. Proportion of visits to *Digitalis purpurea* by *Bombus hortorum, B. pascuorum* and other insect species in (a) woodland habitat and (b) non-woodland habitat. Median denoted by horizontal bar, quartiles by box and range by vertical lines. Per-patch mean (range) total numbers of insect visits were 66.8 (24 - 134) for woodland and 101.4 (15 - 241) for non-woodland habitat (n = 10 *D. purpurea* patches per habitat). Other species consisted of *Apis mellifera, Bombus lapidarius, B. terrestris / B. lucorum* (pooled), Megachilidae, Syrphidae, *Vespula* spp. and other Vespidae.

DISCUSSION

Our findings show that *B. hortorum* was the principal insect visitor to *D. purpurea* in the study population, with a significant preference for this plant, followed by *B. pascuorum*, which lacked such a preference. *B. pascuorum* has the second longest tongue among common U.K. bumblebees (Prys-Jones and Corbet 1991). The predominance of *B. hortorum* and *B. pascuorum* among *Bombus* visitors to *D. purpurea*, which was also noted by Grindeland et al. (2005), is therefore likely to have arisen because *D. purpurea*, through having long corollas, is preferentially attractive to long-tongued bumblebee species. Because it was the principal insect visitor, we infer that *B. hortorum* is likely to be the main pollinator of *D. purpurea* in the study population. This inference is subject to two qualifications. First, visitation does not equate to pollination. However, observations that *B. hortorum* workers completely enter *D. purpurea* flowers and emerge coated in pollen (Benton 2006; unpublished observations), and that *D. purpurea* pollen occurs in their pollen-loads (Brian 1951; Kleijn and Raemakers 2008), together with the experimental evidence showing that bagged *D. purpurea* flowers set no seed (Nazir et al. 2008), leave little doubt that *B. hortorum* and other *Bombus* visitors pollinate *D. purpurea*. Second, *D. purpurea* flowers may also be visited by nocturnal insects such as moths. This possibility, which we did not investigate, seems unlikely given that *D. purpurea* flowers lack features typical of flowers pollinated by nocturnal moths (Proctor et al. 1996). But since floral features are not necessarily reliable guides to pollinator taxon (Ollerton et al. 2009), the possibility remains that nocturnal insects also contribute to *D. purpurea* pollination.

With regard to diurnal pollinators, the relationship between *B. hortorum* and *D. purpurea* therefore provides a potential example of a high degree of specialization in a plant-pollinator system. Because *B. hortorum* visits a wide range of other plant species with long corollas (Benton 2006; Kleijn and Raemakers 2008), any association of *D. purpurea* with it is clearly asymmetric (Vázquez and Aizen

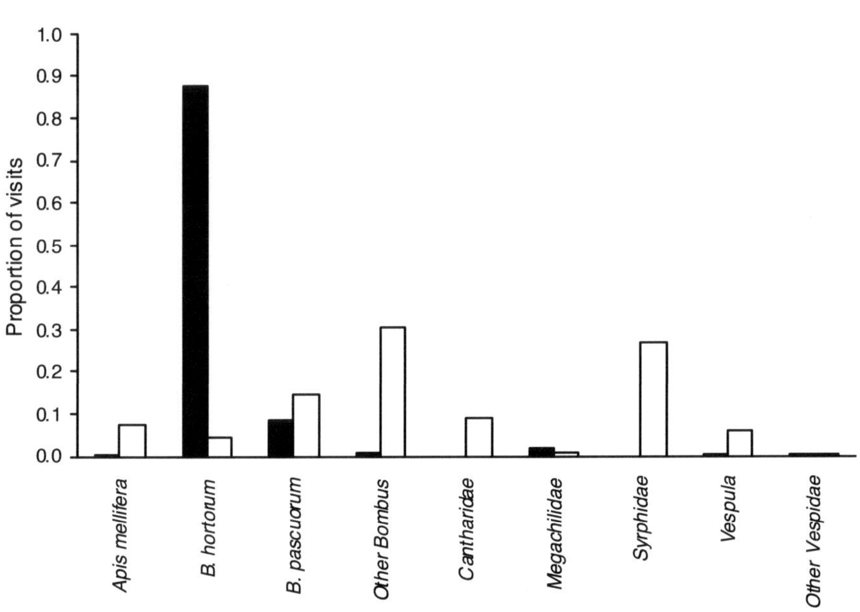

FIGURE. 2. Comparison of proportion of visits by insect taxa to flowers of *Digitalis purpurea* (black columns) and to flowers of other plant species on control transects within 15 m of the focal patches of *D. purpurea* (white columns). Data from woodland and non-woodland habitat have been pooled. Total numbers of insect visits were 1682 to *D. purpurea* and 1133 to flowers on control transects. Figure is illustrative, with statistical analysis (chi-squared test) being conducted on corrected numbers of visits as described in 'Materials and methods' and reported in 'Results'. Other *Bombus* consisted of *B. hypnorum, B. lapidarius, B. pratorum, B. terrestris / B. lucorum* (pooled) and *B. (Psithyrus)* spp.

2004). Studies elsewhere have found that *D. purpurea* is also visited by other *Bombus* species (Brian 1957; Percival and Morgan 1965; Fussell and Corbet 1993; Grindeland et al. 2005; Nazir et al. 2008), including native *Bombus* species recently brought into contact with introduced *D. purpurea* populations (Best and Bierzychudek 1982). These studies, combined with our results, suggest that the extent to which *D. purpurea* has a specialized association with *B. hortorum* or other long-tongued bumblebees must vary geographically. Nonetheless, even spatially-limited asymmetric specialization of *D. purpurea* on one or a few *Bombus* species could be an important phenomenon given that *D. purpurea* appears obligately dependent on insect pollination for reproduction (Hill et al. 2004; Nazir et al. 2008) and that, among bumblebees as a whole, *B. hortorum* and other long-tongued species appear disproportionately prone to decline (see 'Introduction'). Hence our findings also show that *B. hortorum* and *B. pascuorum* or their functional equivalents may play a major role in the viability of *D. purpurea* populations. More broadly, if other species of flowering plant with long corollas likewise exhibit asymmetric, specialized relationships with one or a few long-tongued *Bombus* species, they too would be vulnerable to effects of declines in these bees.

Acknowledgements

We thank Silke and Simon Broadbent and Barbara and Jim Crowland for their logistical support, Jane Memmott for helpful discussion, anonymous referees for comments and land owners for access permission.

References

Benton T (2006) Bumblebees. Collins, London.

Best LS, Bierzychudek P (1982) Pollinator foraging on Foxglove (*Digitalis purpurea*): a test of a new model. Evolution 36:70-79.

Blamey M, Grey-Wilson C (1989) The illustrated flora of Britain and Northern Europe. Hodder & Stoughton, London.

Bommarco R, Lundin O, Smith HG, Rundlöf M (2012) Drastic historic shifts in bumble-bee community composition in Sweden. Proceedings of the Royal Society B 279:309-315.

Brian AD (1951) The pollen collected by bumble-bees. Journal of Animal Ecology 20:191-194.

Brian AD (1957) Differences in the flowers visited by four species of bumblebee and their causes. Journal of Animal Ecology 26:71-98.

Edwards M, Jenner M (2005) Field guide to the bumblebees of Great Britain and Ireland. Ocelli Limited.

Fussell M, Corbet SA (1993) Bumblebee (Hym., Apidae) forage plants in the United Kingdom. Entomologist's Monthly Magazine 129:1-14.

Goulson D (2010) Bumblebees: behaviour, ecology, and conservation, 2nd edn. Oxford University Press, Oxford.

Goulson D, Kaden JC, Lepais O, Lye GC, Darvill B (2011) Population structure, dispersal and colonization history of the garden bumblebee *Bombus hortorum* in the Western Isles of Scotland. Conservation Genetics 12:867-879.

Grindeland JM, Sletvold N, Ims RA (2005) Effects of floral display size and plant density on pollinator visitation rate in a natural population of *Digitalis purpurea*. Functional Ecology 19:383-390.

Hill MO, Preston CD, Roy DB (2004) PLANTATT: attributes of British and Irish plants: status, size, life history, geography and habitats. Centre for Ecology and Hydrology, Huntingdon.

Kleijn D, Raemakers I (2008) A retrospective analysis of pollen host plant use by stable and declining bumble bee species. Ecology 89:1811-1823.

Memmott J, Waser NM, Price MV (2004) Tolerance of pollination networks to species extinctions. Proceedings of the Royal Society B 271:2605-2611.

Nazir R, Reshi Z, Wafai BA (2008) Reproductive ecology of medicinally important Kashmir Himalayan species of *Digitalis* L. Plant Species Biology 23:59-70.

Ollerton J, Johnson SD, Hingston AB (2006) Geographical variation in diversity and specificity of pollination systems. In: Waser NM, Ollerton J (eds) Plant-pollinator interactions: from specialization to generalization. University of Chicago Press, Chicago, pp 283-308.

Ollerton J, Alarcón R, Waser NM, Price MV, Watts S, Cranmer L, Hingston A, Peter CI, Rotenberry J (2009) A global test of the pollination syndrome hypothesis. Annals of Botany 103:1471-1480.

Percival M, Morgan P (1965) Observations on the floral biology of *Digitalis* species. New Phytologist 64:1-22.

Proctor M, Yeo P, Lack A (1996) The natural history of pollination. HarperCollins, London.

Prys-Jones OE, Corbet SA (1991) Bumblebees, 2nd edn. Richmond Publishing Co. Ltd, Slough.

Vázquez DP, Aizen MA (2004) Asymmetric specialization: a pervasive feature of plant-pollinator interactions. Ecology 85:1251-1257.

Waser NM, Ollerton J, eds (2006) Plant-pollinator interactions: from specialization to generalization. University of Chicago Press, Chicago.

14

COMPARATIVE BIOACOUSTICAL STUDIES ON FLIGHT AND BUZZING OF NEOTROPICAL BEES

Andreas Burkart[1*], Klaus Lunau[1] and Clemens Schlindwein[2]

[1] *Institute of Sensory Ecology, Department Biology, Heinrich-Heine-University Düsseldorf, 40225 Düsseldorf, Germany*
[2] *Departamento de Botânica, Instituto de Ciêncais Biológicas, Universidade Federal de Minas Gerais Av. Antônio Carlos, 6627 - Pampulha - Caixa Postal 486, 31270-901, Belo Horizonte, MG, Brazil*

Abstract—The presence of bees is typically accompanied by the humming sound of their flight. Bees of several tribes are also capable of pollen collecting by vibration, known as buzzing behaviour, which produces a buzzing sound, different from the flight sound. An open question is whether bee species have species-specific buzzing patterns or frequencies dependent of the bees' morphology or are capable to adjust their individual buzzing sound to optimize pollen return. The investigations to approach this issue were performed in northeastern Brazil near Recife in the state of Pernambuco. We present a new field method using a commercially available portable system able to record the sound of bees during flight and buzzing at flowers. Further, we describe computer linguistical algorithms to analyse the frequency of the recorded sound sequences. With this method, we recorded the flight and buzzing sequences of 59 individual bees out of 12 species visiting the flowers of *Solanum stramoniifolium* and *S. paniculatum*. Our findings demonstrate a typical frequency range for the sounds produced by the bees of a species. Our statistical analysis shows a strong correlation of bee size and flight frequency and demonstrates that bee species use different frequency patterns.

Keywords: *pollen,* Solanum, Xylocopa, *buzz pollination, poricidal anthers, bioacoustic*

INTRODUCTION

Flowers of numerous species in 72 plant families hide pollen in poricidal anthers, like, for example, those of Melastomataceae, Leguminosae-Cassiinae and Solanaceae including the huge genus *Solanum* (Buchmann & Hurley 1978; Buchmann 1983). This pollen is easily accessible only to bees that are capable of shaking the pollen grains out of the apical pores of the anthers by vibrations (King *et al.* 1996). This phenomenon, described as buzz pollination, is widespread among bees, but restricted to certain clades. Females of carpenter bees (*Xylocopa* – Xylocopini), bumble bees (Bombini), and neotropical orchid bees (Euglossini), Centridini and Augochlorini, among others, are able to vibrate flowers to collect pollen, but not the majority of the common tropical stingless honeybees (Meliponini) or the honeybees (Apini) (Hurd & Linsley 1976; Buchmann 1983; Houston & Thorp 1984; Neff & Simpson 1988; Thorp 2000; Harter et al. 2002; Teppner 2005). Among the fraction of the bee fauna that is capable of buzzing, flowers with poricidal anthers are an abundant pollen source (Harter et al. 2002). Vibrating the flowers, the females produce an audible buzzing sound using the flight musculatur with decoupled wings (Michener 1962; Wille 1963; King et al. 1996). Because of the changing resonance properties of the thorax with decoupled wings, the sound of buzzing usually has an about two times higher frequency than the sound of flight (King & Buchmann 2003). In other insect taxa, body size and wingbeat frequency is correlated in a manner that smaller insects beat their wings faster than larger ones, when flying (Byrne et al. 1988). It is not known, if this is also true for flower vibrating bees, whose flight musculature is decoupled from the wings when collecting pollen. According to the results of King & Buchmann (1996) on flowers of *Solanum laciniatum* the optimal frequency of pollen dispensing is not in the range bees are able to achieve by buzzing. Despite this fact, pollen collecting bee females might be able to adjust buzzing frequency in the given range to the kind of flower or to change their buzzing according to pollen release during buzzing behaviour. Earlier studies by Harter et al. (2002), performed with several plant species including *Solanum,* showed no significant relationship in bee size related to flower diameter, raising the question whether buzz pollination is independent of the bee size.

For our buzzing study we chose two shrubby species of *Solanum* Subgenus *Leptostemonum,* common weeds in northeastern Brazil: *S. stramoniifolium* Jacq. and *S. paniculatum* L. (Levin et al. 2006). The pollination of both species had been already investigated earlier (Forni-Martins et al. 1998; Bezerra & Machado 2003; Silva et al. 2004). These studies have shown that their flowers are mainly pollinated by several buzz pollinating medium to large sized bees of the genera *Xylocopa, Centris, Bombus* and *Melipona.*

Using a sensitive microphone to record the sound of buzzing or flying of flower vibrating bees in the field and software for computer linguistic analysis, we asked: What are the characteristics of the buzzing sounds of bees collecting pollen in flowers of the two species of *Solanum*? Do females

*Corresponding author; email: andreas.burkart@uni-duesseldorf.de

adjust buzzing patterns at the flowers of these species? Is there a specific frequency for individual bee species during flight or buzzing and are these buzzing frequencies correlated to the body size of the bee females?

METHODS

Study site and plant species

The observations and recordings were performed in the municipality of Camaragibe, near Recife, Pernambuco, northeastern Brazil (S 07°58.230' W 035°00.179'). The temperature during the observation period from November 2010 to January 2011 was almost constant with an average maximum temperature of 29.6 ± 0.7 °C and an average minimum night temperature of 25.2 ± 1.1 °C. The relative humidity was on average 69 ± 5 % but increased to 85 % due to heavy rainfalls during the last weeks in January. The study sites are characterized by the presence of houses and small farms with numerous ornamental plants and tropical fruit crops, surrounded by diverse secondary Atlantic Rainforest.

Solanum paniculatum and *S. stramoniifolium* are ruderal plants or weeds and grew in open areas in the vicinity of tropical rainforest and within an orchard of soursop (*Annona muricata* L., Annonaceae). *Solanum paniculatum* is an up to 2 m tall shrub with spiny stems (Knapp & Jarvis 1990). The main stem is usually unbranched and carries an apical inflorescence with light purple to whitish pentamerous flowers. The plants were very common and grew as dominant weeds in large groups with high density in several open areas. *Solanum stramoniifolium* is an overall spiny shrub with a circumference of up to 2 m and up to 2 m height (Silva et al. 2004). Different to *S. paniculatum*, the inflorescences were axillary over the whole plant and usually carried only one or two white flowers at the same time. The plants of *S. stramoniifolium* were also common but occurred less aggregated than the other species.

Flower visitors and sound recordings

At the beginning of the study we collected the flower visitors and determined the pollinator spectrum of both plant species. Bees were caught at the flowers with entomological nets and killed by ethyl acetate. All bees were pinned, dried, labelled and then identified in the laboratory. The specimens were included in the Entomological Collection of the Universidade Federal de Minas Gerais, Belo Horizonte, Brazil.

Sounds of the flower visiting bees were recorded in the field during the flower visits. The microphone was carefully placed as near as possible to the buzzing bees at the flowers. While buzzing, most of the bees did not react to disturbances in the surroundings. This allowed us to place the microphone as close as 1 cm to the bees. Generally, the microphone recordings were done in a distance of 1 to 5 cm. When recording small and silent bees, we tried to move the microphone as close as possible. When departing from the flower we recorded the flight sound of the bee during takeoff. Every bee was followed as long as possible to ensure the recording of at least one evaluable sequence of flight and buzzing. As microphone and recording device an Android smartphone (Legend, HTC, Taoyuan, Taiwan) was used running the application Hertz 1.1.1 producing *.wav data at a sample rate of 16 000 Hz or 22 050 Hz. By Soundforkes and computer generated tones of a given frequency the range of the microphone was tested from 50 Hz up to 1 kHz with positive results and therefore stated as sensitive in the range of bee sounds. After recording the bees were collected for identification. From the bee individuals the intertegular distance as a standard measure for bee size was determined (Cane 1987).

Sound analyses

For analysis the soundfiles were divided in flight samples and single buzzings at flowers, which contained solely the sound of a single buzzing. All soundfiles were processed using Audacity 1.2.6 (Audacity, GNU General Public License). Every file was then labeled to assign it to the specific bee which produced the sound to build up a database of bee sounds. This database was analysed using the toolset of the computer phonetics programm Praat 5.2.15 (Boersma 2002). The pitch (ac) function searched in every sound sample for the basic frequency (Boersma 1993). The parameters of the function were adapted using the earlier results of other authors (King & Buchmann 2003; Nunes-Silva et al. 2010) to output the basic frequency of the bees wingbeat and buzzings. For further analysis the toolset of Praat was used to output average, maximum, minimum and standard deviation of the basic frequency and the length of each sound sample. Samples which contained too much disturbing noise were filtered out by a threshold for unregular high standard deviation of the basic frequency. (The script used with Praat is provided in Appendix 1)

All data processing and statistical tests were performed using Microsoft Excel 2002 (Microsoft Corp., Redmond, USA) and Graphpad Prism 4.0 (GraphPad Software Inc., La Jolla, USA). Statistical analysis was done with the Students t-test. Differences between data sets were regarded as significant with $P < 0.05$.

RESULTS

During the study period we observed bees of 11 species buzzing the flowers of *S. stramoniifolium* and of 12 species those of *S. paniculatum*. Only the large carpenter bees *Xylocopa (Neoxylocopa) frontalis* and *X. (N.) suspecta* visited the flowers of both *Solanum* species (Tab. 1).

During the field work several hundred sound recordings were performed. Here, we evaluated 59 recordings for which the associated bee was caught. The pitch analysis performed in Praat and further processed (Tab. 2), shows buzzing and flight frequencies and buzzing durations for several observed species. Within tribes similarity can be found, as seen in *Centris* and *Euglossa* which have high buzzing frequencies, while *Xylocopa, Eulaema* and *Augochloropsis* buzz with lower frequencies (Tab. 2).

The frequency of wingbeat is negatively correlated (Slope = -17.72 ± 1.73, $r^2 = 0.66$) with the intertegular distance of the bees (Fig. 1), whereas the buzzing frequency has also a

TAB. 1. Bee species found buzzing on both *Solanum* species. *Solanum stramoniifolium* (*S. s.*), *S. paniculatum* (*S. p.*). Nomenclature of bees after Moure et al. (2007).

bee species	family, tribe	visited species S. s.	visited species S. p.
Xylocopa frontalis	Apidae, Xylocopini	X	X
Xylocopa suspecta	Apidae, Xylocopini	X	X
Xylocopa muscaria	Apidae, Xylocopini		X
Bombus brevivillus	Apidae, Bombini	X	
Centris aenea	Apidae, Centridini		X
Centris flavifrons	Apidae, Centridini		X
Centris tarsata	Apidae, Centridini		X
Centris sponsa	Apidae, Centridini		X
Centris spilopoda	Apidae, Centridini		X
Euglossa cordata	Apidae, Euglossini	X	
Eulaema cingulata	Apidae, Euglossini	X	
Eufriesea surinamensis	Apidae, Euglossini	X	
Eulaema nigrita	Apidae, Euglossini	X	
Melipona scutellaris	Apidae, Meliponini	X	
Pseudaugochlora graminea	Halictidae, Augochlorini	X	
Augochloropsis sp. 1	Halictidae, Augochlorini		X
Augochloropsis sp. 2	Halictidae, Augochlorini	X	
Exomalopsis sp. 1	Apidae, Exomalopsini		X
Exomalopsis sp. 2	Apidae, Exomalopsini		X
Thygater sp.	Apidae, Eucerini	X	
Centris sp.	Apidae, Centridini		X

negative trend (Slope = -5.66 ± 1.86, r^2 = 0.14). Large bees were found to fly and buzz with a lower frequency than smaller bees. This correlation was not observed in individuals of *Augochloropsis* which buzz at about the same frequency as their flight frequency (Fig. 2).

In a further analysis, the buzzing frequency of females of *X. frontalis* and *X. suspecta* on the flowers of the *Solanum* species were compared (Fig. 3). A significant difference was found for those of *X. frontalis*, buzzing flowers of *S. stramoniifolium* with a slightly higher frequency (224.2 ± 3.9 Hz) than those of *S. paniculatum* (211.0 ± 2.9 Hz). The buzzing frequencies of individuals of *X. suspecta* were not significantly different between the two *Solanum* species. Measurements of the size of the flowers of *S. paniculatum* and *S. stramoniifolium* showed similar average size in diameter (S.s. 23.7 mm; S.p. 25.5 mm) and stamen length (*S.s.* 7.0 mm; *S.p.* 6.7 mm).

The frequency pattern of buzzings of different species revealed that the buzzing frequency is reducing during each buzzing of the investigated bee species, with exception of those of *Pseudaugochlora graminea* that showed no clear pattern (Fig 4).

DISCUSSION

In this study, a method of sound recording and analysing was tested with two *Solanum* species and several occuring bee species to answer different questions about the buzzing frequency. The pollen removal from poricidal anthers by buzzing is a mechanical phenomenon which consists of several parameters on the bee side including force, acceleration or frequency and other parameters on the plant side such as the mass of the flower or the size of the pores (Buchmann & Hurley 1978; King & Lengoc 1993). In this work, the parameter of the frequency of female bee´s vibration was chosen as it is probably the only parameter that is modifiable by the bee itself and differs from species to species (King & Buchmann 2003; Nunes-Silva et al. 2010).

TAB. 2. Mean frequency of buzzing, duration of single buzzings, flight frequency, and intertegular span of recorded bee species sorted by the latter parameter. In total, 59 samples of individuals from different species on flowers of *S. stramoniifolium* and *S. paniculatum* are tabulated. All data shown in mean ± SD, n is given in brackets.

bee species	buzzing frequency [Hz]	buzzing duration [s]	flight frequency [Hz]	intertegular span [mm]
Xylocopa frontalis (18)	218.3 ± 12.4 (103)	0.59 ± 0.23	106.5 ± 3.5 (69)	8.81 ± 0.41
Xylocopa suspecta (21)	251.0 ± 10.7 (141)	0.51 ± 0.22	116.0 ± 5.4 (55)	6.67 ± 0.46
Centris flavifrons (2)	294.5 ± 18.6 (4)	0.53 ± 0.13	151.2 ± 12.3 (3)	6.12 ± 0.28
Eulaema nigrita (2)	220.9 ± 5.6 (11)	0.97 ± 0.29	126.9 ± 2.5 (7)	5.68 ± 0.11
Eulaema cingulata (1)	206.9 ± 13.8 (4)	1.04 ± 0.32	124.3 ± 3.6 (1)	5.68
Eufriesea surinamensis (1)	239.0 ± 10.1 (3)	0.44 ± 0.26	no data	5.20
Centris aenea (2)	324.1 ± 12.1 (5)	0.68 ± 0.19	221.4 ± 10.0 (4)	5.08 ± 0.17
Bombus brevivillus (1)	267.7 ± 8.9 (3)	0.55 ± 0.22	125.1 ± 2.3 (1)	4.00
Euglossa cordata (2)	294.2 ± 10.3 (7)	0.65 ± 0.08	251.8 ± 3.8 (2)	3.64 ± 0.06
Centris tarsata (2)	311.4 ± 7.0 (4)	0.76 ± 0.16	249.6 ± 3.3 (5)	3.24 ± 0.06
Melipona scutellaris (2)	282.2 ± 27.4 (2)	1.05 ± 0.63	229.0 ± 11.5 (4)	3.00 ± 0.06
Pseudaugochlora graminea (3)	221.9 ± 22.8 (20)	0.84 ± 0.55	190.2 ± 6.8 (3)	2.19 ± 0.37
Augochloropsis sp.2 (1)	172.7 ± 6.8 (10)	0.50 ± 0.26	170.2 ± 4.9 (1)	2.08
Augochloropsis sp.1 (1)	210.0 ± 6.0 (8)	0.83 ± 0.61	191.3 ± 13.4 (1)	1.76

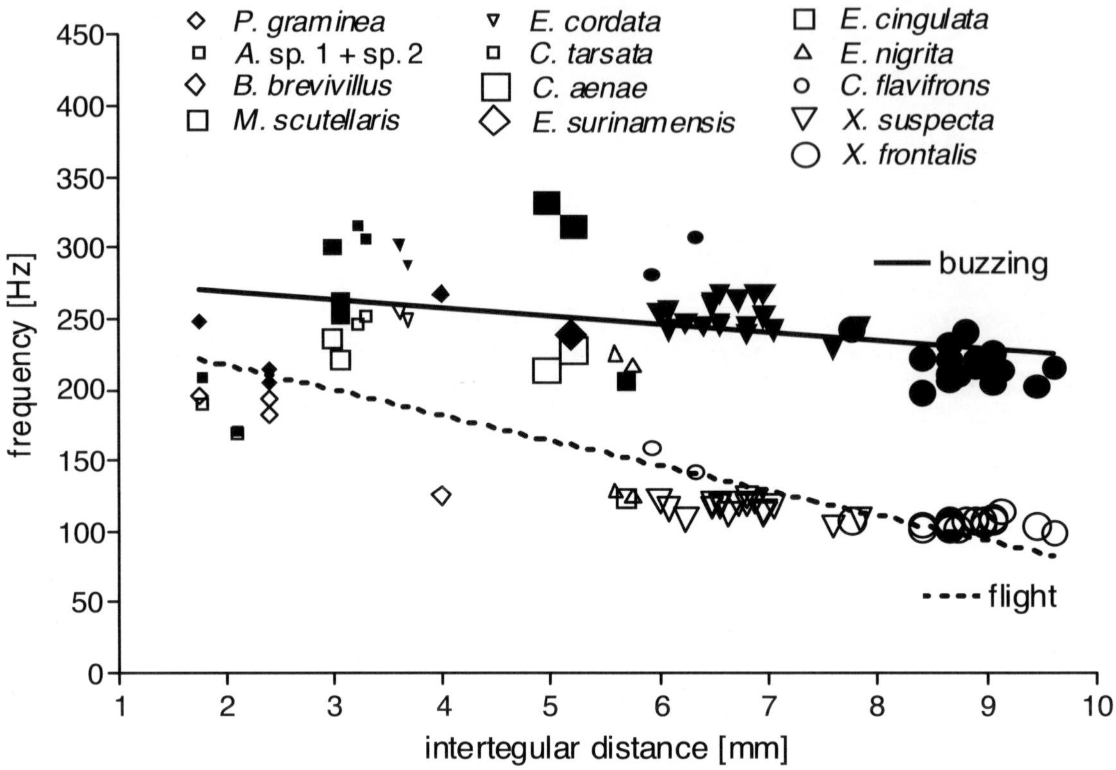

FIG. 1. Correlation of intertegular distance as a standard for bee size and the frequency of buzzing (solid symbols) and the frequency of flight (open symbols). Each symbol stands for a collected individual. All individuals of a species were assigned the same symbol. Linear regression is indicated by the lines, solid for buzzing frequency and dotted for flight frequency.

Earlier measurements of vibratory effects in buzzing and flight were often done using an accelerometer or a contact free laser accelerometer (King & Ferguson 1994; King & Buchmann 2003; Hrncir et al. 2004). Both systems face problems due to their limited in-field capability. A sensitive microphone is able to record the sound field, produced by a flying or buzzing bee (Hrncir et al. 2004), which can be used to analyse the length of a flower visit (Ehrlich & Lunau 2009). In addition, data about the frequency of the flight musculature movement can be made accessible by the sound of the bee. The buzzing sound is of complex structure and composed of harmonics and other noises, generated by different oscillations of parts of the bee body (King & Buchmann 2003). To overcome this, we used computer linguistical analysis after the recording to determine the basic frequency of the sound (Boersma 1993; 2002).

The described system delivered good results as an in field setup. This is an advantage over other methods which need to fix the bee to a laser accelerometer or measure the frequency with stroboscobic methods inside a laboratory. Supporting the observations by King & Buchmann (2003) and Hrncir et al. (2004) we found different frequency ranges in different bee species. The flight frequency seems to be in a range which is anatomically fixed and correlated to the intertegular distance and thus to the size of the bee. These findings follow the measurements of Byrne et al. (1988) who showed that smaller insects have higher wingbeat frequencies. However, the buzzing frequency seems to be linked to the wingbeat frequency (Fig. 1) and is – mainly for large bees – about two times higher, as stated earlier by King & Buchmann (2003). A possible explanation was seen in the second harmonics, which allow a system like the flight apparatus to oscillate at twice the natural frequency, using little energy. In this view, the two-times higher buzzing frequency is an adaptation to save energy costs for collecting pollen at high vibration frequency.

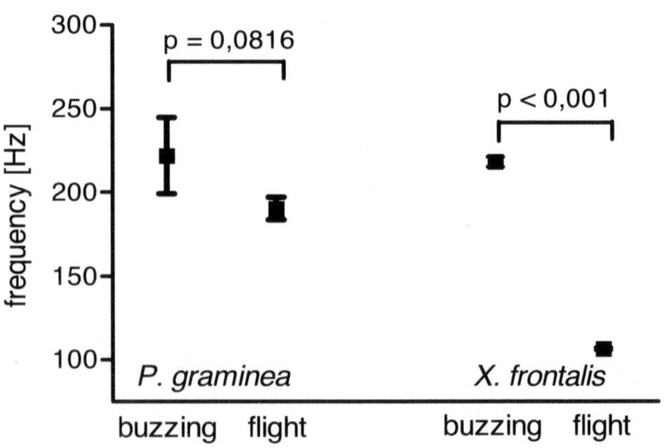

FIG. 2. Comparison of the buzzing and flight frequencies of the small *Pseudaugochlora graminea* (Halictidae) (n = 3) and the big *Xylocopa frontalis* (Apidae) (n = 18) (mean ± SD).

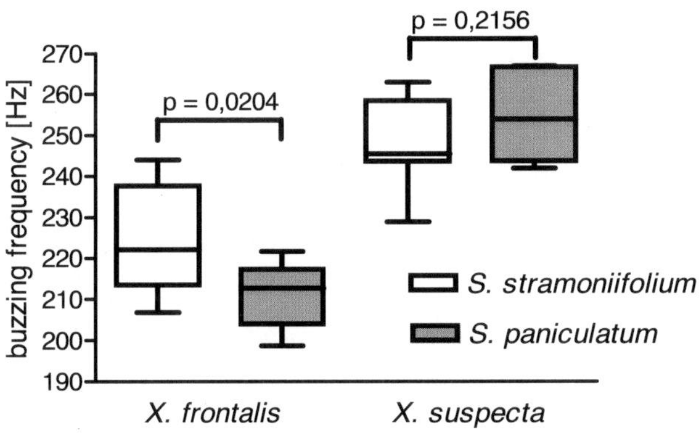

FIG. 3. Box wisker plot showing buzzing frequency of indivudals of *Xylocopa frontalis* on the flowers of *Solanum stramoniifolium* (n = 10) and *S. paniculatum* (n = 8) and of individuals of *X. suspecta* on the flowers of *S. stramoniifolium* (n = 12) and *S. paniculatum* (n = 8).

This correlation did not apply to medium to small sized bees which have an intertegular distance of less than 4 mm. These smaller bees, especially *Euglossa cordata, Centris tarsata* and *Melipona scutellaris*, have a high flight frequency, ranging from 229 to 252 Hz. Other than in larger bees, their buzzing frequency, ranging from 282 to 311 Hz, is far less than the double of the flight frequency. Furthermore in several recorded small *Augochlorini* individuals, belonging to the family of Halictidae, an atypical buzzing frequency close to the wingbeat frequency was found. When comparing the ratio of buzzing to flight frequency (Fig. 5) we found a close to linear relationship ranging from 2, as found in big bees, to 1 in the smallest ones. As the anatomical reason for these findings is largely unknown we could assume a general difference in the buzzing system of small bees as compared to larger ones. On the other hand, an adapation of small bees to a potentially optimal buzzing frequency could cause this deviant behaviour. This range of the buzzing frequency was found to be between 200 and 300 Hz for most bees. Despite these observations all bees collected during the study have obviously collected pollen, thus the ability of shaking pollen out of poricidal anthers of *S. stramoniifolium* and *S. paniculatum* via buzzing is independent of the size of the bee.

Because sound recording is very sensitive, frequency analysis permitted to resolve the frequency pattern performed during a single buzzing. While the wingbeat frequency during flight is almost constant, which is necessary for flight control, the buzzing frequency has no comparable limitation. However, we found a dynamic pattern ranging for more than 50 Hz during one buzzing which lasts for less than one second. When examining recordings of single buzzings no obvious tendency in frequency shift was seen. To overcome this, we performed trend analysis including all buzzings of single species (Fig. 4.). Observed species were found in average to start with a high and end with a low

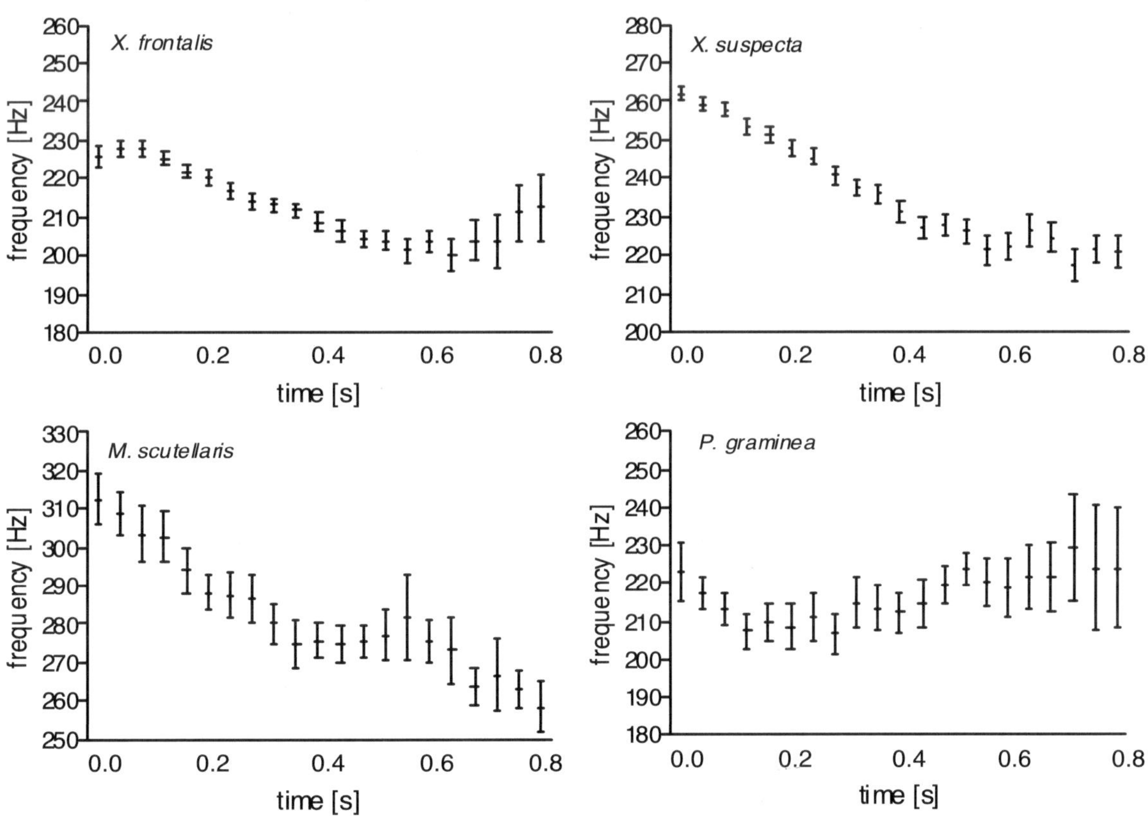

FIG. 4. Buzzing pattern of *Xylocopa frontalis* (n = 103), *X. suspecta* (n = 141), *Melipona scutellaris* (n = 12) and *Pseudaugochlora graminea* (n = 20); average pattern plotted in 0.04 s steps (mean ± SEM).

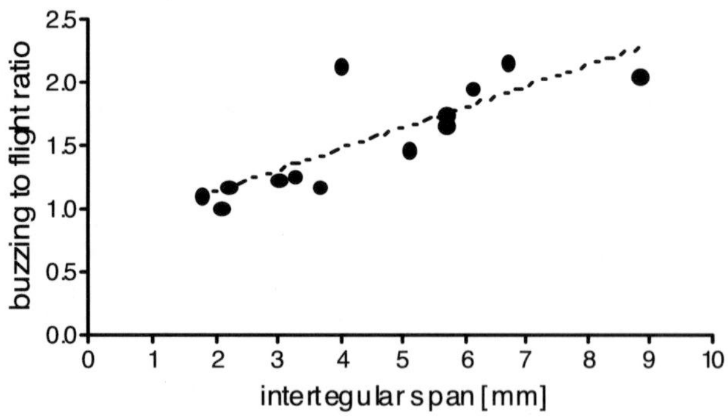

FIG. 5. Buzzing to flight ratio of the average frequencies as found in the bee species listed in Tab.2. The dotted line indicates linear regression ($r^2 = 0.672$).

frequency during a single buzzing, except for *Pseudaugochlora graminea* which showed no clear trend in its buzzing frequency. The buzzing frequency, thus, is not an absolutely fixed parameter, but changes during single buzzings. The possible reasons for frequency shifts are diverse and range from a better pollen removal by applying changing frequencies to fatigue of the flight musculatur due to the energy intensive buzzing behaviour.

We also tested whether there is a difference in the same species of bee buzzing on flowers *of S. stramoniifolium* and *S. paniculatum*. We found a significant difference in the used frequencies of *X. frontalis* but not in *X. suspecta*. This could be caused by behavioural differences or by other unknown reasons. As the flowers of both *Solanum* species in this study were morphologically similar, we encourage further frequency and resonance measurements on flowers with less morphological and size similarities. This could provide an answer to the question whether bees adapt their buzzing frequency to the given flower type.

Taken together, our study demonstrated the successful introduction and evaluation of a novel field and computer linguistical method to investigate aspects of the buzzing behaviour of neotropical bees.

ACKNOWLEDGEMENTS

To the two reviewers for valuable critical comments and suggestions, which improved the manuscript, and Poliana Ojima for the help with bee identification. This work was supported by a PROMOS fellowship of the Heinrich-Heine-Universität Düsseldorf, Germany, to A. B. and a research grant of the Brazilian Research Council - CNPq to C.S..

APPENDICES

Additional supporting information may be found in the online version of this article:

Appendix I. Praat script for analysing bee sounds

REFERENCES

Bezerra ELS, Machado IC (2003) Biologia floral e sistema de polinização de *Solanum stramonifolium* Jacq.(Solanaceae) em remanescente de Mata Atlântica, Pernambuco. Acta Botanica Brasilica 17:247-257.

Boersma PPG (1993) Accurate short-term analysis of the fundamental frequency and the harmonics-to-noise ratio of a sampled sound. IFA Proceedings 17:97-110.

Boersma PPG (2002) Praat, a system for doing phonetics by computer. Glot International 5:341-345.

Buchmann SC (1983) Buzz pollination in angiosperms. In: Eugene CE, Little RJ (eds) Handbook of experimental pollination biology. Van Nostrand Reinhold Company Inc., New York, pp 73-114.

Buchmann SL, Hurley JP (1978) Biophysical model for buzz pollination in angiosperms. Journal of Theoretical Biology 72:639-657.

Byrne DN, Buchmann SL, Spangler HG (1988) Relationship between wing loading, wingbeat frequency and body mass in homopterous insects. Journal of Experimental Biology 135:9-23.

Cane JH (1987) Estimation of bee size using intertegular span (Apoidea). Journal of the Kansas Entomological Society 60:145-147.

Ehrlich I, Lunau K (2009) Untersuchungen zur Pollendetektion von Hummeln. Enomologie heute 21:141-149.

Forni-Martins ER, Marques MCM, Lemes MR (1998) Biologia floral e reprodução de *Solanum paniculatum* L.(Solanaceae) no estado de São Paulo, Brasil. Revista Brasileira de Botanica 21:117-124.

Harter B, Leistikow C, Wilms W, Truylio B, Engels W (2002): Bees collecting pollen from flowers with poricidal anthers in a south Brazilian Araucaria forest: a community study. Journal of Apicultural Research 41:9-16.

Houston TF, Thorp RW (1984) Bionomics of the bee *Stenotritus greavesi* and ethological characteristics of Stenotritidae (Hymenoptera). Records of the Western Australian Museum 11:375-385.

Hrncir M, Jarau S, Zucchi R, Barth FG (2004) Thorax vibrations of a stingless bee (*Melipona seminigra*). I. No influence of visual flow. Journal of Comparative Physiology A 190:539-548.

Hurd PD, Linsley EG (1976) The bee family Oxaeidae with a revision of the North American species (Hymenoptera, Apoidea). Smithsonian Institution Press, Washington.

King MJ, Buchmann SL (2003) Floral sonication by bees: mesosomal vibration by *Bombus* and *Xylocopa*, but not *Apis* (Hymenoptera: Apidae), ejects pollen from poricidal anthers. Journal of the Kansas Entomological Society 76:295-305.

King MJ, Buchmann SL, Spangler H (1996) Activity of asynchronous flight muscle from two bee families during sonication (buzzing). Journal of Experimental Biology 199:2317-2321.

King MJ, Buchmann S (1996) Sonication dispensing of pollen from *Solanum laciniatum* flowers. Functional Ecology 10:449-456.

King MJ, Ferguson AM (1994) Vibratory collection of *Actinidia deliciosa* (kiwifruit) pollen. Annals of Botany 74:479-482.

King MJ, Lengoc L (1993) Vibratory pollen collection dynamics. Transactions-America Society of Agricultural Engineers 36:135-140.

Knapp S, Jarvis CE (1990) The typification of the names of New World *Solatium* species described by Linnaeus. Botanical Journal of the Linnean Society 104:325-367.

Levin RA, Myers NR, Bohs L (2006) Phylogenetic relationships among the "spiny *Solanums*" (*Solanum* subgenus

Leptostemonum, Solanaceae). American Journal of Botany 93:157-169.

Michener CD (1962) An interesting method of pollen collecting by bees from flowers with tubular anthers. Un método interesante de colectar polen en flores con anteras tubulares por parte de las abejas. Revista de Biologia Tropical 10:167-175.

Moure JM, Urban D, Melo GAR (2007) Catalogue of bees (Hymenoptera, Apoidea) in the Neotropical Region, Curitiba. Sociedade Brasilieira de Entomologia I:272-578.

Neff JL, Simpson BB (1988) Vibratile pollen-harvesting by *Megachile mendica* Cresson (Hymenoptera, Megachilidae). Journal of the Kansas Entomological Society 61:242-244.

Nunes-Silva P, Hrncir M, Imperatriz-Fonseca VL (2010) A Polinização por Vibração. Oecologia Australis 14:140-151.

Nunes-Silva P, Hrncir M, Venturieri GC, Imperatriz-Fonseca VL (2010) O potencial das abelhas *Melipona* na polinização por vibração, considerando as características físicas da vibração. Anais do IV Encontro sobre Abelhas 43:76-84.

Silva AC, Kinupp VF, Absy ML, Kerr WE (2004) Pollen morphology and study of the visitors (Hymenoptera, Apidae) of *Solanum stramoniifolium* Jacq. (Solanaceae) in Central Amazon. Acta Botanica Brasilica 18:653-657.

Teppner H (2005) Pollinators of Tomato, *Solanum lycopersicum* (Solanaceae), in Central Europe. Phyton 45:217-235.

Thorp RW (2000) The collection of pollen by bees. Plant Systematics and Evolution 222:211-223.

Wille A (1963) Behavioral adaptations of bees for pollen collecting from *Cassia* flowers. Revista de Biologia Tropical 11:205-206.

15

MUTUALISM OF *STRATIOTES ALOIDES* L. (HYDROCHARITACEAE) AND *HYDRELLIA TARSATA* HALIDAY (DIPTERA: EPHYDRIDAE): TRITROPHIC INTERACTION OF MACROPHYTE, LEAF-MINING DIPTERAN POLLINATOR AND PARASITOID BRACONIDAE

Jakob Katzenberger* and Dietmar Zacharias

Bremen University of Applied Sciences, Applied and Ecological Botany, Faculty 5, Neustadtswall 30, 28199 Bremen, Germany

Abstract—The aquatic macrophyte *Stratiotes aloides* L. is of conservation concern in central Europe due to its high importance for biodiversity in lowland floodplain and ditch ecosystems. However, over the last decades this species has shown population declines for instance in Germany or the Netherlands. *S. aloides* is dioecious with male and female individuals, in mixed or separated stands, often reproducing vegetatively. Generative reproduction is observed less frequently, but of great importance for declining plant populations facing threats of habitat destruction and eutrophication. Precisely which arthropods transfer *S. aloides* pollen was previously unknown. We examined flower visitors of *S. aloides* in the 2011 and 2014 flowering seasons in ditches of a wet grassland ecosystem in Bremen, Northwest Germany. *Hydrellia tarsata* Haliday (Diptera: Ephydridae) was found abundantly in male and female flowers of *S. aloides* in both years. Pollen of *S. aloides* was actively transferred by *H. tarsata* and reproduction of the fly in *S. aloides* leaves was detected by rearing *H. tarsata* from extracted puparia. The mining *Hydrellia* were parasitised by the braconid wasps *Chaenusa "punctulata"* Burghele and *Chorebus "densepunctatus"* Burghele, which also visited *S. aloides* flowers in 2011 and 2014. These results point to a mutualism between *S. aloides* and the ephydrid *H. tarsata*, with both partners benefiting with their own reproduction. This relationship between plant and dipteran pollinator is however complicated in a tritrophic interaction with the braconid parasitoids, which infest the mining stages of the ephydrid flies and could potentially also transfer *S. aloides* pollen.

Keywords: myiophily, pollination, Ephydridae, Braconidae, Dacnusini, plant-insect interaction

INTRODUCTION

Insect pollination is an essential part of the reproductive cycle in numerous angiosperm plant species. The most important pollinators are found among the Hymenoptera, which have a major impact on natural and anthropogenic ecosystems by transporting pollen between flowers of entomophilous plants and thereby facilitate fertilization and generative reproduction (Kevan & Baker 1983). The ubiquitous role of flies (Diptera) as flower visitors has long been noted (Larson et al. 2001), however their importance in plant pollination only recently became evident (Ssymank et al. 2008; Woodcock et al. 2014). Examples of flower pollination by flies (myiophily) are manifold (Larson et al. 2001; Kevan 2002; Ssymank et al. 2008) and come from both terrestrial and freshwater ecosystems (e.g. Mesler & Lu 1993; Lippok & Renner 1997). Biotic pollination commonly occurs in aquatic plant species, as these rely to a large part on insects for pollen transfer and less often on abiotic vectors like wind or water (Ackerman 2000). Although many aquatic plants have the ability to reproduce vegetatively, generative reproduction is an important aspect of their biology and has been retained throughout evolution (Philbrick & Les 1996).

The aquatic macrophyte *Stratiotes aloides* L. belongs to the fully aquatic family Hydrocharitaceae and is a perennial, dioecious plant species (Cook & Urmi-König 1983; Chen et al. 2012). It grows in sexually separated or mixed stands and often reproduces vegetatively, forming dense aggregations in slow-flowing to stagnant freshwater habitats (Cook & Urmi-König 1983). The generative reproduction of *S. aloides* relies on the close proximity of the two sexes (Smolders et al. 1995) but has not been studied comprehensively. Thus, no detailed account of the species' pollination ecology exists and the only descriptions of flower-visiting insects come from early anthecology (Knuth 1899; Daumann 1931). From these early works Diptera emerge as likely pollination vectors (Daumann 1931). More recent observations of the fly *Hydrellia tarsata* Haliday (Diptera: Ephydridae) mining in leaves and visiting flowers of *S. aloides* suggest a possible interaction of the two species for reproduction (Przhiboro 2004; Stuke 2010; Katzenberger et al. 2013).

In general, *S. aloides* is seen as an indicator species for aquatic habitats with high biodiversity value in lowland floodplain and ditch ecosystems and it is therefore of concern to nature conservation in Europe (Suutari et al. 2009; Kunze et al. 2012). *S. aloides* showed population

*Corresponding author; email: jakob.katzenberger@hotmail.de

declines over the last decades in central Europe, for instance in the Netherlands or Germany (Smolders et al. 1996; Sinning 2001; Suutari et al. 2009). The Bremen region in Northwest Germany remains one of the main distribution areas for *S. aloides* (Cordes et al. 2006) and the biology and ecology of the species is under extensive study here (Nüssel & Zacharias 2010; Zantout et al. 2011; Kunze et al. 2012; Katzenberger et al. 2013). Bremen is especially well-suited to studying the generative reproduction of the species as seed-set occurs plentifully (Kunze et al. 2012). The observed negative population trend is mainly attributed to eutrophication, leading to shading and degraded growth conditions (Smolders et al. 2003; Zantout et al. 2011). Another reason suspected to contribute to the decline is an overall lack of generative reproduction by the species and therefore less adaptability to changing environmental factors (Smolders et al. 1996, 2003).

In this study, we address the following questions: (1) Which insects visit flowers of *S. aloides*? (2) Do flower visitors transfer pollen and thereby indicate pollination activity? (3) Is there evidence for stable interactions between *S. aloides* and its flower visitors? With our results we aim to investigate the suggestion from early anthecology that *S. aloides* is likely an example of fly pollination (myiophily) and contribute towards to a detailed account of the pollination ecology for *Stratiotes aloides.*

MATERIALS AND METHODS

Study species

The water soldier, *Stratiotes aloides,* is distributed in Northern, Central and Eastern Europe and also in Central Asia (Cook & Urmi-König 1983). In Germany, *S. aloides* is listed as "vulnerable" in the national red list of flowering plants (Korneck et al. 1996) and considered an important indicator species for conservation management aiming to sustain lowland ditches rich in biodiversity (Kunze et al. 2012). The dense stands of *S. aloides* form a spatially complex and sheltered habitat for diverse generalist macroinvertebrate communities (Tarkowska-Kukuryk 2006) but they also foster specialised species like the dragonfly, *Aeshna viridis* L. (Rantala et al. 2004; Suutari et al. 2009). The plant has a unique life form; during summer it is partially emergent but in autumn it sinks and over-winters under water as a rosette with green leaves (Cook & Urmi-König 1983). Then, in spring, *S. aloides* re-surfaces, old leaves are shed and fresh leaves develop, while during the whole cycle it is usually rooted below in loose substrate (Cook & Urmi-König 1983).

S. aloides is entomophilous and the flowers attract insects with large white petals, osmophores and nectaries (Cook 1998). The flowers emit a foul smell similar to that of rotting meat (Daumann 1931; Cook 1998). Male and female flowers of *S. aloides* look essentially alike only differing in the morphology of the regenerative organs. Male flowers have 5-17 stamens surrounded by 20-30 nectaries while female ones usually show 6 pistils likewise surrounded by 20-30 nectaries (Cook & Urmi-König 1983; Nüssel & Zacharias 2010). The first flowers appear in April and flowering continues until the end of August (Kunze et al. 2012). The only available field study on the generative reproduction of the species found that it required short distances (< 1 km) between male and female stands for successful fertilisation, but did not investigate pollinators (Smolders et al. 1995). Nevertheless, *S. aloides* is parthenocarpic and therefore produces sterile fruits when not pollinated which are distinctly smaller than fertile fruits and quickly start to rot (Cook & Urmi-König 1983; Smolders et al. 1995).

Flower visitation of *S. aloides* by the hoverfly *Eristalis tenax* L. (Diptera: Syrphidae) was already noted by Knuth (1899). In his detailed account on the floral morphology of *S. aloides,* Daumann (1931) also studied the mode of pollination and observed the dipterans *Calliphora erythrocephala* Meigen (Diptera: Calliphoridae), *Calliphora vomitoria* L. (Diptera: Calliphoridae), *Lucilia caesar* L. (Diptera: Calliphoridae), *Sarcophaga carnaria* L. (Diptera: Sarcophagidae), *Odontomyia tigrina* Fabricius (Diptera: Stratiomyidae) and especially abundant *Hydrellia cardamines* Haliday (Diptera: Ephydridae) as flower visitors.

The Braconidae are a diverse group of parasitoid wasps occurring worldwide, with over 40,000 species they form the second largest family among the Hymenoptera (Goulet & Huber 1993). They are often closely attuned to their hosts and highly efficient in controlling host populations and therefore used for biological control (Wharton 1993). Among the Braconidae the subfamilies Alysiinae and Opiinae are endoparasitoids of cyclorrhaphous Diptera, ovipositing in the larva or egg and emerging from the puparium (Wharton 1993). The biology and ecology of the braconid species belonging to the tribe Dacnusini in the subfamily Alysiinae is still largely unknown (Fischer 1999). In general Dacnusini are seen as very host specific and this is especially marked in *Chorebus,* where many species have only been obtained from a single host (Griffiths 1966).

Sample collection and species identification

Samples of flower visitors and mining fauna in leaves of *Stratiotes aloides* were taken in the 2011 and 2014 flowering season (April - July) in the nature reserve "Westliches Hollerland" (53°06'54.1"N, 8°51'30.8"E) in Bremen, Germany. The study area is an approximately 300 ha large and extensively managed agricultural grassland with a dense network of ditches for drainage in the marshland. The samples were obtained from ditches with compact stands of *S. aloides* (both mixed-sex and unisexual separated stands occurred).

Flower visitors were caught with a collapsible net (15 cm diameter, 2 × 2 mm mesh width) on a telescopic pole. The net was lowered over randomly selected open flowers above the water line and was then collapsed, trapping flower visitors inside. Trapped insects were transferred into a vial and killed with ethyl acetate. All samples were taken during the middle of the day between 11 am and 4 pm. Dipteran puparia were collected by randomly gathering 7-10 leaves from several *S. aloides* plants on the water surface and dissecting these in the laboratory. The puparia were reared in petri-dishes on wet cotton wool, kept in a constant temperature of 21°C. Species identification of *Hydrellia tarsata* was achieved by dissection of the male genitalia which corresponded exactly to Collin (1966). Female *Hydrellia* in the samples could not confidently be identified to species level, however, they resembled male specimens of *H. tarsata* morphologically and in some instances copulation was observed on flowers before trapping. Insect dissections and identifications were performed under a Zeiss Stemi 2000-C dissection microscope. Species identification of braconid wasps was made by Dr. HCJ Godfray, Oxford. The samples were compared to museum specimens and will be deposited in the collection of the National Museum of Scotland for future research. Taxonomy and identification of Braconidae is a challenging field and little is known about the tribe Dacnusini to which these wasps belong (Fischer 1999). The specimens collected in this study were determined with every possible care but, for the lack of a functional key, species identifications are termed "probably" or species names are put in quotation marks (HCJ Godfray, pers. comm.).

To determine pollen transfer of individuals, pollen grains were located on male *Hydrellia tarsata* specimens after collection from flowers. Single pollen grains from extremities and body setae, visible under the dissection microscope, were transferred with a dissection needle to a microscope slide coated with basic fuchsin gelatin and thereby stained (Beattie 1971). Under a Zeiss Axio Lab.A1 microscope, pollen grains were observed in 1000 × magnification (with oil) with a Ph1 filter and photographed with a Zeiss AxioCam ERc5s camera. Pollen morphology (exine structure and diameter) from the samples collected from *H. tarsata* individuals was compared to stained pollen from male *S. aloides* flowers under 1000 × magnification to ascertain pollen identity.

Results

Flower visitors of *Stratiotes aloides*

In 21 samples from male and female flowers of *Stratiotes aloides* collected in 2011 and 2014 we found 42 male individuals of *Hydrellia tarsata* and 5 female *Hydrellia* specimens which could not be determined to species level (Tab. 1). *H. tarsata* was abundant in and around flowers of dense stands of *S. aloides* in small ditches (< 3 m) with almost stagnant water in the grassland of the study area (Fig. 1). Flower visiting *Hydrellia* were often observed deep within the flowers, moving actively inside around the flower nectaries and stamens or pistils, sometimes copulating. Even if we approached the flowers closely the insects would not fly away, only a strong movement of the flower would trigger escape.

Sample	Date	Flower	No.	ID
2011-1	27.05.11	♂	1	*Hydrellia tarsata*
2011-2	27.05.11	♀	2	*Hydrellia tarsata*
2011-3	27.05.11	♀	1	*Hydrellia tarsata*
2011-4	27.05.11	♀	2	*Hydrellia tarsata*
2011-5	27.05.11	♀	4	*Hydrellia tarsata*
2011-6	27.05.11	♀	1	*Hydrellia tarsata*
			1	female *Hydrellia* sp.
2011-7	27.05.11	♀	4	*Hydrellia tarsata*
2011-8	28.06.11	♂	1	female *Hydrellia* sp.
2011-9	28.06.11	♀	2	*Hydrellia tarsata*
2011-10	28.06.11	♀	1	*Hydrellia tarsata*
2011-11	28.06.11	♀	1	*Hydrellia tarsata*
			1	female *Hydrellia* sp.
2011-12	28.06.11	♀	1	*Hydrellia tarsata*
2011-13	28.06.11	♂	2	*Hydrellia tarsata*
2011-14	28.06.11	♀	2	*Hydrellia tarsata*
2014-1	29.04.14	♀	6	*Hydrellia tarsata*
			1	female *Hydrellia* sp.
2014-2	30.05.14	♂	4	*Hydrellia tarsata*
2014-3	30.05.14	♂	3	*Hydrellia tarsata*
2014-4	29.06.14	♀	2	*Hydrellia tarsata*
2014-5	29.06.14	♀	2	*Hydrellia tarsata*
2014-6	29.06.14	♀	1	*Hydrellia tarsata*
2014-7	29.06.14	♀	1	female *Hydrellia* sp.

TABLE 1. Abundance of dipteran visitors to male and female *Stratiotes aloides* flowers in our samples from the study area in years 2011 and 2014.

FIGURE 1. *Hydrellia tarsata* flower visitors on female (part "a", sample 2014-1 see Tab. 1) and male (part "b", sample 2014-3 see Tab. 1) flowers of *Stratiotes aloides*. Note the forked, yellow nectaries clustered around regenerative organs offering nectar reward for flower visitors and *H. tarsata* in contact with flower stigmata and anthers.

Sample	Date	Flower	No.	ID
2011-3	27.05.11	♀	1	*Chorebus "densepunctatus"*
2011-4	27.05.11	♀	1	*Chaenusa "punctulata"*
2011-7	27.05.11	♀	1	*Chaenusa "punctulata"*
2011-15	27.05.11	♂	1	*Chorebus "densepunctatus"*
2011-16	28.06.11	♀	1	*Chaenusa "punctulata"*
2011-13	28.06.11	♂	1	*Chaenusa "punctulata"*
2014-1	29.04.14	♀	1	*Chaenusa "punctulata"*
2014-2	30.05.14	♂	1	*Chaenusa "punctulata"*
2014-5	29.06.14	♀	1	*Chaenusa "punctulata"*
2014-7	29.06.14	♀	2	*Chaenusa "punctulata"*

TABLE 2. Abundance of hymenopteran visitors to male and female *Stratiotes aloides* flowers in our samples from the study area in years 2011 and 2014. Species identifications of braconid wasps are put in quotation marks due to the lack of a functional key; see also Material and Methods.

In 8 samples from *S. aloides* flowers we observed 9 specimens of a braconid wasp belonging to the genus *Chaenusa*, probably *punctulata* Burghele (Tab. 2). Another braconid wasp from the genus *Chorebus* (*uliginosus* group sensu Nixon), probably *densepunctatus* Burghele was observed in 2 samples with 2 specimens (Tab. 2). Similar to the flower visiting *Hydrellia*, the braconid wasps were usually deep within the flower, only escaping upon flower movement. A one-time visitation of a *S. aloides* flower by a hoverfly was observed but no other Diptera or Hymenoptera were found. Other rarely observed arthropods in flower samples included three Coleoptera specimens, one type of Collembola, one Lepidoptera specimen and one Araneae specimen. Due to the low abundances those arthropods were not investigated further.

Transport of Stratiotes aloides pollen

Stratiotes aloides pollen could be detected on 4 male specimens of *Hydrellia tarsata*, found in two samples from male *S. aloides* flowers (Fig. 2). Pollen loads were always relatively low with never more than 10 pollen grains attached to body setae and legs of *H. tarsata* (Fig. 2). Exine structure of pollen grains found on *H. tarsata* corresponded to pollen from a male *S. aloides* flower, showing a granular structure and clearly visible bacula (Fig. 2). Pollen extracted from a male *S. aloides* flower had an average diameter of 42.3 µm (± SD 3.1, $N = 5$) while the average diameter of pollen grains found on *H. tarsata* specimens was 39. 8 µm (± SD 1.6, $N = 5$).

FIGURE 2. Male *Hydrellia tarsata* specimen (sample 2014-2, see Tab. 1) caught in a male flower of *Stratiotes aloides* carrying pollen grains of *S. aloides*. Part "a" shows adult fly with pollen (red arrows) on tarsus and tibia (25 × magnification) while part "b" shows dissected genital used for species identification (50 × magnification). Part "c" shows single stained *S. aloides* pollen grain (1000 × magnification, diameter 37 μm) extracted from legs of *H. tarsata* for comparison with stained pollen grain from anthers of *S. aloides* (1000 × magnification, diameter 38 μm) displayed in part "d". Note the granular exine structure and bacula (circular spots) in *S. aloides* pollen grains.

Puparia extracted from leaves of Stratiotes aloides

Dipteran puparia were collected three times in 2014 (29th April, 27th May, 29th June) from mines in *Stratiotes aloides* leaves and subsequently reared in the laboratory. During the first sampling in April *S. aloides* had just resurfaced, most of the leaves were still from the previous year and no flowers were present. In the two later samplings *S. aloides* stands were flowering and most of the old leaves had decayed. The puparia were subcylindrical and semi translucent with a variable degree of brown colour. The length of the puparia ranged between 2-3 mm. In total 10 *Hydrellia* specimens hatched from the puparia, of which two were male and could be identified as *Hydrellia tarsata*. From these puparia three braconid wasps also emerged; two were identified as *Chaenusa*, probably *punctulata*, and one as *Chorebus*, probably *densepunctatus*.

DISCUSSION

The results from our study clearly show that the ephydrid *Hydrellia tarsata* reproduces in the leaves of *Stratiotes aloides*, but also that it is a main candidate to consider as pollination vector for the species. From earlier studies on *S. aloides* we know that fertile seeds are produced plentifully in our study area (Kunze et al. 2012) and thus pollination and fertilization occurs. We found *H. tarsata* abundantly in male and female flowers (Fig. 1) and revealed that it does in fact carry *S. aloides* pollen (Fig. 2). The results were consistent between the two study years and no other flower visitors were found regularly enough to consider relevant pollination activity. It has to be noted that our sample size was limited and the collections of flower visitors were taken from a small study area, chosen for easy accessibility. Also, we did not perform any experiments to demonstrate the pollination activity of *H. tarsata*. Nevertheless, the presented evidence strongly suggests an interaction between *S. aloides* and *H. tarsata*, where the fly reproduces in the plant tissue by means of its phytophageous larvae and the imagines pollinate the dioecious flowers of the macrophyte and thus facilitate its generative reproduction. However, our results indicate that the braconid parasitoids, visiting flowers of *S. aloides* and infesting immature *Hydrellia*, add another level of complexity to this plant-pollinator interaction and could mediate the strength of the relationship between *H. tarsata* and *S. aloides* in a set of tritrophic interactions.

Ecological relationships between flies and flowers are critical to each other's survival in many habitats and flower visitation by small Diptera, including alcalyptrates, warrants more investigation (Ssymank et al. 2008; Woodcock et al. 2014). The Ephydridae in particular are already known as pollinators of aquatic plants such as *Cabomba caroliniana* Gray (Schneider & Jeter 1982) or several species of *Nuphar* (Lippok & Renner 1997; Nagasaki 2007). In two of these studies *Hydrellia* species were confirmed as pollinators of an aquatic macrophyte, one in Europe (Lippok & Renner 1997) and one in North America (Schneider & Jeter 1982). Small Diptera could compensate for their own supposed inefficiency as pollinators by their large numbers and their ubiquity (Ssymank et al. 2008), which seems also to be the case in our study system. The pollen loads evident on *Hydrellia tarsata* were usually low, but flowers were visited by many individuals at a time and *H. tarsata* was abundant in the study area. Two of the only three known records for *H. tarsata* in Northwest Germany were made on, or next to, *S. aloides* stands (Stuke 2010; Katzenberger et al. 2013). Furthermore in the only systematic study on *Stratiotes aloides* floral visitation, Daumann (1931) describes the common species *Hydrellia cardamines* as very abundant. But *Hydrellia* is a taxonomically challenging genus and species identifications are often only possible by dissection of the male genitalia (Deyrup & Deyrup 2008; Stuke 2010), thus it seems possible that old accounts in the literature contain misidentifications. *H. tarsata* probably has been overlooked on *S. aloides* for a long time.

The dipteran reproduction in plant tissues by mining phytophageous larvae has received some attention in the literature, for instance Deonier (1971) lists *S. aloides* as host for *Hydrellia griseola* Fallen and *Hydrellia mutata* Zetterstedt. That the rarer and less studied *Hydrellia tarsata* also mines in *S. aloides* was observed in Russia only recently (Przhiboro 2004) and our study confirmed that the relationship occurs in Germany.

Considering all of this information, the interaction between *S. aloides* and *H. tarsata* is likely a mutualistic one, with both partners benefiting in reproduction. Insect pollination is usually based on a mutual benefit and flies receive primary (nectar/pollen feeding) or secondary (e.g. brood and/or mating places) rewards from flowers, while the plants get cross-pollinated for generative reproduction (Woodcock et al. 2014). Our results clearly show that *H. tarsata* utilises a secondary reward by brooding in leaves, but the high ratio of male to female flies we found on flowers and the observation of copulation suggest that *S. aloides* could also be a mating site for *H. tarsata*. Feeding on nectar as a form of primary reward in flowers is known for *Hydrellia* species as well (Deonier 1971; Schneider & Jeter 1982; Lippok & Renner 1997). Nevertheless the mining *H. tarsata* damage and consume plant tissue during their reproduction in the leaves of *S. aloides*, but the interaction between plant and pollinator will persist if the plant has a benefit in propagules produced (Woodcock et al. 2014). More research on the pollination efficiency of *H. tarsata* could help to understand how much *S. aloides* plants actually benefit from this interaction.

In a tritrophic interaction plants interact with carnivores to indirectly defend against herbivory, a phenomenon also described for braconid wasps attracted by plant volatiles (Heil 2008). The braconids *Chaenusa "punctulata"* and *Chorebus "densepunctatus"* found in our samples from *S. aloides* flowers and infesting *Hydrellia* puparia belong to the very host specific tribe Dacnusini in the subfamily Alysiinae. This suggests that the parasitoids we found are the main limiting factor for *Hydrellia tarsata* in our study area. Thus, the dacnusine wasps could function as natural enemies for *H. tarsata* and thereby also affect the interaction with *S. aloides.* The two species *Chaenusa punctulata* and *Chorebus densepunctatus* were both described in Romania, from material collected in extensive swamps and wetlands near Bucharest (Burghele 1959, 1960). *Chorebus densepunctatus* was, in fact, reared by Burghele (1960) in large numbers from *Hydrellia* puparia, with the host identified as *Hydrellia griseola.* Even earlier, dacnusine wasps determined as *Chorebus uliginosus* Haliday were reared by Ruschka & Thienemann (1913) in West Germany, also from puparia extracted from *S. aloides* leaf mines. Similarly, the host was identified as *Hydrellia griseola* (Ruschka & Thienemann 1913). Despite being scattered and disputable, the literature still reflects that *S. aloides* is a great study object to increase the knowledge about ecological relations between *Hydrellia* and its parasitoids.

The reason why the Dacnusini visit *S. aloides* flowers seems to be primarily to locate their *Hydrellia* hosts and to perform oviposition. Such a strategy has already been supposed by Schneider & Jeter (1982), who similarly found *Chaenusa* sp. together with a *Hydrellia* species on flowers of the aquatic macrophyte *Cabomba caroliniana.* The authors also showed that the braconid parasitoids actively transfer pollen (Schneider & Jeter 1982). Nectar feeding and pollen transfer by parasitoid hymenopterans also occurs in the High Arctic, but their contribution to plant pollination is likely minimal (Kevan 1973). To ascertain which role the Dacnusini play in our study system further study of their ecological relationships is required, especially of pollination activity and induced mortality.

Acknowledgements

We thank HCJ Godfray for identification of braconid wasps and his help with the literature on Dacnusini. Furthermore, H Klugkist from the "Senator für Umwelt, Bau und Verkehr" in Bremen was very helpful and thankfully provided permission for entering and taking samples in the study area. J de Satgé and T Langbehn are thanked for their comments on an earlier draft of the manuscript.

References

Ackerman JD (2000) Abiotic pollen and pollination: Ecological, functional, and evolutionary perspectives. Plant Systematics and Evolution 222:167-185.

Beattie A (1971) A technique for the study of insect-borne pollen. The Pan-Pacific Entomologist 47:82.

Burghele A (1959) New Rumanian species of Dacnusini (Hym., Braconidae) and some ecological observations upon them. Entomologist's Monthly Magazine 95:121-126.

Burghele A (1960) Zwei neue Arten von Dacnusini (Hymenoptera, Braconidae) nebst einer Liste der aus Rumänien bekannten Arten. Zeitschrift der Arbeitsgemeinschaft der österreichischen Entomologen 12:95-100.

Chen L-Y, Chen J-M, Gituru RW, Wang Q-F (2012) Generic phylogeny, historical biogeography and character evolution of the cosmopolitan aquatic plant family Hydrocharitaceae. BMC Evolutionary Biology 12:30.

Collin JE (1966) A contribution to the knowledge of the male genitalia of species of *Hydrellia* (Diptera, Ephydridae). Bolletino del Museo Civico di Storia Naturale di Venezia 16:7-18.

Cook CDK (1998) Hydrocharitaceae. In: Flowering Plants • Monocotyledons. Springer, Berlin Heidelberg, pp 234-248.

Cook CDK, Urmi-König K (1983) A revision of the genus *Stratiotes* (Hydrocharitaceae). Aquatic Botany 16:213-249.

Cordes H, Feder J, Hellberg F, Wittig B, Metzing D (2006) Atlas der Farn- und Blütenpflanzen des Weser-Elbe-Gebietes. Hauschild, Bremen.

Daumann E (1931) Zur Morphologie und Ökologie der Blüte von *Stratiotes aloides* L. Planta 14:766-776.

Deonier DL (1971) A systematic and ecological study of Nearctic *Hydrellia* (Diptera: Ephydridae). Smithsonian Contributions to Zoology 68:1-147.

Deyrup M, Deyrup L (2008) Flower visitation by adult shore flies at an inland site in Florida (Diptera: Ephydridae). Florida Entomologist 91:504-507.

Fischer M (1999) Einiges über Kieferwespen (Hymenoptera, Braconidae, Alysiinae). Linzer biologische Beiträge 31:5-56.

Goulet H, Huber JT (1993) Hymenoptera of the world: an identification guide to families. Research Branch, Agriculture Canada.

Griffiths GCD (1966) The Alysiinae parasites of Agromyzidae. Beiträge zur Entomologie 8:775-963.

Heil M (2008) Indirect defence via tritrophic interactions. New Phytologist 178:41-61.

Katzenberger J, Langbehn T, Zacharias D (2013) Erstnachweis von *Hydrellia tarsata* (Diptera: Ephydridae) für Bremen in Blüten von *Stratiotes aloides*. Abhandlungen des Naturwissenschaftlichen Vereins zu Bremen 47:193-195.

Kevan PG (1973) Parasitoid wasps as flower visitors in the Canadian high arctic. Anzeiger für Schädlingskunde, Pflanzen- und Umweltschutz 46:3-7.

Kevan PG (2002) Flowers, pollination, and the associated diversity of flies. Biodiversity 3:16-18.

Kevan PG, Baker HG (1983) Insects as flower visitors and pollinators. Annual Review of Entomology 28:407-453.

Knuth P (1899) Handbuch der Blütenbiologie II. Engelmann, Leipzig.

Korneck D, Schnittler M, Vollmer I (1996) Rote Liste der Farn- und Blütenpflanzen (Pteridophyta et Spermatophyta) Deutschlands. Schriftenreihe für Vegetationskunde 28:21-187.

Kunze K, Jordan R, Kesel R, Kundel W, Nagler A, Schirmer M, Zacharias D (2012) Erprobung von Managementmaßnahmen zum Erhalt der Krebsschere (*Stratiotes aloides*) als Leitart für die ökologisch wertvollen Graben-Grünland-Gebiete der Kulturlandschaft Nordwestdeutschlands. Natur und Landschaft 87:362-369.

Larson B, Kevan P, Inouye D (2001) Flies and flowers: taxonomic diversity of anthophiles and pollinators. The Canadian Entomologist 133:439-465.

Lippok B, Renner SS (1997) Pollination of *Nuphar* (Nymphaeaceae) in Europe: flies and bees rather than *Donacia* beetles. Plant Systematics and Evolution 207:273-283.

Mesler M, Lu K (1993) Pollination biology of *Asarum hartwegii* (Aristolochiaceae): an evaluation of Vogel's mushroom-fly hypothesis. Madroño 40:117-125.

Nagasaki O (2007) Pollination of the yellow water lily *Nuphar subintegerrima* (Nymphaeaceae) by the shore fly *Notiphila* (*Notiphila*) *maritima* (Diptera: Ephydridae). Plant Species Biology 22:227-230.

Nüssel K, Zacharias D (2010) Zur Morphologie des Makrophyten *Stratiotes aloides* L. (Krebsschere). Abhandlungen des Naturwissenschaftlichen Vereins zu Bremen 46:435-444.

Philbrick CT, Les DH (1996) Evolution of aquatic angiosperm reproductive systems. BioScience 46:813-826.

Przhiboro AA (2004) New data on *Hydrellia tarsata* Haliday (Diptera: Ephydridae). Zoosystematica Rossica 13:239-240.

Rantala MJ, Ilmonen J, Koskimäki J, Suhonen J, Tynkkynen. K (2004) The macrophyte, *Stratiotes aloides*, protects larvae of dragonfly *Aeshna viridis* against fish predation. Aquatic Ecology 38:77-82.

Ruschka F, Thienemann A (1913) Zur Kenntnis der Wasser-Hymenopteren. Zeitschrift für wissenschaftliche Insektenbiologie 9:82-87.

Schneider EL, Jeter JM (1982) Morphological studies of the Nymphaeaceae. XII. The floral biology of *Cabomba caroliniana*. American Journal of Botany 69:1410-1419.

Sinning F (2001) Zur Verbreitung der Krebsschere (*Stratiotes aloides*) in der Wesermarsch. Bremer Beiträge für Naturkunde und Naturschutz 5:65-70.

Smolders AJP, den Hartog C, Roelofs JGM (1995) Observations on fruiting and seed-set of *Stratiotes aloides* L. in the Netherlands. Aquatic Botany 51:259-268.

Smolders AJP, Lamers LPM, den Hartog C, Roelofs JGM (2003) Mechanisms involved in the decline of *Stratiotes aloides* L. in The Netherlands: sulphate as a key variable. Hydrobiologia 506-509:603-610.

Smolders AJP, Roelofs JGM, den Hartog C (1996) Possible causes for the decline of the water soldier (*Stratiotes aloides* L.) in the Netherlands. Archiv für Hydrobiologie 136:327-342.

Ssymank A, Kearns CA, Pape T, Thompson FC (2008) Pollinating Flies (Diptera): A major contribution to plant diversity and agricultural production. Biodiversity 9:86-89.

Stuke JH (2010) Die Uferfliegen Niedersachsens und Bremens (Diptera: Ephydridae) - Teil II. Entomologische Zeitschrift 120:195-222.

Suutari E, Salmela J, Paasivirta L, Rantala MJ, Tynkkynen K, Luojumaki M, Suhonen J (2009) Macroarthropod species richness and conservation priorities in *Stratiotes aloides* (L.) lakes. Journal of Insect Conservation 13:413-419.

Tarkowska-Kukuryk M (2006) Water soldier *Stratiotes aloides* L. (Hydrocharitaceae) as a substratum for macroinvertebrates in a shallow eutrophic lake. Polish Journal of Ecology 54:441-451.

Wharton R (1993) Bionomics of the Braconidae. Annual Review of Entomology 38:121-143.

Woodcock TS, Larson BMH, Kevan PG, Inouye DW, Lunau K (2014) Flies and flowers II: Floral attractants and rewards. Journal of Pollination Ecology 12:63-94.

Zantout N, Wilfert P, Smolders AJP, Weber G, Zacharias D (2011) Effects of sediment pore water qualities on the decline of *Stratiotes aloides* L. stands in Bremen, Germany. Fundamental and Applied Limnology / Archiv für Hydrobiologie 179:131-136.

16

THE CONTRIBUTION OF HONEY BEES, FLIES AND WASPS TO AVOCADO (*PERSEA AMERICANA*) POLLINATION IN SOUTHERN MEXICO

J. Pérez-Balam[1], J.J.G Quezada-Euán[1], R. Alfaro-Bates[1], S. Medina[2], L. McKendrick[3], A. Soro[3,4], R.J. Paxton[3,4]

[1]*Departamento de Apicultura, Campus Ciencias Biológicas y Agropecuarias, Universidad Autónoma de Yucatán, Km. 15.5. carr. Mérida-Xmatkuil CP 97000, Mérida, Yucatán, México*

[2]*Facultad de Matemáticas, Campus Ciencias Exactas, Universidad Autónoma de Yucatán, Mérida, Yucatán, México*

[3]*School of Biological Sciences, Queen's University Belfast, MBC, 97 Lisburn Road, Belfast BT9 7BL, United Kingdom*

[4]*Institute fur Biology Martin- Luther-UniversitätHalle-Wittenberg, HoherWeg 8, D-06099 Halle (Saale), Germany*

Abstract—Although avocado is native to Mexico, there are no comparative measures in this country on the performance of its flower visitors as pollinators. The contribution of honey bees, flies and wasps to the pollination of avocado from tropical Mexico was assessed by comparing abundance, speed of flower visitation, quantity of pollen carried per individual and pollen deposited on virgin flowers after single visits. The values of abundance and frequency of flower visitation with pollen deposition were combined to obtain a measure of pollinator performance (*PP*). The most abundant insects on avocado were flies (mean ± SE: 15. 2 ± 6.2), followed by honey bees (9.4 ± 6.3) and wasps (4.2 ± 3.1) (ANOVA F = 91.71, d.f. = 2,78; $P < 0.001$). Honey bees and wasps visited similar number of flowers (8.2 ± 3.1 and 7.5 ± 2.6 respectively), and more than flies (4.1 ± 1.2) in the same time period (F = 17.63; d.f. = 2,33; $P < 0.01$). Although flies carried far more avocado pollen on their bodies (44.9 ± 16.8 grains) compared with honey bees and wasps, (21.3 ± 6.2 and 23.8 ± 8.11 grains, respectively; H = 26.522, df = 2, $P = 0.001$), the number of pollen grains deposited on a stigma after a single visit was similar for the three taxa (2-5). There was evidence for a significant and similarly positive *PP* of both honey bees and flies as avocado pollinators over wasps, given their abundance, potential for pollen transport and deposition of pollen on stigmas.

Keywords: pollinator, Hymenoptera, Diptera, pollen deposition, fruit set

INTRODUCTION

Avocado (*Persea americana*) is a tree of the family Lauraceae native to southern Mexico and Central America. In its native environment, an array of insect taxa visit avocado flowers, including bees, wasps (Hymenoptera) and flies (Diptera) (Free 1993; Castañeda-Vildózola et al. 1999; Can-Alonzo et al. 2005). However, the relative contribution of different insect taxa to avocado pollination has never been evaluated in its native Mexican range.

Avocado trees have bisexual flowers that show synchronously protogynous dichogamy (Free 1993). The flowers alternate between the female or pistillate stage I, and the male or staminate stage II on consecutive days. The flowers produce nectar in both stages (Davenport 1986). There are two types of avocado cultivars: those classified as type A have flowers in stage I in the morning of their first opening and then the same flowers re-open again in stage II of the following afternoon. Cultivars classified as type B have flowers in stage I in the afternoon of their first opening and stage II on the following morning (Lahav & Gazit 1991). Although there is still debate as to whether insects are needed for avocado pollination (Davenport 1998), it has been shown that insect visits significantly increase fruit set in most cultivars worldwide (Vithanage 1990; Ish-Am & Eisenkowitch 1998a).

In extensive avocado orchards, it is common practice to introduce honeybee colonies (*Apis mellifera*) to promote pollination. However, honey bees show erratic rates of visitation of avocado flowers and tend to reduce their visits when there are other blossoms in the surroundings, which may negatively affect fruit-set in the absence of other insect visitors (Ish-Am & Eisikowitch 1993, 1998a; Ish-Am et al. 1999). It has even been suggested that honey bees do not visit avocado flowers assiduously because of the repellent properties of the high mineral concentrations and the alcohol perseitol in its nectar (Afik et al. 2006; Afik et al. 2011). Because avocado is native to the Neotropics, some of its floral traits could make it attractive to its natural pollinators but possibly less so for honey bees that are exotic to this geographic region (Ish-Am et al. 1999; Afik et al. 2011).Yet there is surprisingly little information on the behaviour of avocado's insect visitors in Mexico and Central America, and practically no comparative measures of their efficiency as pollinators.

Mexico is the world's largest producer of avocado with ca. 40% of the total production and avocado represents a major export crop for the country (Sánchez-Colin & Arriaga

*Corresponding author; email: qeuan@uady.mx

1994). In Mexico, extensive avocado plantations are heavily sprayed and the surrounding forests have been largely logged, so the diversity and numbers of insects have been greatly reduced within plantations (Castañeda-Vildózola et al. 1999). Thus, it is important to evaluate the relative contribution of native insects to avocado pollination in order to propose appropriate and sustainable management programs for this economically important crop in Mexico. Ensuring the productivity of cash crops such as avocado is of particular importance in the tropics in order to improve the environmental performance of agriculture in these areas (Foley et al. 2011).

Measuring pollinator performance is difficult but such measures predominantly assess pollinator behaviour and estimates of stigmatic pollen deposition (Freitas & Paxton 1998; Ne'eman et al. 2010). In this study, the individual visitation activity of honey bees, flies and wasps was evaluated per unit time and their abundance on avocado flowers. The values of visitation and abundance were also combined with the quantity of pollen deposited on virgin flowers after a single visit to compare the relative performance of the different taxa for avocado pollination.

MATERIALS AND METHODS

The study was conducted in an avocado orchard in the locality of Hunucma, 89° 99' 29" N and 21° 07'59" W in the Yucatan Peninsula. The climate in this area is tropical with a range of precipitation between 800 and 1500 mm annually. This orchard held a total of 240 trees, mainly of the 'West Indian'race, planted in alternate rows of type A and type B, 12 trees per row. The age of the trees was on average 5 years. The orchard was routinely sprayed with insecticides and was surrounded by mango and citrus cultivars.

The experiments were conducted from February to April 2010 which encompass the flowering period of avocado in the Yucatan (Can-Alonzo et al. 2005). Insects were collected at flowers in three consecutive weeks and identified in the Laboratory of Entomology of the Campus of Biological Sciences in UADY. Wasp specimens belonged to the species *Brachygastra mellifica* and *Polybia occidentalis*. Flies were mainly of the species *Chrysomya megacephala* and *Ornidia obesa* and a few *Musca* sp. In the analyses we only included individuals of *B. mellifica* and *C. megacephala* as these were the most abundant species in the wasp and fly taxa, respectively.

Relative contribution of open and hand cross-pollination to avocado fruit set

The relative contribution of insects to avocado pollination was estimated by comparing open pollination and hand cross-pollination versus zero insect visitation using bagging methods (Kearns & Inouye 1993; Dafni et al. 2005).

Twenty trees of type 'A' were selected for the experiments. Three panicles with only flower buds were selected at random from each of tree and covered with cheesecloth bags before anthesis. When the flowers started to open, a group of flowers on each of the three panicles were left unbagged across the experiment and their fruit set represented the effect of open pollination. Another group of flowers was cross-pollinated with two drops of a fresh solution obtained by macerating the anthers of 20 mature type B flowers in 1 ml of distilled water. The two drops contained an average of 800 pollen grains. A third group of flowers on each panicle remained bagged during the whole experiment and represented the effect of zero insect visitation. Flowers were marked with different coloured thread depending on the treatment that they received and fruit set was recorded 2 weeks after the experiments.

Contribution of different insect taxa to avocado pollination

This part of the study was conducted in March 2010, in order to carry the experiments during the flowering peak of both, type A and B trees. The relative contribution of three insect taxa to avocado pollination was evaluated using three approaches. During 5 min we counted the number of individuals of each taxon that were present on 100 female flowers of 10 type "A" trees at 8 am, 10 am and 12 pm in two occasions separated by one week. These data were used to calculate the average abundance of each taxa per min (A). For periods of 3 min, we observed 12 individuals per taxon to determine the average number of female flowers visited per min. An estimation of flower coverage (FC) was obtained multiplying the abundance of each insect type (A) by the number of flowers visited per individual per min (NF).

Thereafter, estimates of pollinator performance in picking up and transferring avocado pollen grains to avocado stigmas were obtained. Insects of the three taxa visiting flowers in phase I were collected and their body was washed twice using the protocol of Freitas (1997). The average number of avocado pollen grains per individual (P) was obtained by counting two aliquots of the washed off grains in a haemacytometer (Dafni et al. 2005). In the case of honey bees, the rear legs were removed before washing to avoid sampling of corbicular pollen loads. The avocado pollen grain has distinctive shape, form and size and thus, is easy to differentiate from those of other plants that were found on the bodies of the three insect types (see Appendix I). An index of the number of avocado pollen grains removed and transported per taxon, PT (for 'pollen transportation'), was then calculated as $PT = A \times P$. Three specimens from each taxa were also analysed under the microscope to determine the body regions on which pollen grains were more frequent.

Finally, the number of avocado pollen grains deposited (PD) on virgin flowers of type A in stage I by the three taxa of pollinator was recorded. For this, we used flowers from previously bagged inflorescences and attached them to a wire stick on the day that they first opened. The stick was then taken close to avocado trees for an insect to visit the flower (Appendices II and III). The stigmas of the visited flowers were fixed in Fuchsin-stained glycerine jelly and the avocado pollen grains attached to the stigma surface were counted under a microscope using the protocols by Kearns and Inouye (1993).

Pollinator performance

The parameters of flower coverage by each insect taxa and the number of grains deposited in a single visit were combined to obtain an estimate of pollinator performance (*PP*) (Olsen 1997; Ne'eman et al. 2010) as follows:

$$PP = FC \times PD$$

Where:

FC = Flower coverage (a product of the abundance of a given taxon on flowers at stage I(A) and the number of flowers visited per unit time, NF); and

PD = Number of pollen grains deposited per single visit.

Statistical analyses

The abundance of each insect taxon was compared using a two-way ANOVA with tree and insect taxon as main effects on the log transformed data. In the case of numbers of flowers visited per individual, we compared insect taxa by means of a one-way ANOVA. For analysis of both abundance and number of flowers visited by insect taxa, we subsequently used Duncan's multiple comparison tests to compare taxon means.

Numbers of avocado pollen grains carried by each insect taxon were compared by means of a Kruskal-Wallis test followed by a Dunn multiple comparison test. The number of pollen grains deposited on the stigma and the individual values of pollinator performance were also compared between insect taxa by means of Kruskal-Wallis and Dunn multiple comparison tests. All statistical analyses were conducted using the SAS statistical package 9.2 (SAS Institute Inc. 2008).

RESULTS

Relative contribution of open and cross-pollination to avocado fruit set

It was found that only a small number of flowers in the bagged inflorescences set fruit (Tab. 1). On the other hand, both the open and cross-pollination treatments had similar percentages of fruit set that were higher than the treatment with permanently bagged flowers (Tab. 1). However, as hand self-pollination trials were not performed, it was not possible to differentiate the relative contribution of self and cross-pollination on fruit set.

Contribution of different insect taxa to avocado pollination

The abundance of honey bees, flies and wasps at avocado flowers (*A*) differed significantly ($F = 91.71$, d.f. = 2,78; $P < 0.001$). The most abundant insects per 100 flowers in stage I of anthesis were flies (15.2 ± 6.2), followed by honey bees (9.4 ± 6.3) and wasps (4.2 ± 3.1).

Honey bees and wasps visited a similar number of flowers per min ($NF = 8.2 \pm 3.1$ and 7.5 ± 2.6, respectively). However, flies visited on average only half the number of flowers visited by the other two taxa in the same period (Tab. 2) ($F = 17.63$; d.f. = 2,33; $P < 0.001$).

TABLE 1. Fruit set two weeks after anthesis for three pollination treatments on avocado trees of type A.

	N flowers	Formed fruit	Fruit set (%)
Open pollination	436	257	58.9
Cross-pollination	457	253	55.3
Bagged (Zero insect pollination)	312	26	8.3

Significant differences between insect taxa in the number of avocado pollen grains that each insect type carried on its body were found ($H = 26.522$, df = 2, $P = 0.001$). On average, flies carried over twice the number of avocado pollen grains on their bodies compared with honey bees or wasps (Tab. 2). Therefore, flies, due to the large number of avocado pollen grains on their bodies compared to both honey bees and wasps and their high abundance on the crop, had the largest pollen transportation value (*PT*; Tab. 2).

Pollinator performance

Flowers on sticks were quickly visited once taken close to the trees, usually in less than 5 min. Pollen deposition on the stigma after a single visit (*PD*) showed no differences between insect taxa ($H = 5.23$, df = 2, $P = 0.072$; Tab 2). There were differences regarding the areas on which avocado pollen grains were observed on the body of each insect type. The areas on which pollen grains were more abundant on the honey bee were the first two pair of legs, the ventral part of the thorax and the last abdominal segment. Flies, in contrast, carried pollen mainly in the ventral region between their thorax and abdomen and the last two abdominal segments In the case of wasps, pollen grains were present in the mandibles, the base of the first pair of legs and the ventral part of the thorax. Possibly all the body regions where pollen was found in the three insect taxa are good sites for transfer of pollen to stigmas as there were no differences in pollen deposition (Appendices I and II).

When the values of *PP* per individual were compared between the three taxa, no significant differences were detected between flies and honey bees, but wasps had lower values compared with the other two taxa ($H = 9.533$, df = 2, $P = 0.01$; Tab. 2).

DISCUSSION

In a cultivated avocado orchard in southern Mexico we found that three major insect groups were attracted in significant numbers to flowers, namely non-native bees (*A. mellifera*) and flies (*C. megacephala*) and native wasps (*B. mellifica*). Similar values were found for the pollinator performance of both, honey bees and flies; but wasps were less important for avocado pollination on this particular site, possible a result of their lower numbers.

Table 2. Pollinator behaviour and performance of non-native bees (*Apis mellifera*), flies (*Chrysomya megacephala*) and native wasps (*Brachigastra mellifica*) visiting avocado flowers.

Insect taxon	Abundance on flowers stage I per min (*A*)	N flowers of stage I visited per min (*NF*)	Flower coverage: *FC* (*A* x *NF*)	Pollen grains on body (*P*)*	Pollen grains moved by taxon: *PT* (*A* x *P*)	Pollen grains deposited by taxon (*PD*)**	Pollinator Performance (*FC* x *PD*)
Apis mellifera honey bee	9.4 ± 6.3^{B}	8.2 ± 3.1^{A}	77	21.4 ± 6.2^{B}	201	2.8 ± 2.6^{A}	218 ± 18^{A}
C. megacephala fly	15.2 ± 6.2^{A}	4.1 ± 1.2^{C}	62	44.9 ± 16.8^{A}	683	3.7 ± 3.2^{A}	229 ± 18^{A}
B. mellifica wasp	4.2 ± 3.1^{C}	7.5 ± 2.6^{B}	31	23.8 ± 8.11^{B}	100	5.0 ± 3.9^{A}	158 ± 14^{B}

*N for bees =19; flies = 24; wasps = 14; **N for bees= 14; flies = 15 and wasps= 9.
Different upper case letters within the same column signify significant differences in *post hoc* tests ($P<0.05$)

We also corroborated the idea that pollen vectors have a beneficial effect on fruit set of avocado cultivars in southern Mexico, as bagged flowers set fewer fruit. However, it is not possible to clearly establish whether it is self or cross-pollination that flower visitors undertake, as hand self-pollination trials to separate both were not performed. Yet other studies have demonstrated that autogamy in avocado flowers does not lead to fertilization (Free 1993; Ish-Am 2005) so it seems that pollen vectors in avocado are good pollinators.

In accordance to their visitation rate, behaviour and anatomical features, it has been suggested that stingless bees (Hymenoptera: Meliponini) and the honey wasp (*Brachygastra mellifica*) could be the primary and original avocado pollinators in central and western Mexico (Wysoki et al. 2002). Flies have also been reported as frequent visitors of avocado flowers and they have been considered as potential pollinators in its native range (Vischer & Sherman 1998). On avocado flowers in tropical Yucatan, we also found that the fly *C. megacephala* was highly abundant and that the most frequent wasp was *B. mellifica* which indicate that they are present across the different regions where avocado is cultivated. These latter two species along with the honey bee visited both type A and B trees (pers. obs.) and had avocado pollen grains on body areas that may effectively contact the stigma. Indeed, the body regions on which the pollen was found on the honey bees significantly touch the stigma of the avocado flower (Ish-Am & Eisikowitch 1993), and the same could be true for flies and wasps.

It is interesting that none of the studied insects was best for all the indicators of pollinator ability that we evaluated. They did not carry equal amounts of avocado pollen, and they were not equally abundant nor visited similar number of flowers per unit time. For instance, honey bees were capable of visiting a larger number of flowers compared to flies and wasps. Their advantage with respect to flies was through the higher numbers of flowers that they visited per time unit and, with respect to wasps, it was through the higher density of honey bees. However, honey bees carried significantly less pollen on their bodies compared with flies. In contrast, flies were more abundant on flowers and seemed more capable of moving larger numbers of pollen grains, but they visited less flowers per unit time. Regardless these differences, the indices of pollen deposition (*PD*) on virgin stigmas were similar for all three taxa. Yet when the values of *PD* were combined with *FC*, it was evident that the honey bees and flies were equally important for avocado pollination and more important compared with wasps. Nevertheless, when considered together, flies and wasps could have had a wider *FC* and overall importance compared with the honey bees. Such a combined effect of flies and wasps is relevant if and when honey bees are attracted to nearby blossoms other than avocado.

Unlike most bee-pollinated flowers where floral rewards are usually hidden (Kevan and Baker 1983), avocado flowers have exposed nectar and easily collected pollen (Visscher and Sherman 1998) that attract different insect taxa (Roubik 1995; Castañeda-Vildózola et al. 1999). Moreover, the morphological features of avocado flowers do not seem to fit a specific pollinator (Wysoki et al. 2002), thus, it is hardly surprising that different insects could effectively pollinate it, as our results indicate.

Our measure of pollinator performance also contradict the view that flies may have low pollinator efficiency in avocado (Ish-am et al.1999), in spite of being highly abundant in various regions of Mexico (Castañeda-Vildózola et al. 1999; Ish-am et al. 1999). Flies have been suggested to carry large numbers of avocado pollen and make contact with stamens and stigmas (Wysoki et al. 2002), a view that our results support. Thus, amongst avocado insect visitors, flies may play an important role in pollen transfer due to their very high abundance in orchards. In Australia, Vithanage (1990) found that Diptera visited avocado in large numbers. Vithanage (1990) even suggested that, in agreement to the open structure of the flower and the accessibility of the nectar to insects with lapping mouthparts such as flies, avocado could be a fly-pollinated plant in its natural environment.

The importance of Diptera as pollinators cannot be neglected. They are one of the most important groups of pollinating insects, second only to the Hymenoptera (Kearns 2001; Ssymank et al. 2008). Flies are regular visitors of at least 100 cultivated plant species including economically important crops (Heath 1982). Although flies are usually regarded as generalist flower visitors (Faegri and Van der Pijl 1979), they are present in many ecosystems and their role in pollen transfer could be considerable, mainly due to their large numbers (Ssymank et al. 2008), as our results show with Calliphoridae in Yucatecan avocado orchards.

The use of insecticides could have been an important factor that reduced the diversity of insects visiting avocado flowers in our study site. In non-commercially cultivated/sprayed orchards from other regions of Mexico, over 100 insect species have been documented visiting avocado (Ish-Am et al. 1999). We only found two species of wasp and 5 species of fly belonging to the families Calliphoridae (1 species) Syrphidae (2 species) and Tachinidae (2 species). Surprisingly, meliponine bees, which are considered good avocado pollinators (Castañeda-Vildózola et al. 1999; Can-Alonzo et al. 2005), were not seen in the orchard during our experiments possibly due to the continuous use of insecticides, too.

In the light of the erratic visits of honey bees to avocado flowers and to ensure the presence of alternative pollinators in orchards, we consider essential to reduce and carefully plan the use of insecticides especially during the blossom period. It is also important to preserve forest patches around avocado orchards that provide nesting sites and food to ensure long term populations of alternative pollinators.

In multiple pollinator systems, the relative importance of a given pollinator may change as its populations fluctuate spatially and temporarily (Thompson and Thompson 1992; Kearns 2001; Ssymank et al. 2008). Therefore, although our preliminary findings revealed a positive performance of insects other than the honey bee on avocado pollination in southern Mexico, it is still necessary to further evaluate the contribution and combined effects of various insect visitors over seasons and cultivars in order to ensure sustainable levels of fruit production of this economically important crop across Mexico.

Acknowledgements

We are thankful to the editors and anonymous reviewers for their comments and suggestions that substantially improved our manuscript. We thank the CONACYT-EU project FONCICYT (MUTUAL 94293) and CONACYT 103341 for funding. LMcK thanks DARD-NI for support. We are also very thankful to Estela Galicia, Guillermo Pinto, Claudia Castillo, Liliana Guerrero and Ligia Martin for their invaluable help during experimental set-up and field work. We are also very grateful to Sr. Horacio Gomez who let us stay and work in his orchard, even at weekends. Franklin Rocha helped during pollen counting in the lab. Pablo Manrique Saide kindly identified the fly specimens and J. González the wasps.

Appendices

Additional supporting information may be found in the online version of this article:

Appendix I. Avocado pollen grains and other types of pollen found on the body of the three insect taxa.

Appendix II. *Brachygastra mellifica* lateral (left) and ventral view (right). Two white avocado pollen grains can be seen on the base of the first left leg (white arrow).

Appendix III. *Chrysomya megacephala* (Diptera) (A); and honey bee (*Apis mellifera*) (B) visiting virgin avocado flowers on a stick.

References

Afik O, Dag A, Kerem Z, Shafir S (2006) Analyses of avocado (*Persea americana*) nectar properties and their perception by honey bees (*Apis mellifera*). Journal of Chemical Ecology 32:1949-1963.

Afik O, Delaplane KS, Moo-Valle H, Quezada-Euán JJG (2011) Nectar minerals as regulators of flower visitation in stingless bees and nectar hoarding wasps. Memorias VII Seminario Mesoamericano sobre abejas nativas, Cuetzalan México, pp. 237.

Can-Alonzo C, Quezada-Euán JJG, Xiu-Ancona P, Moo-Valle H, Valdovinos-Nunez GR, Medina-Peralta S (2005) Pollination of 'criollo' avocados (*Persea americana*) and the behaviour of associated bees in subtropical Mexico. Journal of Apicultural Research 44: 3-8.

Castañeda-Vildózola A, Equihua-Martinez A, Valdés-Carrasco J, Barrientos-Priego AF, Ish-Am G, Gazit S (1999) Insectos polinizadores del aguacatero en los estados de Mexico y Michoacan, Mexico. Revista Chapingo Serie Horticultura 5: 129-136.

Dafni A, Kevan P G, Husband B C (2005) Practical pollination biology. Enviroquest Ltd., Cambridge.

Davenport TL (1986) Avocado flowering. Horticultural Reviews 8: 257-289.

Faegri K, van der Pijl L (1979) The principles of pollination ecology. Pergamon Press, Oxford.

Foley JA, Ramankutty N, Brauman KA, Cassidy ES, Gerber JA et al. (2011) Solutions for a cultivated planet. Nature 478: 337-342.

Free JB (1993) Insect Pollination of Crops, 2nd edn. Academic Press, London.

Freitas BM (1997) Number and distribution of cashew (*Anacardium occidentale*) pollen grains on the bodies of its pollinators, *Apis mellifera* and *Centris tarsata*. Journal of Apicultural Research 36: 15-22.

Freitas BM, Paxton RJ (1998) A comparison of two pollinators: the introduced honey bee *Apis mellifera* and an indigenous bee *Centris tarsata* on cashew *Anacardium occidentale* in its native range of NE Brazil. Journal of Applied Ecology 35: 109-121.

Heath ACG (1982) Beneficial aspects of blowflies (Diptera: Calliphoridae). New Zealand Entomologist 7: 343-348.

Ish-Am G (2005) Avocado pollination - a review. New Zealand and Australia avocado grower's conference '05.Tauranga New Zealand.Session 7.

Ish-Am G, Eisikowitch D (1993) The behavior of honey bees (*Apis mellifera*) visiting avocado (*Persea americana*) flowers and their contribution to its pollination. Journal of Apicultural Research 32 : 175-186.

Ish-Am G, Eisikowitch D (1998a) Quantitative approach to avocado pollination. Proceedings Third World Avocado Congress 1998, pp. 46-51.

Ish-Am G, Eisikowitch D (1998b) Low attractiveness of avocado (*Persea americana* Mill.) flowers to honeybees (*Apis mellifera* L.) limits fruit set in Israel. The Journal of Horticultural Science and Biotechnology73: 195-204.

Ish-Am G, Barrientos Priego AF, Castaneda-Vildozola A, Gazit S (1999) Avocado (*Persea americana* Mill.) pollinators in its region of origin. Revista Chapingo Serie Horticultura 5: 137-143.

Kearns CA (2001) North American dipteran pollinators: assessing their value and conservation. Conservation Ecology 5(1):5. http://www.consecol.org/vol5/iss1/art5. (accessed July 2011).

Kearns CA, Inouye DW (1993) Techniques for pollination biologists. University Press of Colorado, Niwot.

Kevan PG, Baker HG (1983) Insects as flower visitors and pollinators. Annual Review of Entomology28: 407-453.

Lahav E, Gazit S (1991) World listing of avocado cultivars according to flowering type. Fruits 49: 299-313.

Ne'eman G, Jürgens A, Newstrom-Lloyd L, Potts SG, Dafni A (2010) A framework for comparing pollinator performance: effectiveness and efficiency. Biological Reviews 85: 435-451.

Olsen KM (1997) Pollination effectiveness and pollinator importance in a population of *Hetheroteca subaxillaris* (Asteraceae). Oecologia 109: 114-121.

Roubik DW (1995) Pollination of cultivated plants in the tropics. Agricultural Services Bulletin FAO, Rome.

SAS Institute Inc (2008) SAS/STAT® 9.2 User's Guide. Cary, NC: SAS Institute Inc.

Sánchez-Colin S, Arriaga MR (1994) Situación actual del cultivo del aguacate en México. California avocado society 1994 yearbook 78: 61-74.

Ssymank A, Kearns CA, Pape T, Thompson FC (2008) Pollinating flies (Diptera): a major contribution to plant diversity and agricultural production. Biodiversity 9: 86-89.

Thomson J, Thomson B (1992) Pollen presentation and viability schedules in animal-pollinated plants consequences for reproductive success. In:Wyatt, R (ed) Ecology and evolution of plant reproduction: new approaches. Chapman and Hall, New York, NY,pp 1-24.

Vischer PK, Sherman G (1998) Insect visitors to avocado flowers. Subtropical fruit news 6: 7-10.

Vithanage V (1990) The role of the European honeybee (*Apis mellifera* L.) in avocado pollination. The Journal of Horticultural Science 65: 81-86.

Wysoki M, van den Berg MA, Ish-Am G, Gazit S, Peña JE, Waite G K (2002) Pests and pollinators of avocado. In: Peña JE, Sharp JL, Wysoki M (eds). Tropical fruit pests and pollinators: biology, economic importance, natural enemies and control. Cabi Publishing International, NY, pp 223-293.

17

INSECT POLLINATORS BOOST THE MARKET PRICE OF CULTURALLY IMPORTANT CROPS: HOLLY, MISTLETOE AND THE SPIRIT OF CHRISTMAS

Jeff Ollerton[1,*], James R. Rouquette [1], and Tom D. Breeze[2]

[1]*Faculty of Arts, Science and Technology, University of Northampton, Avenue Campus, Northampton, NN2 6JD, UK*
[2]*Centre for Agri-Environmental Research, School of Agriculture, Policy, and Development, University of Reading, Whiteknights, PO Box 237, Reading, RG6 6AR, UK*

Abstract—Animal pollination is a vital ecological process in both natural and agricultural ecosystems. Economic valuation studies have demonstrated that pollination services may underpin a significant proportion of global crop market outputs. However these assessments are probably under-estimates because they have rarely included non-food crops, for which very few data are currently available. In particular, culturally symbolic plants have received no attention. Here we show that pollinators have considerable economic benefits to sales of European holly and mistletoe, two seasonal cultural crops that are almost wholly dependent upon insect pollinators for the production of ornamental berries. Analyses of a time series of auction records spanning 11 years indicates that wholesale prices paid for holly with berries are twice those paid for the crop without berries, whilst mistletoe with berries sells for almost three times the price of the crop lacking fruit. The benefits of this insect pollination ecosystem service to this market are therefore considerable. These findings demonstrate that pollinators can play a significant role in adding value to crops that provide resources other than food. In the future such crops should be included in assessments of regional and global value of animal pollination to increase the accuracy of assessments of the value of pollination as an ecosystem service. Our results also support arguments for continued efforts to conserve pollinator diversity and abundance in agro-ecological systems, not least for their contribution as providers of ecosystem services.

Keywords: Ilex, Viscum, cultural services, pollination value, ecosystem services, natural capital, insect pollination

INTRODUCTION

"The shepherd, now no more afraid, since custom doth the chance bestow,
Steps up to kiss the giggling maid beneath the branch of mistletoe
That 'neath each cottage beam is seen, With pearl-like berries shining gay"

John Clare *The Shepherd's Calendar* (1827)

"The general complaint of the retail purchasers this season seems to be that holly is much too dear, and that there are fewer berries than ever…"

Anon. (1872)

An estimated 87.5% of the 352,000 species of flowering plants are biotically pollinated, the majority by insects (Ollerton et al. 2011), whilst 75% of the 115 most productive crop plants, accounting for 35% of worldwide crop production, are likewise dependent to some extent upon pollinators for seed and fruit production (Klein et al. 2007). Estimates of the economic benefits of pollinators to agriculture has focussed primarily on food crops and includes global, national, and regional estimates, as well as estimates for specific crops; e.g. the additional output added by animal pollination to agriculture is estimated at €153 billion per year, almost 10% of the economic value of food crop production (Gallai et al. 2009).

Not all agricultural plant production is edible, with many species cultivated for fibres, construction materials, pharmaceuticals, and ornamentals contributing significantly to the world's economy. In the latter category are included a number of culturally significant plants. However with the exception of cotton (Pires et al. 2014) and some industrial crops such as oilseed rape, the benefits of pollination services to non-food crops has hardly been studied (Klein et al. 2007; Ollerton et al. 2011). Furthermore, no previous research has considered the impacts of pollination services on markets for culturally symbolic species.

European holly (*Ilex aquifolium* – Fig. 1A) and mistletoe (*Viscum album* – Fig. 1B) have similar distributions across most of Europe and into southern Scandinavia (though mistletoe is a naturalised introduction to Ireland – Nelson 2008). In some areas they are semi-domesticated and have a strong symbolic cultural association with both Christian and Pagan winter festivities in many north European countries (Mabey 1996; Peterken & Lloyd 1967; Zuber 2004), and other parts of the world following European colonisation (e.g. North America – see below). For example Christian folklore interprets holly wreaths as a

*Corresponding author: Jeff.Ollerton@northampton.ac.uk

FIGURE I: (A) Mistletoe and (B) holly berries growing on female plants *in situ*, and the same species packaged displayed for retail (C & D) Photos by J. Ollerton.

symbol of Christ's blood (berries) amidst his crown of thorns (the sharp leaves) while older Celtic lore describes holly wreaths as a ward against evil (Varner 2006). Mistletoe's usage is linked with pre-Christian winter solstice customs and with Norse mythology. Its use as an ancient symbol of fertility underwent a resurgence of interest in the 18th and 19th centuries and consolidated into the kissing tradition with which it is most commonly associated nowadays (Mabey 1996). As a result of these past associations, holly and mistletoe are commercially valuable non-food crops that are harvested and traded for their seasonal ornamental value. Although the traditions and folklore surrounding holly and mistletoe originate in Europe, their cultural value as ornamental plants has spread to many parts of the world, with other species of holly and mistletoe being used beyond the European range of *I. aquifolium* and *V. album*. For example, the mistletoe *Phoradendron leucarpum* shares a similar ecology to *V. album* and performs a comparable cultural and ornamental role in the USA.

Holly is a small to medium-sized multi-stemmed tree, whilst mistletoe has a hemi-parasitic lifestyle. Both species possess evergreen foliage, a significant factor in their use as home decoration in the northern winter. The flowers of holly are thought to mainly be pollinated by a range of bees (Hymenoptera – including honey bees and unmanaged species of *Bombus, Andrena* and *Osmia*) and hoverflies (Syrphidae). Mistletoe is pollinated by a number of taxa of flies (Diptera – Zuber 2004). The pollination ecology of neither species has been well studied, however, and a limited amount of wind pollination and/or parthenocarpic fruit production may be possible (Peterken & Lloyd 1967; Zuber 2004). Both species produce fleshy berries that are inedible to humans but regularly consumed by birds (Peterken & Lloyd 1967; Zuber 2004). Importantly, holly and mistletoe are dioecious species with separate male and female individuals. As such, self-pollination, a significant source of uncertainty in valuing pollination services (Melathopolous et al. 2015), is impossible, making dependence upon insect pollination likely to be near 100%. It is also highly unlikely that the benefits of pollination services could be supplanted by changes to cultivation techniques or additional inputs (Melathopolous et al. 2014).

A significant fraction of the mistletoe and holly that is produced in Britain is traded through an annual set of auctions in Tenbury Wells, Worcestershire (Sanderson & Prendergast 2002). There are many ways in which the value of ecosystem services such as animal pollination can be assessed (Breeze et al. 2016). In this study we focused only on the financial value of these culturally important crops, whilst acknowledging that their cultural value could also be assessed using different approaches. To do this we used a data set comprising 11 years of auction records to address the question of the extent to which insect pollinators add commercial value to these crops via the production of berries on female plants.

Materials and Methods

Holly and mistletoe auction reports for the period 2005 to 2015 were provided by Nick Champion, Auctioneer, of Tenbury Wells and are summarised in the electronic supplementary material. Each report consists of a summary of the general prices paid by buyers for the two crops, together with total number of lots (not split by crop and only available systematically for post-2006 auctions). In each year, three auctions were conducted in late November, and early and mid December, giving a total of 33 wholesale auctions on which to base the analyses. Records up to and including 2009 were provided as price per lb, which were then converted to price per kg to match the post-2009 records.

Each auction comprised 550 to 1450 lots of ~10 kg (holly) or 10 to 25 kg (mistletoe) each. The records include both maximum price and average price paid; in each year average and maximum prices were highly correlated ($N = 33$ auctions: $R^2 = 0.87$ for holly with berries; $R^2 = 0.73$ for mistletoe with berries – see Appendix 1) therefore only average values were used in the subsequent analyses. Data from the 3 auctions in each year were then averaged to give mean price paid per year for the two crops for each of the 11 years.

In each year, average price paid for 1st quality holly with berries was compared with that of 1st quality plain holly; 2nd quality holly was not considered as the findings were similar for this category, with a clear difference in the relative value of berried and plain 2nd quality holly (see Results and Discussion and electronic supplementary material). For mistletoe, the comparison was between average price of 1st quality and 2nd quality material; although factors such as leaf colour and woodiness are important, number of berries is a major factor determining mistletoe quality (Rachel Farmer, Nick Champion Auctions, pers. comm.). We therefore assume that 2nd quality mistletoe has few if any berries, in comparison to 1st quality material.

Results and Discussion

As the 19th century quote at the beginning of this paper demonstrates, there has long been known to be a link between the quantity of berries on decorative crops such as holly, and their market and symbolic values. Analyses of over a decade of auction records shows that average prices per kilogram for the two crops have fluctuated significantly and in parallel ($N = 11$ years, $R^2 = 0.85$). Nonetheless, auction lots of mistletoe and holly which possess berries command a price premium which is typically 1.2 to 3.7 (holly) and 1.9 to 4.3 (mistletoe) times the price for material without berries (Fig. 2).

Year-to-year variations notwithstanding, averaged over the whole eleven years, the price of holly without berries was £1.05 per kg, whilst holly with berries was double the price, selling for £2.09 per kg. Similarly mistletoe without berries sold for £0.49 per kg, and mistletoe with berries was almost three times as valuable at £1.36 per kg. These wholesale prices are comparable to or greater than the price/kg of high value fruit crops such as strawberries (£2.52/kg, class 1) and apples (£0.64/kg, class 1) over the same period (Defra 2016). Final retail prices of mistletoe and holly can be at least an order of magnitude higher, e.g. £10 to £20 per kg is not exceptional (Intermistletoe 2016; Fig. 1 C & D).

By influencing the supply of a crop with berries, variations in pollination services can also have a strong impact on inter-annual variation in prices, lowering average and premium prices in years where the berry crop is plentiful relative to demand, potentially increasing consumer welfare (Gallai et al. 2009). Indeed an auction report for 2006 notes that "The bumper year for berries has resulted in a very large supply of berried holly and mistletoe and prices were down on last year". Longer term time series data would be required to determine the precise extent of such benefits, as well as the impacts on producer profits (lower per kg profit may be compensated with higher bulk sales), however better management of these price fluctuations is likely to be beneficial for both producers and consumers. As far as we are aware differences in prices for different quality categories of these crops do not reflect different production practices (e.g. inputs of fertilisers, pesticides or managed pollinators) because holly is not commercially grown for harvest and mistletoe is a by-product of apple orchards or collected from the wild (Sanderson & Prendergast 2002). Leaf quality can be affected by herbivores and by growing conditions, but the large price differentials are due mainly to the presence of the berries.

No statistics on the annual crop of holly and mistletoe in Britain are officially collected by the Department for the Environment, Food and Rural Affairs (Defra), and it is known that a large quantity is imported from northern France (Sanderson & Prendergast 2002) and has been since at least the latter half of the 19th century (Anon. 1872). Therefore it is impossible to accurately assess the total market for these crops or to quantify the economic benefits of pollination services for the British market as a whole. However it is clear that they contribute to a small but significant element of seasonal spending: in 2014 the British public spent £623 million on Christmas decorations, averaging around £23.35 per household, which was 3% of their total Christmas spending (similar averages and proportions were recorded for 2015 - Centre for Retail Research 2016). This does not include the decorations used by shops, businesses and other commercial properties, so the true value is probably much higher.

Locally, the economic and cultural significance of holly and mistletoe can be considerably higher. Tenbury Wells has proclaimed itself to be the "mistletoe capital" of the UK and has obtained UK parliamentary endorsement for a National Mistletoe Day on 1st December each year. The town runs a series of annual cultural events based around the plant, culminating in the Tenbury Mistletoe Festival and the crowning of the Mistletoe Queen with a wreath of berried mistletoe (The Tenbury Mistletoe Association 2015).

There are on-going concerns about declines in pollinator abundance and diversity in Britian and elsewhere in the

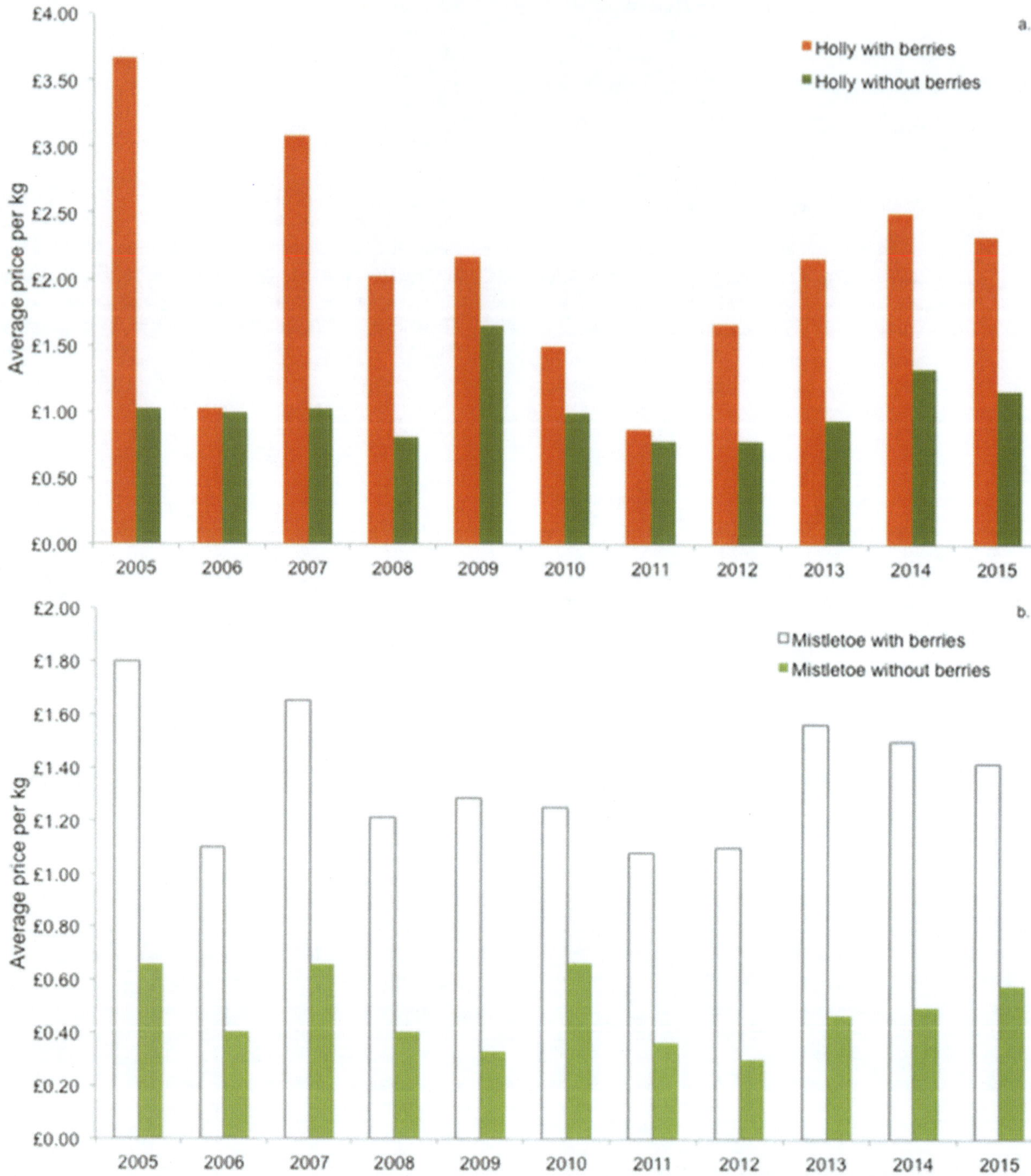

FIGURE 2: Average auction prices (£ per kg) between 2005 and 2015 of (A) holly and (B) mistletoe, with and without berries.

world in the face of habitat loss and degradation, and changes in agricultural practices (Biesmeijer et al. 2006; Potts et al. 2010; Ollerton et al. 2014). A wide range of these pollinators support agricultural outputs and their role is becoming more important as the area of animal-pollinated crops is increasing (Aizen et al. 2008) which means that a fuller understanding of the ecology and evolution of plant-pollinator interactions could play a role in securing global food supplies, e.g. through focussed selective breeding of varieties (Bailes et al. 2015). In the case of holly and mistletoe, loss of pollinators would have both ecological and economic impacts because as well as facilitating the reproduction and dispersal of these plants, the berries of both species are an important food source for birds in winter and (as we have demonstrated) add considerably to the market price of these crops.

This unique case study demonstrates the potential benefits of pollination services to markets for minority non-food crops. Furthermore, it serves to emphasise the need to examine values beyond economics alone by demonstrating direct links between cultural, symbolic value and pollination services. As such, this work highlights the requirement to consider the wide range of non-food crops that also benefit from animal pollinators, directly for crop yield and indirectly for crop reproduction. The study also emphasises the fact that there are large gaps in our knowledge of the contribution of these and other culturally significant plants to local and national economies, even for a country such as Britain where detailed agricultural statistics are widely collected and readily available. Future research on the role of pollinators in enhancing the value of such crops will require focused data collection on both their ecology and their ecological economics.

ACKNOWLEDGEMENTS

Thanks to Nick Champion and Rachel Farmer for providing the auction reports and answering questions, and two anonymous reviewers and Jane Stout for their comments which improved the final manuscript.

APPENDICES

Additional supporting information may be found in the online version of this article:

APPENDIX I. Auction prices data for holly and mistletoe (2005-2015) used in the analyses.

REFERENCES

Aizen MA, Garibaldi LA, Cunningham SA, Klein AM (2008) Long-term global trends in crop yield and production reveal no current pollination shortage but increasing pollinator dependency. Current Biology 18:1572-1575.

Anon. (1872) Christmas in Covent Garden. The Morning Post (London, England) Issue 31352, 25th December, p8.

Bailes E, Ollerton J, Pattrick J, Glover BJ (2015) How can an understanding of plant-pollinator interactions contribute to global food security? Current Opinion in Plant Biology 26:72-79.

Biesmeijer JC, Roberts SPM, Reemer M, Ohlemüller R, Edwards M, Peeters T, Schaffers AP, Potts SG, Kleukers R, Thomas CD, Settele J, Kunin WE (2006) Parallel declines in pollinators and insect-pollinated plants in Britain and the Netherlands. Science 313:351-354.

Breeze TD, Gallai N, Garibaldi LA, Li XS (2016) Economic measures of pollination services: shortcomings and future directions. Trends in Ecology and Evolution 31: 927-939.

Centre for Retail Research (2016) http://www.retailresearch.org/shoppingforxmas.php Accessed 1st November 2016.

Defra (2016) Dataset of weekly prices https://www.gov.uk/government/statistics/wholesale-fruit-and-vegetable-prices. Accessed 1st November 2016.

Gallai N, Sallesc J-M, Setteled J, Vaissière BE (2009) Economic valuation of the vulnerability of world agriculture confronted with pollinator decline. Ecological Economics 68:810–821.

Intermistletoe (2016) http://www.intermistletoe.co.uk/index.shtml. Accessed 1st November 2016.

Klein A-M, Vaissière BE, Cane JH, Steffan-Dewenter I, Cunningham SA, Kremen C, Tscharntke T (2007) Importance of pollinators in changing landscapes for world crops. Proceedings of the Royal Society series B 274:303–313.

Mabey R (1996) Flora Brittanica. Sinclair-Stevenson, London.

Melathopolous AP, Cutler GC, Tyedmers P (2015) Where is the value in valuing pollination ecosystem services to agriculture? Ecological Economics 109:59–70.

Melathopoulos AP, Tyedmers P, Cutler GC (2014) Contextualising pollination benefits: effect of insecticide and fungicide use on fruit set and weight from bee pollination in lowbush blueberry. Annals of Applied Botany 165:387-394.

Nelson EC (2008) Mistletoe (*Viscum album* L.) in Ireland: a review of records and status. The Irish Naturalists' Journal 29: 87-94.

Ollerton J, Winfree R, Tarrant S (2011) How many flowering plants are pollinated by animals? Oikos 120:321-326.

Ollerton J, Erenler H, Edwards M, Crockett R (2014) Extinctions of aculeate pollinators in Britain and the role of large-scale agricultural changes. Science 346:1360-1362.

Peterken GF, Lloyd PS (1967) Biological Flora of the British Isles: *Ilex aquifolium* L. Journal of Ecology 55:841-858.

Pires VC, Silveira FA, Sujii ER, Torezani KRS, Rodrigues WA, Albuquerque FA, Rodrigues SMM, Salomão AN, Pires CSS (2014) Importance of bee pollination for cotton production in conventional and organic farms in Brazil. Journal of Pollination Ecology 13:151-160.

Potts SG, Biesmeijer JC, Kremen C, Neumann P, Schweiger O, Kunin WE (2010) Global pollinator declines: trends, impacts and drivers. Trends in Ecology and Evolution 25:345-353.

Sanderson H, Prendergast HDV (2002) Commercial uses of wild and traditionally managed plants in England and Scotland. Unpublished report; Centre for Economic Botany, Royal Botanic Gardens, Kew, London.

The Tenbury Mistletoe Association (2015) http://www.tenburymistletoe.org/index.html. Accessed 1st November 2016.

Varner GR (2006) The Mythic Forest, the Green man and the Spirit of Nature; Algora Publishing, New York.

Zuber D (2004) Biological flora of Central Europe: *Viscum album* L. Flora 199:181–203.

18

BUMBLE BEE FORAGER ABUNDANCE ON LOWLAND HEATHS IS PREDICATED BY SPECIFIC FLORAL AVAILABILITY RATHER THAN THE PRESENCE OF HONEY BEE FORAGERS: EVIDENCE FOR FORAGE RESOURCE PARTITIONING

Elizabeth Franklin[1,2]*, Tadhg Carroll[1], Kate Rickard[1], Deborah Blake[1], and Anita Diaz[1]

[1]*Department of Life and Environmental Sciences, Faculty of Science and Technology, Bournemouth University, Poole, BH12 5BB*
[2]*School of Environmental Sciences, University of Guelph, Guelph, Ontario, Canada, N1G 2W1*

Abstract—Honey bees are being scrutinized for their potential impact upon wild bees. In lowland heath mosaics, a simple but resource rich habitat for pollinators, there is a higher probability of niche overlap for bumble bees and honey bees due to the requirement of similar resources and limited floral diversity. This study assesses i) if there is any evidence of forage competition between bumble bees and honey bees and ii) asks to what extent the number of bumble bee foragers in a lowland heath mosaic over the summer months is affected by floral resource availability in different heath types (wet/dry). Bumble bee and honey bee counts were conducted at 30 wet heath and 30 dry heath 20 m × 20 m sites, in the Poole Basin, UK. The relationships between bumble bee and honey bee forager observations and ericaceous forage availability throughout the summer were evaluated using GLMMs considering presence and abundance of honey bees and specific floral availability as factors.

Only weak correlations of honey bee forager abundance on bumble bee forager abundance were detected. Instead, the most important factors relating to bumble bee numbers were the abundance of specific floral resources within the heath type (wet/dry). Bumble bees and honey bees showed resource use consistent with resource partitioning with bumble bees predominantly using wet heaths and honey bees using dry heaths. These findings provide evidence of the importance of maintaining complex habitat mosaics within broader habitats to promote coexistence between bumble bees and honey bees.

Keywords: Bumble bees, Honey bees, Competition, Lowland Heaths, Coexistence

INTRODUCTION

The honey bee (*Apis mellifera*) has long been suspected of competing for resources with wild bees as its populations are artificially modified by humans and their impact per colony is large (See: Goulson 2003b; Paini 2004; Mallinger, Grains-Day & Gratton 2017). This is important as many wild pollinators, such as bumble bees, are already in decline, due to many factors including agricultural intensification and urbanisation (Ghazoul 2005; Vanbergan 2013; Potts et al. 2010; Dicks 2012; Montero-Castano & Vila 2012). Bumble bees have been shown to use habitats in response to different factors including food availability (McFrederick & LeBuhn 2006; Ahrne et al. 2009; Bates et al. 2011; Moron et al. 2014) nest site availability (Osborne et al. 2008; Knight et al. 2009) and competition (Goulson 2003a; Goulson 2003b; Paini 2004). Several studies suggest that the introduction of honey bees disrupts the foraging of native wild bees through nectar depletion, reduction of foraging rates and increasing competition (Schaffer et al. 1983; Kato et al. 1999; Gross 2001; Paini 2004). Conversely, other studies suggest that honey bees have little or no effect on native bees from nectar depletion (Roubik 1983; Horskins & Turner 1999) and that changes in abundance of wild bees often correspond to changes in habitat induced by human activity with simple or homogeneous habitats having greater pollinator competition (Huryn 1997; Steffan-Dwenter & Kuhn 2003; Herbertsson et al. 2016; Lindström et al. 2016). It is important to consider the impact of the honey bee on wild bees as a reduction in bee diversity can lead to reduced crop yields in agricultural areas where honey bees are not efficient pollinators (Garibaldi et al. 2013) and a reduction in wild plant biodiversity (Biesmeijer et al. 2006).

Heathlands have high floral density and contribute one third of total nectar provision in the UK (not including Northern Ireland) (Baude et al. 2016). In the UK, heathlands are dominated by three ericaceous species (also called "ericoids"), *Calluna vulgaris, Erica tetralix* and *Erica cinerea* and as such are comparatively floristically simple (Rodwell 1998). Heathlands consist of a mosaic of different habitat types including dry, wet and mire heath habitats (JNCC 1990). In addition to this, the abundance of available forage in heathlands changes dynamically throughout the summer,

*Corresponding author: Elizabeth.franklin@uoguelph.ca

starting before May with (*Ulex europaeus*) then moving through the ericoids, *Erica cinerea, Erica tetralix* and *Erica ciliaris,* smaller species of *Ulex* and finishing with *Calluna vulgaris* into September (Forup & Memmot 2005). Ericoids are among the top four largest contributors to national nectar production in the UK, with 50% of the nation's nectar production coming from *Calluna vulgaris* and *Erica cinerea* along with the non-ericaceous *Cirsium palustre* and *Trifolium repens* (Baude et al. 2016). *Calluna vulgaris* has a predicted nectar yield of 147.5 Kg/ha/yr, *Erica cinerea* 160 Kg/ha/yr and *Erica tetralix* 160 Kg/ha/yr potentially making all three very valuable to pollinators (Baude et al. 2016). The long resource availability and presence of rich resources makes heathlands potentially very important to bumble bees. Heathlands also provide a highly economically valuable resource for honey bee keepers around Europe (Descamps et al. 2015). However, potential for niche overlap and competition on heathlands between honey bees and wild bees may be enhanced by the limited floral diversity available at the beginning and the end of the summer.

Bumble bees, especially short tongued bumble bees, have a high likelihood of niche overlap and foraging competition with honey bees due to having similar tongue lengths which dictate the resources that the bees can access (Hawkins 1969; Goulson et al. 2008). This study addresses if there is any evidence of foraging competition between bumble bees and honey bees for forage on lowland heath and explores the possibility of floral resource partitioning by monitoring the abundance of bumble bee foragers on lowland heath mosaics over the summer months in response to i) honey bee forager abundance and ii) floral resource availability in different heath types (wet/dry). If honey bees are competing with bumble bees for forage we expect to see a strong negative correlation between bumble bees and honey bees that cannot be explained by specific floral resource availability.

MATERIALS AND METHODS

Sixty heathland sites were surveyed 17th-23rd July, 17th-26th August and 8th- 19th September 2015 across the Poole Basin, Dorset, UK (FigI, for Grid reference see SI). Site locations were selected at random within heathland areas with the restriction that each site consisted of a 20 × 20 m quadrat patch of homogeneous vegetation that was classified either as dry or wet heath (JNCC 1990). Sites were on average separated by 4494m (with 95% confidence limits of 419-8231m and an absolute minimum of 30m) and showed no spatial autocorrelation found with a Moran's I test for honey bee abundance (Z = -0.005, P = 0,385) or bumble bee abundance (Z= -0.010, P= 0.729). Thirty dry heaths were classified as having at least 25% ericaceous or *Ulex* cover with the presence of *Calluna vulgaris, Erica cinerea, Ulex gallii* and or *Ulex minor* (D1 or E1.8 Phase I habitat survey) (JNCC 1990). Thirty wet heaths were classified as having at least 25% ericaceous cover with the presence of *Molinia caerulea,* some *Sphagnum* and *Erica tetralix* (D2 or E1.7 Phase I habitat survey) (JNCC 1990).

To calculate the floral abundance of ericoid species in terms of actual available percentage cover of ericoid floral resource, five 2 × 2 m quadrats were sampled haphazardly (but ensuring no overlap) within the 20 × 20 m quadrat in July, August and September. A record was made of the percentage cover of each ericoid species in full flower and percentage cover of each heather species in partial flower. Actual available % cover of floral resource was calculated as % cover of heathers in full flower + 0.5 (% cover of heathers in partial flower). The foraging interactions between pollinators and ericaceous floral resources were recorded each month on a 10 minute roughly 'W' shaped walk (to reduce resampling) within the 20 × 20 m quadrat to evaluate foraging preferences and site use. All foraging interactions were all observed under good conditions for pollinators, warm (>15°C) and with low wind (Beaufort scale score 0-3). Pollinators were initially identified as *B. terrestris/lucorum* group, *B. lapidarius, B. pratorum, B. pascuorum* or *Apis mellifera.* Subsequently, due to some low observation counts, the *Bombus* species were considered together as one group. Most were *B. terrestris/lucorum* which together with *B. lapidarius* and *B. pratorum* have similar, relatively short, tongue lengths (Goulson et al. 2008). The remaining species, *B. pascuorum* has a medium tongue length (Goulson et al. 2008) but occurred very infrequently in our dataset.

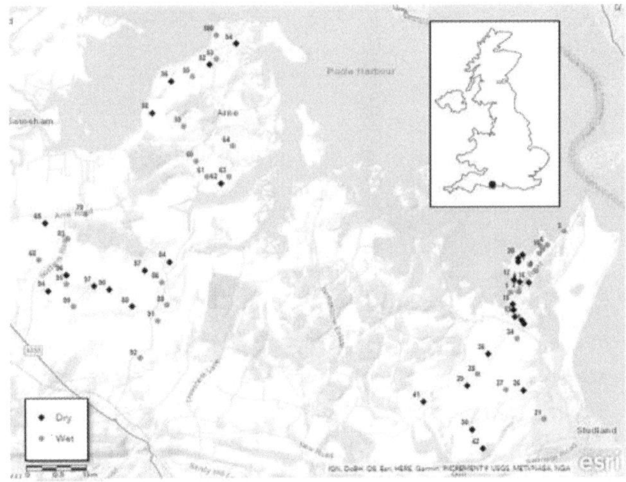

FIGURE I. Locations of wet and dry heath field sites (grey circles wet sites and black diamonds dry sites) with an inset map identifying the location of the sites within the UK (Grid references in SI.)

Analysis

To assess the relationships between bumble bee presence/abundance, specific floral availability and the presence/abundance of honey bees a series of generalised linear mixed models were conducted within the glmmTMB (Magnussun et al. 2018) package in R (2017).

Relationships with the presence of bumble bees were tested with a binomial glmm model with month and site as random intercept factors and the presence of honey bees and the % floral cover of *C. vulgaris, E. cinerea, E. tetralix* and *E. cilliaris* as factors. Relationships with the presence of honey bees were also tested with a binomial glmm model with month and site as random intercept factors and the presence of bumble bees and the % floral cover of *C. vulgaris, E. cinerea, E. tetralix* and *E. cilliaris* as factors. Initial models contained all interactions but these were removed due to their limited effect (S2).

Relationships with the abundance of bumble bees in all sites and those shared with honey bees were tested with a zero inflated, Poisson, glmm model with month and site as random intercept factors and the abundance of honey bees and the % floral cover of *C. vulgaris, E. cinerea, E. tetralix* and *E. cilliaris* as factors. Relationships with the abundance of honey bees were also tested with a zero inflated, Poisson, glmm model with month and site as random intercept factors and the abundance of bumble bees and the % floral cover of *C. vulgaris, E. cinerea, E. tetralix* and *E. cilliaris* as factors. Initial models contained all interactions but these were removed due to their limited effect (S3 and S4).

Wet heaths are characterized by the presence of *E. tetralix* and dry heaths with the presence of *C. vulgaris* and *E. cinerea.* As such heath type was co-linear with these relative abundances and not included in the models.

RESULTS

The presence of bumble bees at a site was only significant positively correlated with the percentage floral cover of *E. cinerea.* Honey bee presence was only significantly correlated with the percentage floral cover of *C. vulgaris* (Table I, Fig 2A).

Bumble bee abundance was positively and significantly correlated with all heather percentage floral cover. Bumble bee abundance showed a significant but weakly negative correlation with honey bee abundance (Table 2, Fig 2A). Honey bee abundance was positively and significantly correlated with the percentage floral cover of *C. vulgaris* and *E. cinereal* and significantly and weakly negatively correlated with bumble bee abundance (Table 2. Fig 2A, Fig 3).

When both bumble bees and honey bees are present at a site, bumble bee abundance was only positively correlated with the percentage cover of *E. cinerea* and *E. tetralix.* Honey bee abundance was significantly and positively correlated with the percentage cover of C. *vulgaris* and *E. cinerea.* and significantly and weakly negatively correlated with bumble bee abundance (Table 3, Fig 2B).

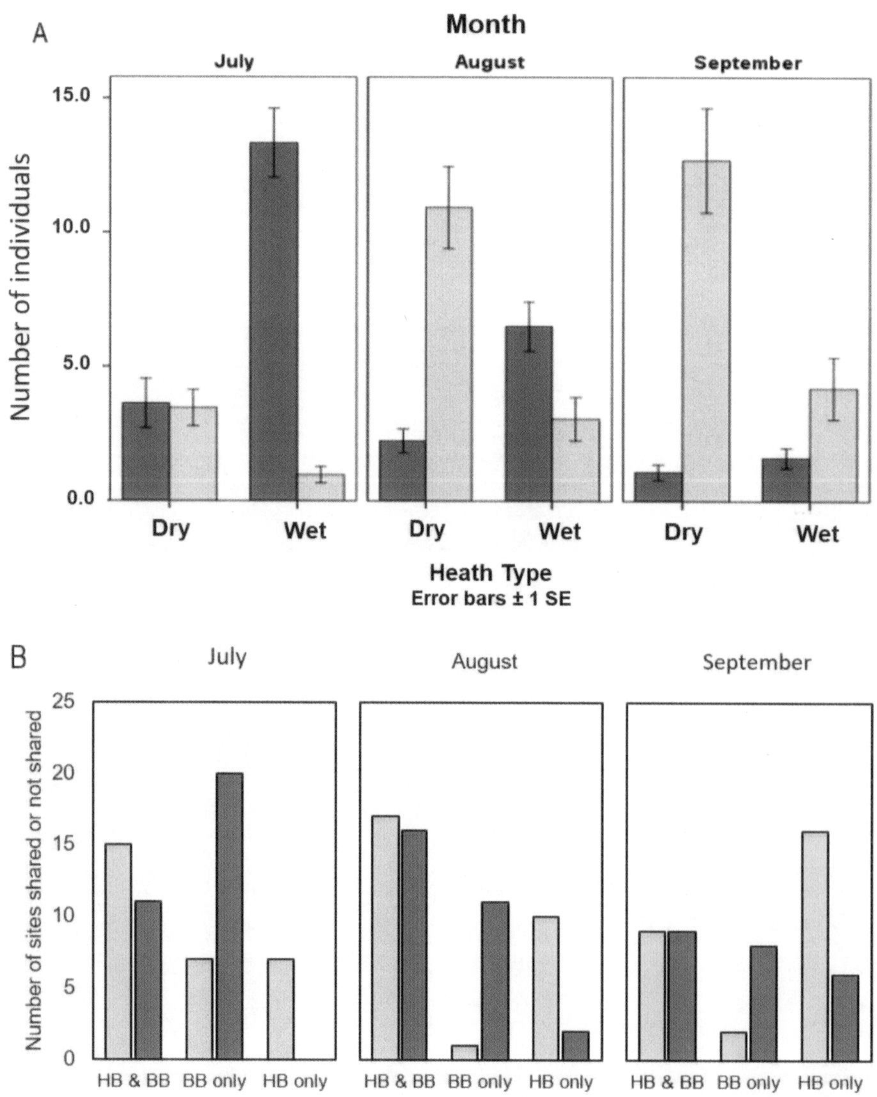

Figure 2. (A) The mean number of bumble bee (dark grey) and honey bee (light grey) observations on wet and dry heaths in July, August and September. Wet heath sites are dominated by *E. tetralix* and dry sites by *C.* vulgaris (Fig. 3). Whiskers display ±1SE.

(B) The number of sites shared by honey bees and bumble bees (HB & BB) and those with only bumble bees (BB only) or only honey bees (HB only) out of a total of 30 sites each month. Dark grey bars represent wet heath sites and dry heaths are light grey.

TABLE 1. (A) The presence of bumble bees in response to specific heather availability and presence of honey bees. (B) The presence of honey bees in response to specific heather availability and presence of bumble bees. Significance is marked with * at $P \leq 0.05$, ** at $P \leq 0.01$ and *** at $P \leq 0.001$.

Factor	Estimate (log)	Standard error	Z- Value	P
(A) bumble bee presence $N = 180$				
% floral cover *C. vulgaris*	-0.013	0.011	-1.185	0.236
% floral cover *E. cinerea*	0.130	0.073	1.782	0.0747
% floral cover *E. tetralix*	0.176	0.070	2.500	0.0124*
% floral cover *E. cilliaris*	0.086	1.185	0.464	0.6427
Presence of honey bees	0.0152	0.5944	0.026	0.9796
(B) honey bee presence $N = 180$				
% floral cover *C. vulgaris*	0.064	0.022	2.882	0.004**
% floral cover *E. cinerea*	0.022	0.041	0.530	0.596
% floral cover *E. tetralix*	-0.024	0.019	-1.214	0.225
% floral cover *E. cilliaris*	-0.472	0.272	-1.736	0.083
Presence of bumble bees	0.059	0.693	0.086	0.932

TABLE 2. (A) The abundance of bumble bees in response to specific heather % cover and abundance of honey bees. (B) The abundance of honey bees in response to specific heather % cover and abundance of bumble bees. Significance is marked with * at $P \leq 0.05$, ** at $P \leq 0.01$ and *** at $P \leq 0.001$.

Factor	Estimate (log)	Standard error	Z- Value	P
(A) bumble bee abundance $N = 180$				
% floral cover *C. vulgaris*	0.0126	0.005	2.620	0.009**
% floral cover *E. cinerea*	0.027	0.006	4.507	<0.001***
% floral cover *E. tetralix*	0.223	0.003	6.881	<0.001***
% floral cover *E. cilliaris*	0.040	0.010	3.935	<0.001***
Abundance of honey bees	-0.062	0.014	-4.375	<0.001***
(B) honey bee abundance $N = 180$				
% floral cover *C. vulgaris*	0.020	0.003	6.809	<0.001***
% floral cover *E. cinerea*	0.032	0.009	3.558	<0.001***
% floral cover *E. tetralix*	0.006	0.006	1.031	0.303
% floral cover *E. cilliaris*	-0.420	0.132	-3.190	0.001
Abundance of bumble bees	-0.103	0.021	-5.025	<0.001***

TABLE 3. (A) The abundance of bumble bees in response to specific heather % cover and abundance of honey bees when sites are shared. (B) The abundance of honey bees in response to specific heather % cover and abundance of bumble bees when sites are shared. Significance is marked with * at $P \leq 0.05$, ** at $P \leq 0.01$ and *** at $P \leq 0.001$.

Factor	Estimate (log)	Standard error	Z- Value	P
(A) bumble bee abundance – when sites are shared $N = 76$				
% floral cover *C. vulgaris*	0.006	0.004	1.46	0.143
% floral cover *E. cinerea*	0.21	0.001	2.418	0.016*
% floral cover *E. tetralix*	0.025	0.004	6.034	<0.001***
% floral cover *E. cilliaris*	0.077	0.074	1.039	0.299
Abundance of honey bees	-0.020	0.015	-1.401	0.161
(B) honey bee abundance– when sites are shared $N = 76$				
% floral cover *C. vulgaris*	0.019	0.003	5.461	<0.001***
% floral cover *E. cinerea*	0.026	0.010	2.731	0.006**
% floral cover *E. tetralix*	0.002	0.006	0.370	0.711
% floral cover *E. cilliaris*	-0.285	0.183	-7.557	0.119
Abundance of bumble bees	-0.057	0.024	-2.397	0.017*

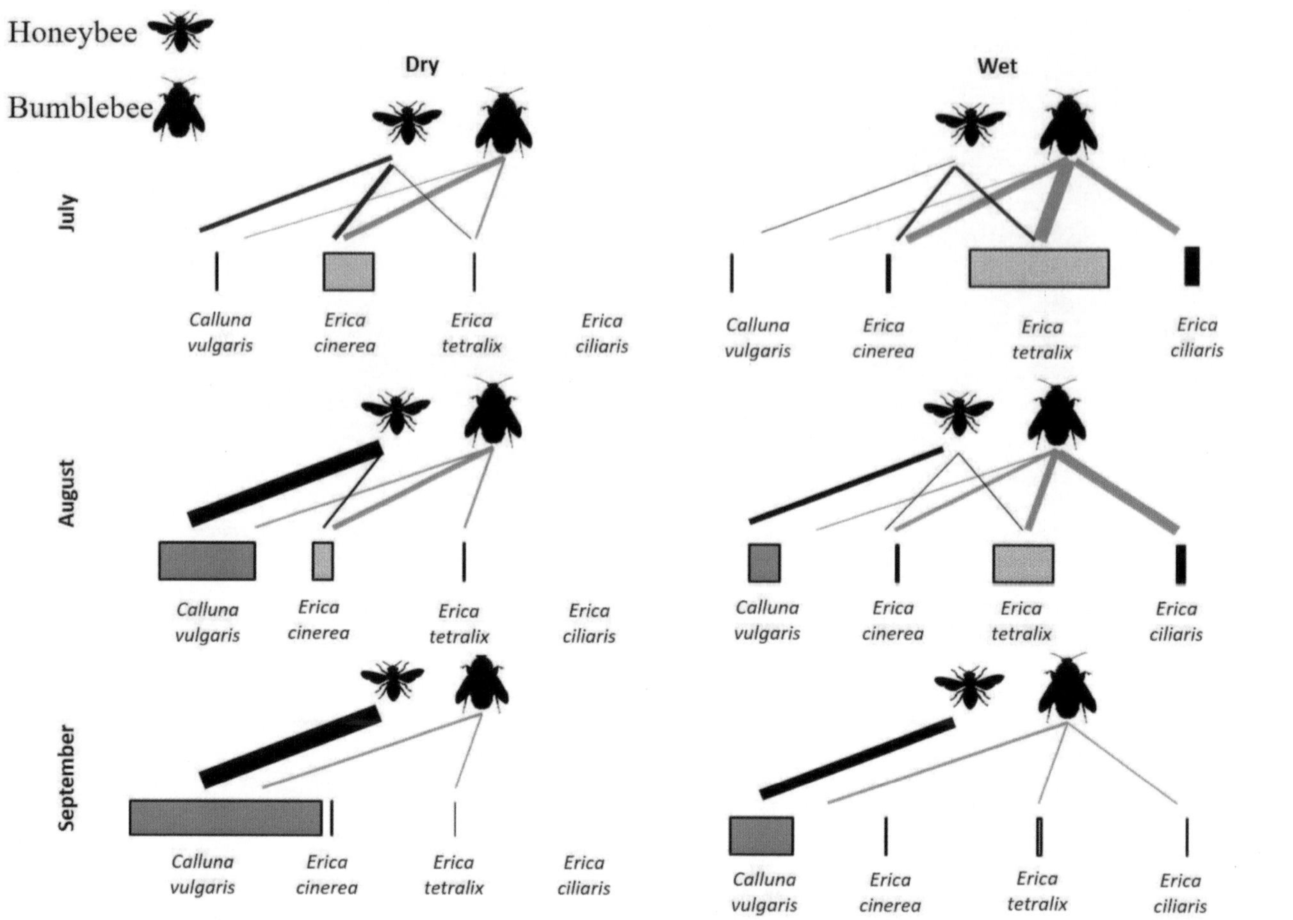

FIGURE 3. Relative mean forage availability (% cover) of each heather species on dry and wet heath types and mean forager observations on each heather species. Boxes are to the same scale with 1 unit representing 1% floral cover. Lines are to the same scale with each unit representing forager observation honey bees in black and bumble bees in grey.

DISCUSSION

The effect of honey bee forager abundance and any evidence of honey bee competition

This study detected no evidence of forager exclusion between bumble bees and honey bees for forage on lowland heath and demonstrates that where bumble bees and honey bees coexist in lowland heaths, their forager distribution is not highly related to each other's presence but the presence of the specific available forage. This indicates that honey bees are only likely to greatly impact bumble bee forager numbers when they are in very high density, such as close to a honey bee hive (Klemens & Volkmar 2006). This study also put honey bees 'in the spot light' and found that their abundance is also correlated with specific forage plants, mainly *C. vulgaris* and *E. cinerea,* characteristic of dry heaths. They also showed a weak but significant negative correlation with an increase bumble bee abundance even in shared sites but showed no evidence of being excluded by bumble bees. Although, the findings from this study suggest that it is unlikely that there is competition between bumble bees and honey bees for forage on lowland heath sites we cannot conclusively exclude any effect of competition (in either direction) as we have not considered colony fitness in our measures.

The bumble bee foragers in this study did not change their preferred forage plants in response to the presence of honey bee foragers as we found no significant interactions between heather availability and presence of honey bees. Instead bumble bees consistently focus their foraging on the *Erica* species regardless of number of honey bees. This contrasts with the findings of other studies where the presence of honey bees has resulted in a foraging pattern shift of the wild bees (Klemens & Volkmar 2006; Ishii et al. 2008; Magrach et al. 2007). It is probable that on lowland heath mosaics, where the forage heterogeneity exists over longer time periods, that stable resource portioning has the potential to exist.

It is in relatively homogeneous habitats, such as agricultural land, that negative impacts of honey bees on bumble bees have predominantly been reported (Lindström et al. 2016; Herbertsson et al. 2016). In such simple habitats there is limited alternative forage on which bumble bees can escape the competition with honey bees and where heterogeneity exists it is short lived with the presence of mass flowering crops. Our findings concur with those found in a study on woodlands, which also found that when heterogeneity of resources were available honey bees and wild bees demonstrated resource partitioning (Magrach et al. 2017). These authors found that competition and diet switches occurred when the crop the honey bees were feeding on stopped flowering and the honey bees moved into semi natural areas to find forage (Magrach et al. 2007). This study also supports the findings of Herbertsson et al. (2016) working in open grassland and road verge landscapes, who conclude that lower bumble bee forager abundance was correlated with honey bee abundance only where habitats were homogeneous; when habitats were heterogeneous, no negative effect of honey bee presence was detected. This study additionally provides an explanation for previously detected negative associations between bumble bees and honey bees (Forup & Memmot 2005). Forup and Memmot's (2015) study focused on dry heaths and suggests that their detected negative relationship between honey bees and bumble bees may be a result of honey bees moving into areas not suitable for the bumble bees. The present study confirms that hypothesis with a greater number of bumble bee forager being associated with *Erica* species in wet areas and honey bee foragers on *Calluna* in dry. This highlights the importance of looking beyond a single habitat when considering inter species competition for both potential habitat partitioning, as seen in this study, or in the breakdown of resource partitioning as seen in Magrach et al. (2017).

Lowland heath flowering phenology also correlates with the peak abundances of bumble bees and honey bees (Forup & Memmot 2005). Species in the genus *Erica* flower earlier and are the preferential forage of bumble bees, where *Calluna vulgaris* flowers later in the season and is the preferential forage for honey bees. In this way the potential for competition between these bees on lowland heaths is limited further.

Evidence for resource partitioning and implications for management

This study has demonstrated the crucial role of habitat heterogeneity though the availability of both wet and dry heaths for enabling bumble bees and honey bees to coexist through potentially portioning their use of floral resources. However, to support a diversity of bees within the landscape, connectivity between these spatially separated heath types needs to be maintained within foraging distances. For bumble bees this is likely ~ 1000 m (Walther-Hellwig & Frankl 2000; Knight et al. 2005) and other wild bees this is likely ~ 250 m (Gathmann & Tscharntke 2002). The temporal availability of resource heterogeneity is also important to avoid times of intense competition when resources are limited (Magrach et al. 2007). In agricultural systems where there is a fluctuating availability of mass flowering resources, the creation of floral resources with long flowering phenology may provide the long-term heterogeneity required to support stable resource partitioning as well as pollinator diversity (Albrecht et al. 2007). Habitat heterogeneity is also important for the other habitat requirements of pollinations such as nest site availability and overwintering sites (Osborne et al. 2008; Knight et al. 2009) that were not addressed in this study. In terms of forage, the results show that wet heaths provide better forage for bumble bees and findings from Moquet et al. (2015) suggest that wet heaths provide important early season (April and May) forage for bumble bees. However, wet heaths are likely to become saturated by water and be unfavourable for bumble bee nests of any of the subterranean nesting bumble bees (*Bombus terrestris, B. lucorum, B. lapidarius* and *B. pratorum*) or the ground surface nesting species like *B. pascuorum* and *B. jonellus* (Prys-Jones & Corbet 1987) and consequently dry heaths may be needed for nesting areas. In addition, if there were only wet heaths the bumble bees come into more competition with honey bees for resources as the favoured honey bee forage plants (*Calluna vulgaris* and *Erica cinera*) would not be as abundant on wet heaths. Management to support the value of heathland as both a wildlife habitat and an economic resource for beekeepers, should therefore aim to maintain connectivity and floral diversity over the duration of the season.

Conclusions

This study found that bumble bee forager abundance is highly related to the floral availability of their favoured *Erica* forage on heathlands rather than honey bee forager abundance which indicates that bumble bees and honey bees may limit competition by resource partitioning. This study highlights that it is important to i) assess habitats for their within habitat heterogeneity when considering inter species interactions to detect resource partitioning ii) consider both sides of a potential competitive relationship and iii) maintain high habitat heterogeneity and forage diversity throughout the season in order to maintain coexistence between species with high potential niche overlap.

Author Contributions: Dr Franklin and Dr Diaz conceived the ideas, conducted analyses and designed methodology along with Tadhg Carroll, Kate Rickard and Deborah Blake collected and collated the data; Dr Franklin and Dr Diaz led the writing of the manuscript. All authors contributed critically to the drafts and gave final approval for publication.

ACKNOWLEDGEMENTS

We would like to thank the National Trust and the RSPB for the permission to survey on their land. We are particularly grateful for the specific support of David Brown and Michelle Brown. We would also like to thank the following Bournemouth University 2015 Purbeck Student Environment Research Team (SERT) members and supporting research staff for their assistance with mapping, site setup and floral surveys: Dr Andy Ford, Dr Pippa Gillingham, Gitte Kragh, Tom Dando, Suzanne Gibbon, Grace Burger, Jackie Jobes, Belinda George, Kinzie Watts, Emma-Louise Gill, Sam Hodgkinson, Cara Doyle, Michaela Pozzetti, Jessica Wiffen, Ollie Roberts, Frances Hale, Ash Pyke, Natalie Hewitt, Laura Hart, Glenn Saxby, Scarlett Robinson, Honey Hollingsworth, Alexandra Boliver-Brown, Hamish Lillywhite and Rebecca Gane.

APPENDICES

Additional supporting information may be found in the online version of this article:

APPENDIX I. Grid references of sample site locations, information on glmm models.

APPENDIX II. Full observational data set.

REFERENCES

Albrecht M, Duelli P, Müller C, Kleijn D, Schmid B (2007) The Swiss agri-environment scheme enhances pollinator diversity and plant reproductive success in nearby intensively managed farmland. Journal of Applied Ecology 44:813-22.

Ahrne K, Bengtsson J, Elmqvist T (2009) Bumble bees (*Bombus* spp) along a gradient of increasing urbanization. PLoS One 15:4:e5574. doi.org/10.1371/journal.pone.0005574.

Bates AJ, Sadler JP, Fairbrass AJ, Falk SJ, Hale JD, Matthews TJ (2011) Changing bee and hoverfly pollinator assemblages along an urban-rural gradient. PloS one 12:6:e23459. doi.org/10.1371/journal.pone.0023459.

Baude M, Kunin WE, Boatman ND, Conyers S, Davies N, Gillespie MA, Morton RD, Smart SM Memmott J (2016) Historical nectar assessment reveals the fall and rise of floral resources in Britain. Nature 530:85-88. doi.org/10.1038/nature16532.

Biesmeijer JC, Roberts SP, Reemer M, Ohlemüller R, Edwards M, Peeters T, Schaffers AP, Potts SG, Kleukers R, Thomas CD, Settele J (2006). Parallel declines in pollinators and insect-pollinated plants in Britain and the Netherlands. Science 313:351-354.

Descamps C, Moquet L, Migon M, Jacquemart A-L. (2015) Diversity of the Insect Visitors on *Calluna vulgaris* (Ericaceae) in Southern France Heathlands. Journal of Insect Science 15:130. doi.org/10.1093/jisesa/iev116.

Dicks LV, Abrahams A, Atkinson J, Biesmeijer J, Bourn N, Brown C, Brown MJ, Carvell C, Connolly C, Cresswell JE, Croft P (2013) Identifying key knowledge needs for evidence-based conservation of wild insect pollinators: a collaborative cross-sectoral exercise. Insect Conservation and Diversity 6:435-46. doi.org/10.1111/j.1752-4598.2012.00221.x.

Forup ML, Memmott J (2005) The relationship between the abundances of bumble bees and honey bees in a native habitat. Ecological Entomology 30:47-57. doi.org/10.1111/j.0307-6946.2005.00660.x.

Gathmann A, Tscharntke T (2002) Foraging ranges of solitary bees. Journal of Animal Ecology 71:757-764. doi.org/10.1046/j.1365-2656.2002.00641.x.

Ghazoul J (2005) Buzziness as usual? Questioning the global pollination crisis. Trends in Ecology and Evolution 31:367-373. doi.org/10.1016/j.tree.2005.04.026.

Goulson D (2003a) Bumble bees: Their Behaviour and Ecology. Oxford University Press, Oxford, UK.

Goulson D (2003b) Effects of introduced bees on native ecosystems. Annual Review of Ecology Evolution and Systematics. 34:1-26. doi.org/10.1146/annurev.ecolsys.34.011802.132355.

Goulson D, Lye GC, Darvill B (2008) Diet breadth, coexistence and rarity in bumble bees. Biodiversity Conservation 17:3269-3288. doi.org/10.1007/s10531-008-9428-y.

Gross CL (2001) The effect of introduced honey bees on native bee visitation and fruit-set in Dillwynia juniperina (Fabaceae) in a fragmented ecosystem. Biodiversity Conservation 102:89-95. doi.org/10.1016/S0006-3207(01)00088-X.

Hawkins RP (1969) Length of tongue in a honey bee in relation to the pollination of red clover. Journal of Agricultural Science 73:489-493. doi.org/10.1017/S0021859600020128.

Herbertsson L, Lindström SA, Rundlöf M, Bommarco R, Smith HG (2016) Competition between managed honey bees and wild bumble bees depends on landscape context. Basic and Applied Ecology 17:609-616. doi.org/10.1016/j.baae.2016.05.001.

Horskins K, Turner VB (1999) Resource use and foraging patterns of honey bees, *Apis mellifera*, and native insects on flowers of Eucalyptus costata. Austral Ecology 24:221-227. doi.org/10.1046/j.1442-9993.1999.00965.x

Huryn VM (1997) Ecological impacts of introduced honey bees. The Quarterly Review of Biology 72:275-297. doi.org/10.1086/419860.

Ishii HS, Kadoya T, Kikuchi R, Suda SI, Washitani I (2008) Habitat and flower resource partitioning by an exotic and three native bumble bees in central Hokkaido, Japan. Biological Conservation 141:2597-607. doi.org/10.1016/j.biocon.2008.07.029.

JNCC (1990) Unit EF. Handbook for Phase I habitat survey a technique for environmental audit

Kato M, Shibata A, Yasui T, Nagamasu H (1999) Impact of introduced honey bees, *Apis mellifera*, upon native bee communities in the Bonin (Ogasawara) Islands. Researches on Population Ecology 41:217-228. doi.org/10.1007%2Fs101440050025.

Klemens E, Volkmar W (2006) Increased density of honey bee colonies affects foraging bumble bees. Apidologie 37:517-532. doi.org/10.1051/apido:2006035.

Knight ME, Martin AP, Bishop S, Osborne JL, Hale RJ, Sanderson RA, Goulson D (2005) An interspecific comparison of foraging range and nest density of four bumble bee (*Bombus*) species. Molecular Ecology 14:1811-1820. doi.org/10.1111/j.1365-294X.2005.02540.x.

Knight ME, Osborne JL, Sanderson RA, Hale RJ, Martin AP, Goulson D (2009) Bumble bee nest density and the scale of available forage in arable landscapes. Insect Conservation and Diversity 2:116-124. doi.org/10.1111/j.1752-4598.2009.00049.x.

Lindström SA, Herbertsson L, Rundlöf M, Bommarco R, Smith HG (2016) Experimental evidence that honey bees depress wild insect densities in a flowering crop. Proceedings of the Royal Society B 283:20161641.

Magrach A, González-Varo JP, Boiffier M, Vilà M, Bartomeus I (2017) Honey bee spillover reshuffles pollinator diets and affects plant reproductive success. Nature Ecology & Evolution 1:1299–1307. doi:10.1038/s41559-017-0249-9.

McFrederick QS, LeBuhn G (2006) Are urban parks refuges for bumble bees *Bombus* spp. (Hymenoptera: Apidae)? Biological Conservation 129:372-382. doi.org/10.1016/j.biocon.2005.11.004

Magnussun et al. (2018) glmmTMB: Generalized Linear Mixed Models using Template Model Builder. [online] URL https://CRAN.R-project.org/package=glmmTMB

Mallinger RE, Gaines-Day HR, Gratton C. (2017) Do managed bees have negative effects on wild bees? A systematic review of the literature. PloS one 12:e0189268.

Montero-Castaño A, Vila M (2012) Impact of landscape alteration and invasions on pollinators: a meta-analysis. Journal of Ecology 100:884-893. doi.org/10.1111/j.1365-2745.2012.01968.x.

Moquet L, Mayer C, Michez D, Wathelet B, Jacquemart AL (2015) Early spring floral foraging resources for pollinators in wet heathlands in Belgium. Journal of Insect Conservation 19:837-848. doi.org/10.1007/s10841-015-9802-5.

Moron D, Skorka P, Lenda M, Rozej-Pabijan E, Wantuch M, Kajzer-Bonk J, Waldemar C. et al. (2014) Railway Embankments as New Habitat for Pollinators in an Agricultural Landscape. PloS one 9:7: e101297. doi.org/10.1371/journal.pone.0101297.

Osborne JL, Martin AP, Shortall CR, Todd AD, Goulson D, Knight ME, Hale RJ, Sanderson RA (2008) Quantifying and comparing bumble bee nest densities in gardens and countryside habitats. Journal of Applied Ecology 45:784-792. doi.org/10.1111/j.1365-2664.2007.01359.x.

Paini DR (2004) Impact of the introduced honey bee (*Apis mellifera*)(Hymenoptera: Apidae) on native bees: a review. Austral Ecology 29:399-407. doi.org/10.1111/j.1442-9993.2004.01376.x.

Potts SG, Biesmeijer JC, Kremen C, Neumann P, Schweiger O, Kunin WE (2010) Global pollinator declines: trends, impacts and drivers. Trends in Ecology and Evolution 25:345-353. doi.org/10.1016/j.tree.2010.01.007.

Prys-Jones OE, Corbet SA (1987) Bumble bees. Cambridge University Press.

R Core Team (2017). R: A language and environment for statistical computing. R Foundation for Statistical Computing, Vienna, Austria. [online] URL https://www.R-project.org/.

Rodwell JS (1998) British Plant Communities: Volume 2, Mires and Heaths. Vol. 2. Cambridge University Press.

Roubik DW (1983) Experimental Community Studies: Time-Series Tests of Competition between African and Neotropical Bees. Ecology 64:971-978. doi.org/10.2307/1937803.

Schaffer WM, Zeh DW, Buchmann SL, Kleinhans S, Schaffer MV, Antrim J (1983) Competition for nectar between introduced honey bees and native North American bees and ants. Ecology 64:564-577. doi.org/10.2307/1939976.

Steffan-Dewenter I, Tscharntke T (2000) Resource overlap and possible competition between honey bees and wild bees in central Europe. Oecologia 122:288-296. doi.org/10.1007/s004420050034.

Steffan-Dewenter I, Kuhn A (2003) Honey bee foraging in differentially structured landscapes. Proceedings of the Royal Society London B 270:569-575. doi.org/10.1098/rspb.2002.2292.

Vanbergen AJ (2013) Threats to an ecosystem service: pressures on pollinators. Frontiers in Ecology and the Environment 11:251-259. doi.org/10.1890/120126.

Walther-Hellwig K, Frankl R. (2000) Foraging habitats and foraging distances of bumble bees, Bombus spp. (Hym., Apidae), in an agricultural landscape. Journal of Applied Entomology 124:299-306. Doi.org/10.1046/j.1439-0418.2000.00484.x.

19

VISITATION FREQUENCIES OF BUMBLEBEES AND SWALLOWTAIL BUTTERFLIES TO FLOWERS AND THE NECTAR SUGAR CONCENTRATION OF *RHODODENDRON KAEMPFERI* AND *R. JAPONICUM* IN MOUNTAINS OF CENTRAL JAPAN

Keigo Takahashi[1] and Takao Itino*[2,3]

[1]*Graduate School of Science and Technology, Shinshu University, Matsumoto, Nagano 390-8621, Japan*
[2]*Department of Biology, Faculty of Science, Shinshu University, Matsumoto, Nagano 390-8621, Japan*
[3]*Institute of Mountain Science, Shinshu University, Matsumoto, Nagano 390-8621, Japan*

Abstract—Relationships between flower visitor composition and floral traits help us to understand floral evolution. Swallowtail butterflies have been assumed to be the main pollinators of *Rhododendron kaempferi* and *R. japonicum* based on their floral shapes, and *R. kaempferi* was reported to be pollinated by butterflies in southern Japan. In the mountains of central and northern Japan, however, bumblebees are abundant and candidate pollinators of *Rhododendron*. We found that visitation frequencies of bumblebees were higher than those of swallowtail butterflies to both of the *Rhododendron* species at three study sites in the mountains of central Japan. *R. japonicum* pollen adhered to the bodies of both floral visitors. Further, the nectar sugar concentration of *R. kaempferi* was in the preference range of bumblebees, whereas that of *R. japonicum* was suited to both bumblebees and butterflies. Taken together, these findings support the hypothesis that bumblebees are potentially also important pollinators of *R. kaempferi* and *R. japonicum* in mountains of central Japan, where they are more abundant than swallowtail butterflies.

Keywords: bumblebee, floral characteristics, nectar, rhododendron, swallowtail butterfly

INTRODUCTION

Relationships between flower visitor composition and floral traits can help us to understand floral evolution (Kevan & Baker 1983; Schemske & Bradshaw 1999). Suites of floral traits have evolved in response to natural selection imposed by different pollinator taxa, and these relationships are referred to as "pollination syndromes" (Johnson & Steiner 2000; Fenster et al. 2004). In recent years, the validity of the traditional pollination syndrome concept on a world scale has been challenged (Ollerton et al. 2009; Rosas-Guerrero et al. 2014). The expected pollinator of some plant species on the basis of their floral traits (shape, colour, scent, size, etc.) is often not realised (Ollerton et al. 2009; Vlašánková et al. 2017). In these cases, although floral traits may have initially evolved in response to the most effective pollinator (Rosas-Guerrero et al. 2014), pollinator transitions can occur (Ollerton et al. 2009), which can result in floral traits that do not predict the current pollinator.

Rhododendron species can be pollinated by various animals and thus exhibit a diversity of floral traits and a range of pollination syndromes. *Rhododendron calendulaceum* (subgenus *Pentanthera*, section *Pentanthera*), which has big orange petals, a long pistil, and long stamens, is butterfly-pollinated (Epps et al. 2015). *R. floccigerum* (subgenus *Hymenanthes*, section *Pontica*), which has red petals with very short pistil and stamens exertion, is bumblebee- and bird-pollinated (Georgian et al. 2015). On the other hand, *R. aureum* (subgenus *Hymenanthes*, section *Pontica*), which has medium-sized yellow flowers with relatively short pistil and stamens (Kudo 1993; Kudo et al. 2011), *R. ponticum* (subgenus *Hymenanthes,* section *Pontica*), which has medium-sized purple flowers (corolla diameter 3.5 to 5 cm) (Stout et al. 2006), and *R. ferrugineum* (subgenus *Rhododendron,* section *Rhododendron*), which has medium-sized pink flowers with relatively short pistil and stamens (Escaravage & Wagner 2004), are all bumblebee-pollinated.

The shrubs *R. kaempferi* (subgenus *Azaleastrum,* section *Tsutsusi*) and *R. japonicum* (subgenus *Pentanthera,* section *Pentanthera*) have medium to large red or red-orange flowers (corolla diameter 4 and 7 cm, respectively), and a long pistil and long stamens that stick out of the corolla. *Rhododendron* flowers with this colour and shape have been thought to be pollinated by large butterflies, because butterflies easily recognize reddish colours and when they alight on the corolla, their ventral abdomen is suitably positioned to brush against the long pistil and stamens (Tanaka 1993, 1997, 2001). Indeed, in the Kirishima Mountains in southern Japan, *R. kaempferi* is frequently visited by the swallowtail butterflies *Byasa alcinous* and *Papilio xuthus* (Yokogawa & Hotta 1995). However,

*Corresponding author: itinot@shinshu-u.ac.jp

swallowtail butterflies are rare in northern and mountainous regions in Japan, whereas bumblebees, the other potential *Rhododendron* pollinator, are abundant. When the body sizes of bumblebees match the size of *Rhododendron* flowers, their ventral abdomen touches the anthers and the stigma (Mejias et al. 2002). Whether *R. kaempferi* and *R. japonicum* are pollinated by large swallowtail butterflies or by bumblebees in these central mountainous regions of Japan remains unknown.

In this paper, we pose three hypotheses. First, we hypothesize that bumblebees, not swallowtail butterflies, are main flower visitors of *R. kaempferi* and *R. japonicum* in the mountains of central Japan. Second, sugar concentration of nectar of *R. kaempferi* and *R. japonicum* is high (suited to bumblebee pollination). Third, that due to similar selection pressures by the same pollinators, flower size of *R. kaempferi* and *R. japonicum* does not vary between populations.

MATERIALS AND METHODS

Study sites

This study was carried out in Nagano Prefecture, central Japan, for 26 days from 29 April to 10 July 2016. This period includes the peak flowering season of both *Rhododendron kaempferi* and *R. japonicum*. We investigated *R. kaempferi* at sites in Ikeda (Ike, 36°23'N, 137°53'E; 633 m a.s.l.) and Norikura (Nor, 36°07'N, 137°37'E; 1,488 m a.s.l.), and *R. japonicum* at sites in Nor and Utsukushigahara (Uts, 36°14'N, 138°06'E; 1,892 m a.s.l.). In Ike, 50 *R. kaempferi* plants grew within an area of 50 m × 50 m at the edge of a mixed forest in which *Pinus densiflora* was the dominant tree species. In Nor, a few dozen *R. kaempferi* plants grew sporadically over a distance of about 1 km along the edge of a mixed forest in which *Quercus crispula* and *Malus sieboldii* were dominant tree species. Also in Nor, 200–300 *R. japonicum* plants grew sporadically along the forest edge. In Uts, 500–600 *R. japonicum* plants grew in pastureland within an area of about 200 m × 800 m. We haphazardly selected 15–20 *R. kaempferi* and *R. japonicum* plants at each study site for the following investigation.

Visitation frequencies and pollen adhesion on flower-visiting insects

Visitation frequencies of insects, calculated as per hour per flower, were counted for a total of 10 hours for the two plant species between 08:00 and 16:00 local time (LT) at each site. We regarded an insect as a flower visitor if it inserted its mouthpart into at least one flower nectar tube in a patch. The number of *R. kaempferi* flowers in a patch was 137.8 ± 94.6 (mean ± S.D., $N = 10$) in Ike and 87.8 ± 5.5 ($N = 4$) in Nor whilst the number of *R. japonicum* flowers in a patch was 396.8 ± 985.4 ($N = 8$) in Nor and 53.4 ± 42.8 ($N = 7$) in Uts. The visitation frequencies of *Bombus* species and Papilionidae species were compared by using a paired *t*-test on log-transformed values of visitation per flower per hour. Additionally, we observed the flower-visiting behaviour of *Papilio maackii* and *Bombus diversus* at a *R. japonicum* patch with 3,200 flowers in Nor for one hour. We recorded the time spent by insect individuals at the patch and the number of flowers visited per foraging bout.

To quantify the number of *R. japonicum* pollen grains that adhered to the bodies of the visiting insects, we placed round seals (8 mm in diameter, adhesive on one side, NICHIBAN ML-151) on the ventral abdomen of *P. maackii* ($N = 1$) and on the dorsal and lateral thorax and the ventral abdomen (i.e., three seals per individual) of *B. diversus* queens ($N = 4$) in Nor on 15 June 2016. We then counted the number of pollen grains adhering to the seals under a stereoscopic microscope (Nikon SMZ1500). In *R. kaempferi*, we captured a queen and a worker of *B. ardens*, and an individual of *P. maackii* in Ike and confirmed pollen grains of *R. kaempferi* attached to their bodies with the naked eye.

Although we could not test pollination efficiency directly (e.g., by single visit experiments, or other ways of measuring pollen deposition), we collected visitation data and floral characteristics data (describe below) as proxies.

Daytime and night-time nectar collection quantities

To determine whether *R. kaempferi* produced more nectar in the daytime or at night, we used 0.5 μL capillary tubes (Drummond Scientific) to remove all the standing crop of nectar from 12 haphazardly selected *R. kaempferi* flowers in Nor at 16:00 LT on 31 May and then covered each flower with a nylon net to prevent insect access. The next morning, at 07:30 LT, we measured the nectar accumulated during the previous night in nine flowers (three flowers fell off the plant during the night). Then we re-covered each of the nine flowers with a nylon net and, at 16:00 LT, measured the nectar volume in six flowers (three flowers fell off the plant during the day) accumulated in the daytime. We compared night-time and daytime nectar collection quantities by the Wilcoxon rank sum test (R version 3.4.0; R core team 2017).

Sugar concentration of nectar

We haphazardly selected 3–20 flowers (each flower from a different plant) per site per species and measured the sugar concentration of the standing crop of nectar at 08:00 LT (three flowers of *R. kaempferi* in Ike, three flowers of *R. kaempferi* in Nor, three flowers of *R. japonicum* in Nor, and 20 flowers of *R. japonicum* in Uts) by using 2μL capillary tubes (Drummond Scientific) and hand-held refractometer (Bellingham and Stanley). To investigate the pattern of the large variation of sugar concentration in *R. japonicum* in Uts, we increased the sample size there. In addition, we haphazardly selected three flowers from each of four randomly selected *R. japonicum* plants in Uts and measured the sugar concentration of the standing crop of nectar between 09:00 and 12:00 LT, and between-plant differences in nectar sugar concentration were tested by using one-way ANOVA (R version 3.4.0; R core team 2017).

Floral size

We haphazardly selected 20 flowers (one from each of 20 plants) per site per plant species (but 15 *R. kaempferi*

FIGURE 1. Flowers of *Rhododendron kaempferi* (A & B) and *R. japonicum* (C & D): (A) Front and (B) side view of a flower of *R. kaempferi*. In (A), there is a reddish nectar guide and a nectar tube in the position indicated by the white line. In (B), the pistil and five stamens can be seen to stick out of the corolla and curve upward. (C) Front and (D) side view of a flower of *R. japonicum*. In (C), the solid white line shows how we measured corolla diameter. The dashed line connects the tips of the two lower petals.

flowers in Nor), and measured corolla diameter, lengths of the shortest and longest stamen (among five stamens), style length, and nectar tube length. Corolla diameter was defined as the distance between the top centre of the upper petal and the midpoint between the apices of the two lower petals (Fig. 1C). Nectar tube refers to the tubular structure at the base of the upper petal (Fig. 1A). Intraspecific and between-site differences in floral sizes were tested by using an unpaired two-tailed *t*-test.

RESULTS

Visitation frequencies and pollen adhesion on flower-visiting insects

Two species of butterfly were observed to visit *Rhododendron kaempferi* in Ike (*Parnassius glacialis* and *Papilio maackii*), whilst none were observed to visit *R. kaempferi* in Nor. *P. maackii* was also observed to visit *R. japonicum* in Nor, whilst no butterflies were observed to visit *R. japonicum* in Uts (Fig. 2). Three species of bumblebee were observed: *R. kaempferi* was visited by *Bombus ardens* in Ike and *B. diversus* in Nor; and *R. japonicum* was visited by *B. ardens*, *B. hypocrita* and *B. diversus* in Nor and *B. hypocrita* in Uts.

Bumblebees were more frequent visitors to *R. kaempferi* and *R. japonicum* than swallowtail butterflies at all study sites although the difference was only statistically significant for *R. japonicum* in Nor (Fig. 2).

Our limited observations of 3,200 flowers over 1 hour suggested that bumblebees stayed longer in a flower patch and visited more flowers per hour than butterflies. One individual of *P. maackii* perched and took nectar from only two flowers within a minute whilst three *B. diversus* queens stayed for 12, 15, and 18 minutes, visiting about 10 flowers per minute.

Queens of *B. diversus* had 44.5 ± 35.5 (mean ± S.E., $N = 4$) pollen grains per seal on the dorsal thorax, 160.5 ± 67.3 on the lateral thorax, and 125.5 ± 47.8 on the ventral abdomen. The swallowtail *P. maackii* had 857 ($N = 1$) pollen grains per seal on its ventral abdomen (Fig. 3).

Nectar collection quantities during the day and night

In Nor, a total of 0.005 ± 0.004 μL (mean ± S.E., $N = 9$) of *R. kaempferi* nectar was extracted after flowers had been bagged overnight (for 15 and a half hours), and a total of 0.289 ± 0.049 μL ($N = 6$) at the end of the following day (8 and a half hours, $W = 0$, $P < 0.001$).

Sugar concentration of nectar

The sugar concentration of *R. kaempferi* nectar was 54.7 ± 4.9% (mean ± S.E., $N = 3$) in Ike and 51.3 ± 0.3% ($N = 3$) in Nor, whereas that of *R. japonicum* nectar was 45.7 ± 0.7% ($N = 3$) in Nor and 30.2 ± 1.9% ($N = 20$) in Uts. In Uts there was a significant difference in nectar

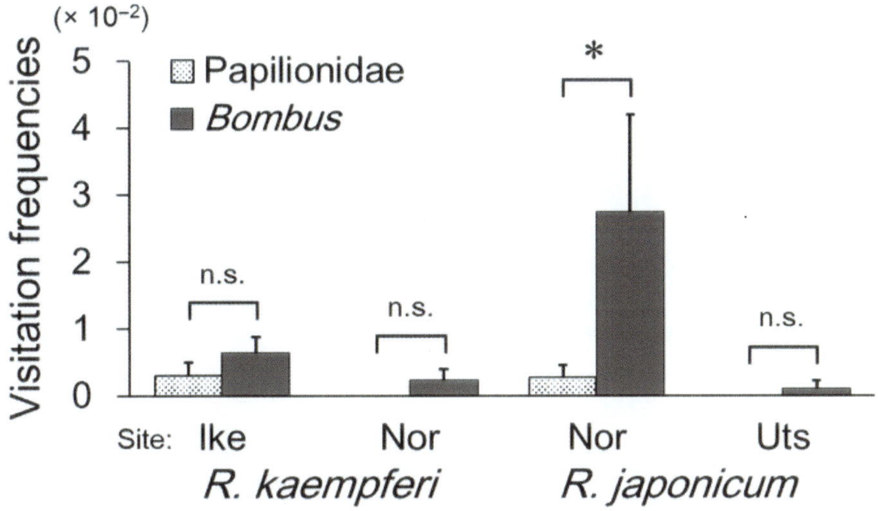

FIGURE 2. Visitation frequencies (number of visits per hour per flower) of Papilionidae spp. and *Bombus* spp. to *Rhododendron kaempferi* and *R. japonicum*.

*$P < 0.05$, paired t-test (log-transformed data)

FIGURE 3. A *Papilio maackii* butterfly that has just visited a *Rhododendron japonicum* flower. Pollen grains are adhering to its abdomen and wings.

concentration between the four plants tested (mean ± S.E. in Plant A: 23.3 ± 4.1%, B: 17.0 ± 3.5%, C: 22.0 ± 1.0%, and D: 33.7 ± 1.2%; d.f. = 3, $F = 6.27$, $P < 0.05$).

Floral size

Regional differences in floral size were not detected in *R. kaempferi* (corolla diameter: $t = 0.69$, $P = 0.49$; minimum stamen length: $t = 0.32$, $P = 0.75$; maximum stamen length: $t = 0.46$, $P = 0.65$; style length: $t = 1.10$, $P = 0.28$; nectar tube length: $t = 1.26$, $P = 0.22$; Tab. 1), but detected in *R. japonicum*: flowers were larger in Nor than in Uts in terms of corolla diameter ($t = 2.78$, $P < 0.01$), maximum stamen length ($t = 2.46$, $P < 0.05$), and style length ($t = 4.38$, $P < 0.0001$; Tab. 1).

DISCUSSION

Bumblebees were more frequent visitors than swallowtail butterflies to *Rhododendron kaempferi* and *R. japonicum* spp., although the difference is only significant in one of four cases. Yokogawa and Hotta (1995) reported that the most frequent flower visitor to *R. kaempferi* was the swallowtail butterfly *Byasa alcinous* in the Kirishima Mountains, Kagoshima Prefecture (31°53'N), southern Japan. The species diversity and abundance of swallowtail butterflies are higher in southern Japan (Kiritani 2006), whereas those of bumblebees are higher in northern and mountainous central Japan (Goulson 2003). Therefore, the main pollinator of *R. kaempferi* may change between mountainous central Japan and southern Japan (over a distance of 777 km) according to local bee and butterfly diversity and abundance.

Although Tanaka (1993, 1997, 2001) inferred that floral shape of *Rhododendron* spp. is adapted to swallowtail butterfly pollination in Japan, pollen grains also adhere to the bodies of the flower-visiting bumblebees, suggesting that they could also be pollinators of *R. kaempferi* and *R. japonicum*. When bumblebees hovered in front of the corolla, they changed their position frequently and pollen grains adhered all parts of their bodies. On the other hand, when the swallowtail butterfly *P. maackii* visited a flower, it landed on the pistil and stamens and sucked nectar with its legs and the ventral side of abdomen being in contact with the stigma and anthers, which extend outward horizontally with slightly curving tips in both *R. kaempferi* and *R. japonicum*. This more precise pollen placement may have implications for the relative efficiency of butterflies vs. bumblebees as pollinators, but this requires further investigation.

The adaptation of nectar production time has been reported to be a component of a pollination syndrome (Cruden & Hermann-Parker 1979; Itino et al. 1991). *R. kaempferi* nectar was mainly produced in the daytime, making it suited to daytime pollinators such as swallowtail

TABLE 1. Sizes (mean ± S.E.; mm) of floral parts of *Rhododendron kaempferi* and *R. japonicum*. *N* = number of flowers measured per species per site. Tests were conducted to compare sizes between the two populations of each species.

Species	*Rhododendron kaempferi*			*Rhododendron japonicum*		
Site	Ike	Nor		Nor	Uts	
Altitude (m)	633	1439 - 1480		1440 - 1488	1829 - 1892	
N	20	15		20	20	
Corolla diameter	42.3 ± 0.8	43.1 ± 1.0	n.s.	69.5 ± 1.9	62.3 ± 1.7	**
Stamen length (Min)	31.3 ± 0.6	30.9 ± 0.8	n.s.	41.5 ± 1.1	39.3 ± 0.9	n.s.
Stamen length (Max)	36.5 ± 0.6	37.0 ± 1.0	n.s.	47.8 ± 1.2	44.2 ± 0.8	*
Style length	36.5 ± 0.7	37.9 ± 1.2	n.s.	53.3 ± 1.2	46.4 ± 1.0	***
Nectar tube length	12.9 ± 0.4	13.7 ± 0.5	n.s.	14.8 ± 0.5	13.7 ± 0.4	n.s.

*$P < 0.05$; **$P < 0.01$; ***$P < 0.0001$; n.s.: not significant, unpaired two-tailed *t*-test.

butterflies and bumblebees, although it would have been better to add a treatment reversing the order between daytime collection and night-time collection to test production in day vs. night independently. We do not know if the same is true for *R. japonicum* and this would be worth further exploration.

The mean nectar sugar concentration of *R. kaempferi* (51–54%) and *R. japonicum* (30–45%) suggests an adaptation to pollination by either bumblebees, which prefer nectar with a high sugar concentration (30–55%, Willmer 2011), or butterflies (butterfly-pollinated flowers sugar concentration ranges between 17–40%, Willmer 2011). However, in Uts, the nectar sugar concentration of *R. japonicum* varied from 19% to 45% between individual plants. Chwil and Weryszko-Chmielewska (2009) also reported that, in Poland, the sugar concentration of *R. japonicum* nectar rises ontogenetically from 19% (on the third day of flowering) to 37% (on the fourth day of flowering). Since our sample sizes were small, and did not take into account flower age, the patterns of variation in the nectar sugar concentration warrant a future research.

Nagano et al. (2014) and Kuriya et al. (2015) showed that flower size and pollinator size were correlated in *Campanula punctata* and *Prunella vulgaris*, respectively, across regions several kilometres apart, reflecting adaptation of flowers to the pollinators distributing each region with different elevations. In contrast to these plant species that are adapted to bumblebee pollination, we expected that flower size of *R. kaempferi* and *R. japonicum* that are visited not only by bumblebees but also by swallowtail butterflies does not vary between populations. As a result, it is not surprising that we found no differences in floral size between sites for *R. kaempferi*. Smaller flowers of *R. japonicum* at one site may just have been a response to local abiotic conditions.

We conclude that both bumblebees and swallowtail butterflies are potential pollinators of *R. kaempferi* and *R. japonicum* in the study sites. Further investigation is required to determine the real effective pollinator (bumblebees or butterflies) of *R. kaempferi* and *R. japonicum*, and to see whether the pollinator-related traits of *R. kaempferi* and *R. japonicum* changes between populations depending on the abundance of and adaptation to the effective pollinator group.

ACKNOWLEDGEMENTS

We thank the Chubu District Forest Office (Forestry Agency) and the Matsumoto Regional Office (Nagano Prefectural Government) for permission to work in the study areas. This research was funded by the Japan Ministry of Education, Culture, Sports, Science and Technology (15H02641 to TI).

REFERENCES

Chwil M, Weryszko-Chmielewska E (2009) Characteristics of nectaries and nectar in flowers of two *Rhododendron* species. Journal of Apicultural Science 53: 17-27.

Cruden RW, Hermann-Parker SM (1979) Butterfly pollination of *Caesalpinia pulcherrima*, with observations on a psychophilous syndrome. The Journal of Ecology 67: 155-168.

Epps MJ, Allison SE, Wolfe LM (2015) Reproduction in flame azalea (*Rhododendron calendulaceum*, Ericaceae): A rare case of insect wing Pollination. The American Naturalist 186: 294-301.

Escaravage N, Wagner J (2004) Pollination effectiveness and pollen dispersal in a *Rhododendron ferrugineum* (Ericaceae) population. Plant Biology 6: 606-615.

Fenster CB, Armbruster WS, Wilson P, Dudash MR, Thomson JD (2004) Pollination syndromes and floral specialization. Annual Review of Ecology, Evolution, and Systematics: 375-403.

Georgian E, Fang Z, Emshwiller E, Pidgeon A (2015) The pollination ecology of *Rhododendron floccigerum* Franchet (Ericaceae) in Weixi, Yunnan Province, China. Journal of Pollination Ecology 16: 72-81.

Goulson D (2003) Bumblebees: their behaviour and ecology. Oxford University Press, New York.

Itino T, Kato M, Hotta M (1991) Pollination ecology of the two wild bananas, *Musa acuminata* subsp. *halabanensis* and *M. salaccensis*: chiropterophily and ornithophily. Biotropica 23: 151-158.

Johnson SD, Steiner KE (2000) Generalization versus specialization in plant pollination systems. Trends in Ecology & Evolution 15: 140-143.

Kevan PG, Baker HG (1983) Insects as flower visitors and pollinators. Annual Review of Entomology 28: 407-453.

Kiritani K (2006) Predicting impacts of global warming on population dynamics and distribution of arthropods in Japan. Population Ecology 48: 5-12.

Kudo G (1993) Relationship between flowering time and fruit set of the entomophilous alpine shrub, *Rhododendron aureum* (Ericaceae), inhabiting snow patches. American Journal of Botany 80: 1300-1304.

Kudo G, Hirao AS, Kawai Y (2011) Pollination efficiency of bumblebee queens and workers in the alpine shrub *Rhododendron aureum*. International Journal of Plant Sciences 172: 70-77.

Kuriya S, Hattori M, Nagano Y, Itino T (2015) Altitudinal flower size variation correlates with local pollinator size in a bumblebee-pollinated herb, *Prunella vulgaris* L. (Lamiaceae). Journal of Evolutionary Biology 28: 1761-1769.

Mejias JA, Arroyo J, Ojeda F (2002) Reproductive ecology of *Rhododendron ponticum* (Ericaceae) in relict Mediterranean populations. Botanical Journal of the Linnean Society 140: 297-311.

Nagano Y, Abe K, Kitazawa T, Hattori M, Hirao AS, Itino T (2014) Changes in pollinator fauna affect altitudinal variation of floral size in a bumblebee-pollinated herb. Ecology and Evolution 4: 3395-3407.

Ollerton J, Alarcón R, Waser NM, Price MV, Watts S, Cranmer L, Hingston A, Peter CI, Rotenberry J (2009) A global test of the pollination syndrome hypothesis. Annals of Botany 103: 1471-1480.

R Core Team (2017) R: A language and environment for statistical computing. R Foundation for Statistical Computing, Vienna, Austria. URL https://www.R-project.org/.

Rosas-Guerrero V, Aguilar R, Martén-Rodríguez S, Ashworth L, Lopezaraiza-Mikel M, Bastida JM, Quesada M (2014) A quantitative review of pollination syndromes: do floral traits predict effective pollinators? Ecology Letters 17: 388-400.

Schemske DW, Bradshaw HD (1999) Pollinator preference and the evolution of floral traits in monkeyflowers (*Mimulus*). Proceedings of the National Academy of Sciences 96: 11910-11915.

Stout JC, Parnell JA, Arroyo J, Crowe TP (2006) Pollination ecology and seed production of *Rhododendron ponticum* in native and exotic habitats. Biodiversity & Conservation 15: 755-777.

Tanaka H (1993) Introduction to floral ecology. Nouson Bunkasha, Tokyo. (In Japanese)

Tanaka H (1997) Ecology of flower-insect interactions. Hoikusha, Osaka. (In Japanese)

Tanaka H (2001) Mutual deception between flowers and insects. Kodansha, Tokyo. (In Japanese)

Vlašánková A, Padyšáková E, Bartoš M, Mengual X, Janečková P, Janeček Š (2017) The nectar spur is not only a simple specialization for long-proboscid pollinators. New Phytologist 215: 1574-1581.

Willmer P (2011) Pollination and floral ecology. Princeton University Press, Princeton.

Yokogawa M, Hotta M (1995) Taxonomical notes on plants of southern Japan II. Variation in the *Rhododendron obtusum* group in the Kirishima Mts., southern Kyushu. Acta Phytotaxonomica et Geobotanica 46: 165-183.

20

THE EFFECT OF THE DISPERSION OF REWARDING AND REWARDLESS FLOWERS ON VISITATION AND CONSTANCY BY BUMBLEBEES (*BOMBUS IMPATIENS*)

Emma Katz[1]* and Carla J. Essenberg[1,2]

[1]*Bates College, Program in Environmental Studies, Bates College, Lewiston, ME 04240, USA*
[2]*Department of Biology, Faculty of Science, Bates College, Lewiston, ME 04240, USA*

Abstract—Deceptive pollination strategies, in which plants rely on animal pollinators but provide no benefits to their pollinators, have evolved many times in angiosperms. However, the conditions that favour deceptive pollination strategies over mutualistic strategies are poorly understood. One factor that may be important for the success of deceptive strategies is the plants' dispersion in relation to co-flowering, rewarding species. We compared pollinator behaviour across two artificial environments, one in which a rewardless species was mixed with a rewarding species and one in which the two species occurred in contiguous patches. Bumblebees both encountered and visited rewardless flowers more often when they were mixed with rewarding species. However, the rate of switching was also higher under those conditions, which could result in higher rates of interspecific pollen transfer. The environmental conditions most favourable to deceptive pollination strategies may vary depending on the vulnerability of the plant species to interspecific pollen transfer.

Keywords: Generalized food deception; magnet species; aggregation; facilitation; competition; foraging behaviour

INTRODUCTION

Deceptive pollination, a strategy in which plants provide no reward to their pollinators but still rely on pollinator visits, has evolved independently many times in the angiosperms, and reversion from rewardlessness to mutualism has also occurred multiple times (Renner 2006; Smithson 2009). An active area of research focuses on the fitness costs and benefits of deceptive strategies (Jersáková et al. 2006), but why deception evolves in some species while others evolve mutualistic strategies has received less attention. Biotic environmental conditions may be important in determining the success of deceptive strategies, as deceptive species rely on heterospecific flowers or other biological conditions to keep pollinators in their vicinity (Renner 2006; Duffy & Johnson 2017). Probably the most common deceptive strategy, known as generalized food deception, is to present flowers that offer signals typical of rewarding plant species without mimicking any particular model (Jersáková et al. 2006). Whether generalized food deception is favoured over mutualism may depend on the abundance of co-flowering, rewarding species as well as the dispersion of rewarding and rewardless plants (Lammi & Kuitunen 1995; Johnson et al. 2003; Internicola et al. 2006; Juillet et al. 2007).

Previous studies suggest that the degree of mixing between rewarding and rewardless plant species can affect pollinator visitation rates to the rewardless species, how often pollinators switch between species, and the rewardless species' reproductive output (Keasar 2000; Internicola et al. 2006; Internicola et al. 2007; also see Hanoteaux et al. 2013; Geslin et al. 2014; de Waal et al. 2015; Bruckman & Campbell 2016, which consider effects of dispersion on pollinator-mediated interactions between rewarding plant species). However, the nature of these effects varies across studies, suggesting that multiple aspects of dispersion influence rewardless species' success (Tab. 1). An experiment manipulating both the degree of mixing and density of the rewardless orchid *Dactylorhiza sambucina* and a dissimilar rewarding species, *Muscari neglectum,* found that increasing aggregation of both species negatively influenced the reproductive success of the rewardless plant (Internicola et al. 2006). In contrast, Keasar (2000) found that rewardless artificial flowers received more visits and switching between species was less common when rewarding and rewardless flowers were separated in spatially distinct, monospecific patches than when they were mixed within a single, large patch, suggesting that aggregation may sometimes be favourable to rewardless species. One explanation for the relatively high visitation to rewardless flowers in Keasar's separated treatment is that the distance between the monospecific patches made switching away from the rewardless species more costly than it was when both species occurred in a single patch, favouring strings of visits to the rewardless species each time a rewardless patch was visited. Consistent with this idea are the contrasting results of

*Corresponding author: emmajkatz1@gmail.com

Table 1. Summary of previous studies exploring effects of dispersion of rewarding and rewardless flowers on aspects of the rewardless flowers' pollination success. The posited mechanism explaining each result is our interpretation and may not reflect the original researchers' text.

Citation	Manipulated Variables	Effect of Aggregation on rewardless flowers	Posited Mechanism
Internicola et al. 2006	Degree of mixing and density of rewardless and rewarding species	Lower reproductive success	Travel cost of switching distant patches favoured thorough sampling in both treatments
Keasar (2000)	Intermixed vs clustered, spatially separated patches	Increased visits, decreased switching	Distance between monospecific patches made switching costly
Internicola et al. (2007)	Mixed vs monospecific patches, both treatments spatially separated	No significant effect	Travel cost of switching patches favoured thorough sampling in both treatments

Internicola et al. (2007), who compared bee behaviour in mixed versus monospecific flower patches, with patches in both treatments spatially separated. In the Internicola et al. (2007) study, mixing had no significant effect on visitation to the rewardless species, perhaps because travel costs associated with moving to a new patch favoured relatively thorough sampling of rewardless flowers in both mixed and monospecific patches.

These results leave open the question of how mixing versus separation of rewarding and rewardless species affects pollinator behaviour when there are no between-patch travel costs. Rewardless flowers may actually receive more visits when mixed with rewarding flowers than when rewarding and rewardless flowers occur in contiguous, pure patches. In both situations, the cost of switching away from the rewardless species should be small, but pollinators would likely encounter the rewardless flowers more frequently when they were mixed with rewarding flowers than when they occurred in monospecific zones of the flower patch.

The aim of the current study was to examine the effect of the mixing of rewarding and rewardless flowers within a continuous patch on visitation rates to the rewardless species as well as the frequency of switching between species. We observed bumblebees (*Bombus impatiens*) foraging on arrays of artificial flowers in which rewarding and rewardless species had similar, but distinguishable colours. This experimental context modelled a deceptive species with imperfect mimicry of a co-flowering, rewarding species.

Materials and Methods

We observed the effect of the dispersion of rewarding and rewardless artificial flowers on the foraging behaviour of bumblebees (*Bombus impatiens*). The two flower "species" differed in colour so that bees could distinguish them, although the colours were similar enough that bees may have needed to learn to discriminate between them. Prior to experimental trials, bees were given experience visiting the rewarding flower type, so our experiment modelled a situation in which bees were only naïve to the rewardless species. In each experimental trial, a bee was randomly assigned to one of two treatments: a floral array in which the rewarding and rewardless flowers formed contiguous patches ("separated") or an array in which the two flower types were mixed together ("mixed"). Each experimental trial was split into two parts, with each part consisting of 15 flower visits. Analyzing these two parts of the experimental trial separately allowed us to assess how bee behaviour changed with experience. We recorded both the number of visits to each flower type and the number of switches between flower types.

Bumble bee husbandry

We used commercially-reared bumble bees (*Bombus impatiens*, Koppert Biological Systems) for this experiment. Bees were given pollen within the colony box and 20% sucrose solution (w/w) from a feeder in a foraging arena. The data we present in this paper came from a total of 41 individual foragers from three *Bombus impatiens* colonies (16, 14, and 11 foragers from each).

Artificial flowers and experimental chamber

Artificial flowers consisted of a small reservoir made from the cap of a 1 mL microcentrifuge tube surrounded by a coloured paper corolla (3.5 cm diameter) reinforced with a plastic ring positioned underneath it (Fig. 1). Each flower sat

FIGURE 1. Artificial flower used in the experiment.

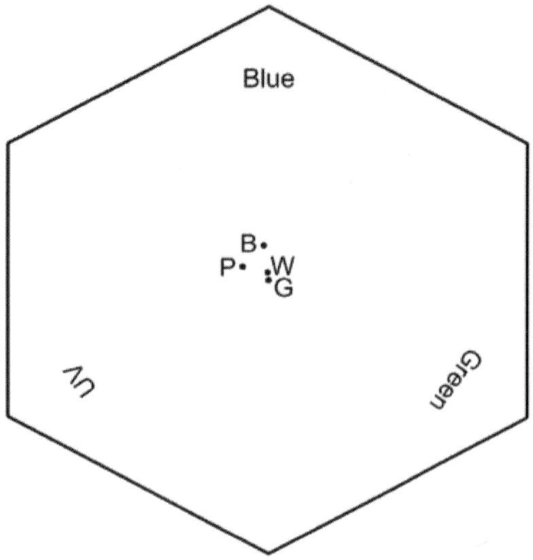

FIGURE 2. Colours of artificial flowers in *B. impatiens* colour space. W = white, B = blue, and P = purple corolla. G = average colour of the green turf background. The diagram was made according to Chittka (1992) using data on the spectral sensitivities of *B. impatiens* receptors from Skorupski and Chittka (2010).

on top of a clear plastic rod that served as a stalk. Flowers with blue corollas contained sucrose rewards, located in the center of the flower, while flowers with purple corollas were empty. Positions of the corolla colours in *B. impatiens* colour space are shown in Fig. 2 (Chittka 1992; Skorupski & Chittka 2010). The blue and purple colours used were relatively similar: their chromatic contrast (D = 0.1059 units) was great enough to be distinguishable to bees but with reduced accuracy compared to more highly contrasting colours (Dyer & Chittka 2004). Therefore, bees should have been capable of distinguishing the two flower types but may have needed to learn to discriminate between them. Our experiment, therefore, is based on a situation in which the rewardless species is an imperfect mimic of the rewarding species.

Both the training and experimental trials took place inside an experimental chamber (dimensions: 70 cm long × 60 cm wide × 50 cm tall), the floor of which was covered with artificial turf. Holes to accommodate flower stalks were drilled into a board forming the floor of the chamber in an 8 × 8 array with 6 cm spacing between holes.

Training Sessions

Prior to individual experimental trials, foragers from each colony were given opportunities to visit white artificial flowers (which match the colour of the maintenance feeder) and then blue flowers. Training sessions took place several times a day during the 3-5 days prior to the beginning of experimental trials for each colony and then were repeated periodically throughout the experiment. Flowers had white corollas for the first 1-2 days, until several foragers learned to use the flowers. We used blue corollas for all subsequent training sessions. Prior to each training session, six flowers were placed in randomly-chosen locations in the 8 × 8 array and each flower was filled with 20 μl 50% sucrose solution (w/w). The entire colony was allowed to forage on the training array for 30 minutes, and flowers were refilled as they emptied. We recorded the identities of bees foraging on the flowers during each training session.

Experimental trials

Foragers were selected for experimental trials and assigned to treatments as follows. Only bees that had been observed foraging on artificial flowers during the training sessions were used. Of those bees, we selected the first bee to emerge from the colony box after the chamber was prepared for an experimental trial. After a bee was selected, it was randomly assigned to either the separated or mixed treatment. We recorded 21 bees in the separated treatment and 20 bees in the mixed treatment

In both treatments, the experimental array consisted of 24 blue (rewarding) flowers and 24 purple (unrewarding) flowers arranged in a continuous patch (45.5 cm × 45.5 cm). Each blue flower contained 1 μl of 50% sucrose solution and each purple flower was empty. Equal numbers of flowers were located within each quarter of the 8 × 8 array, with their locations chosen at random. In the mixed treatment, flowers of both colours were randomly dispersed across the entire array. In the separated treatment, each quarter of the array contained flowers of one type only (i.e., blue or purple), with neighbouring quarters containing different flower colours (Fig. 3). Randomizations of flower locations were generated using the sample() command in R version 3.3.0 (R Core Team 2014).

Each experimental trial was split into two parts. After the bee had probed the reservoirs of 15 different flowers, the light was turned off, all blue flowers that had been visited were refilled, and locations of rewarding and rewardless flowers and zones were re-assigned. The bee was then released to probe another 15 flowers. Flowers were not refilled while the bee was foraging. Revisits to flowers did not count towards the 15 visits. Therefore, the number of flowers probed corresponds to the number that could have been pollinated, had these been real flowers. The division of the experiment into two parts allowed us to compare the bee's initial, naive responses to the rewardless flowers with its responses after limited experience with the rewardless flowers. Making this comparison without taking a break to rearrange and refill the flowers would have been complicated by the bees' tendency to revisit flowers.

Experimental trials were recorded using a video camera, which was located above and just to one side of the array. All behavioural data were acquired by transcribing these video recordings. Two bees were excluded from subsequent analysis because the camera stopped recording very early in one trial. In three other trials, we failed to videotape a full 15 flower visits (not counting revisits) because of a camera failure or other error. We used only the section of the trial that we had on video in our analyses (constituting 7, 13, and 13 flower visits for the three trials).

Data Analysis

We analysed the effects of treatment (separated or mixed) and trial part (1 or 2) on several response variables. To provide information about how floral dispersion would be likely to influence plant pollination success, we analysed the proportion of the bee's flower visits, excluding revisits, that

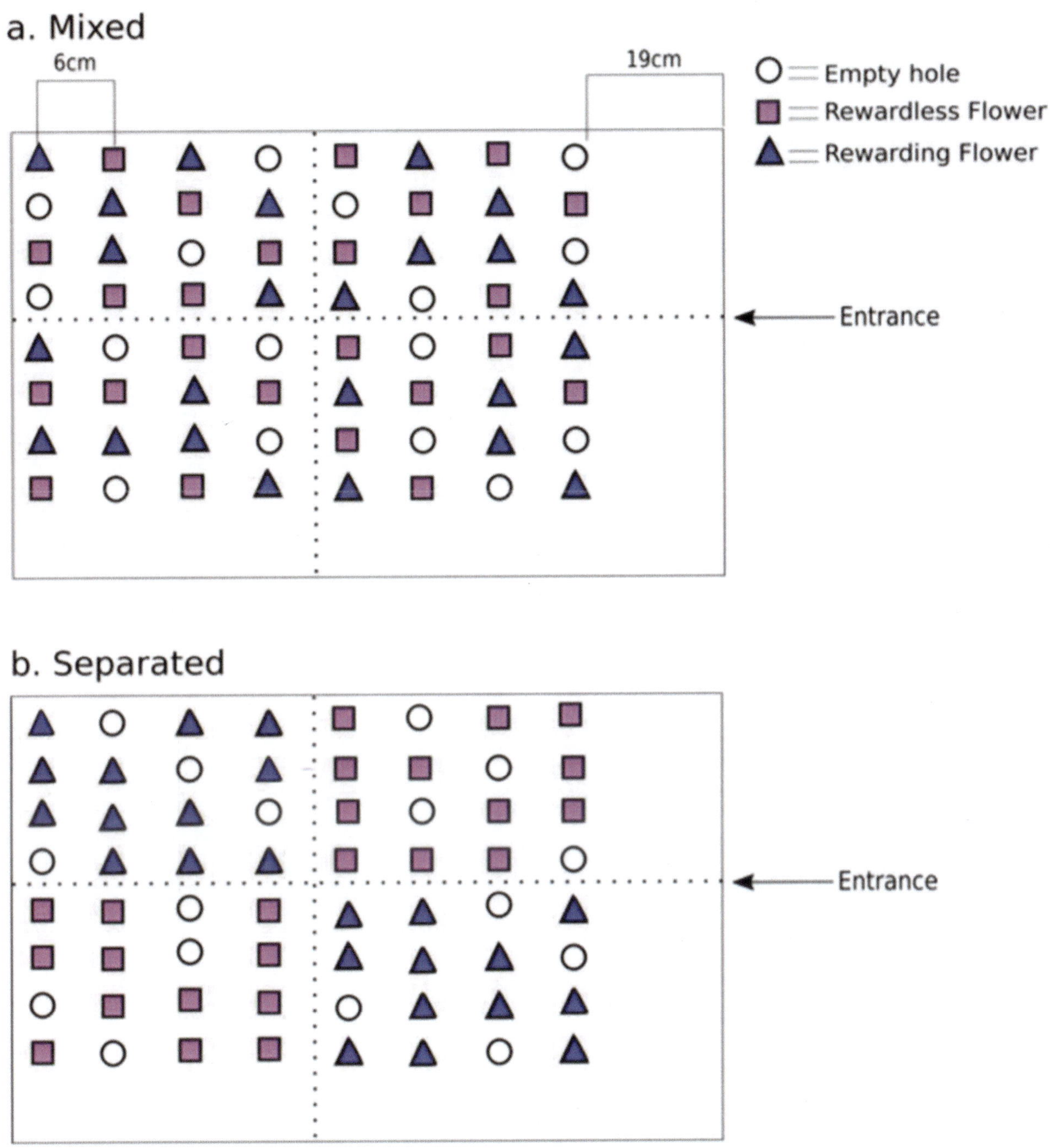

FIGURE 3. Example of spatial arrangements of rewarding and rewardless flowers in the mixed (a) and separated (b) treatments.

were to the rewardless flower type, and the proportion of flower to flower transitions in the bee's visit sequence (including revisits) in which the bee switched flower colours. We counted as visits only occasions when the bee both landed on and came in contact with the centre of the flower (i.e., its head crossed at least part of the reservoir in the flower's centre) because those were occasions when pollination could have occurred, were these real flowers.

To provide further insight into the behavioural mechanisms underlying patterns we observed in flower visitation, we also analysed the proportion of flower encounters that were to the rewardless flower type, excluding encounters of previously visited flowers. A flower encounter was defined as any occasion on which the bee flew within I cm of the edge of the corolla, as seen from above, whether it landed or not. We did not count as encounters occasions when the bee was flying near the top of the chamber rather than close to the flowers or occasions when the bee flew backwards towards a flower, unless she ended up directly over the flower. Bees sometimes flew over the array for extended periods without landing. We were interested in occasions when the bee was actively searching for a flower to visit and therefore counted only the last five flower encounters before the bee landed on a flower.

In analyzing proportions of flower visits and encounters that were to the rewardless flower type, we used linear, mixed effects models in which treatment, trial part, and their interaction were treated as fixed effects. Bee ID nested within colony were treated as random effects; including Bee ID as a random effects factor allowed us to compare parts 1 and 2 of each bee's trial.

We also wished to know whether the bees discriminated against rewardless flowers and, if so, whether they discriminated more strongly as they gained experience. Therefore, we analysed the effects of flower type, treatment (separated or mixed), trial part (1 or 2), and all interactions on the proportion of encounters of each flower type that resulted in the bee visiting the flower. We considered only encounters of flowers that the bee had not visited previously. As before, we used a linear, mixed effects model, with bee ID nested within colony included as random effects.

In all of these analyses, interaction terms were dropped from the model if nonsignificant. Proportion of flower visits was log-transformed to improve fit to the normal distribution. After this transformation, residuals from all of the tests met the assumptions of normality and homoscedasticity. Analyses were carried out using the lmer() function in the lmerTest package in R version 3.4.1 (Kuznetsova et al. 2015; R Core Team 2017).

RESULTS AND DISCUSSION

Foraging bumblebees (*Bombus impatiens*) made a higher proportion of their visits to rewardless flowers when they were mixed with rewarding flowers than when rewarding and rewardless flowers formed contiguous patches (Fig. 4a, Tab. 2). We excluded revisits because of what we perceived as an unnaturally high revisitation rate: 15% of the bees' visits to rewardless flowers and 33% of their visits to rewarding flowers were revisits. A similar treatment effect was observed if revisits were included.

Two different mechanisms could explain a lower rate of visits to rewardless flowers when the two flower types formed single-species patches than when they were mixed together. First, bees could simply encounter rewardless flowers less frequently when the two species occurred in separate patches than when they were mixed with rewarding flowers. Area-restricted search behaviour, in which foragers turned more frequently and made shorter flights after encountering rewards than after unrewarding visits, could allow bees to do a larger proportion of their searching in rewarding than rewardless patches, even if they had no preference for the rewarding species (Pyke 1978; Waddington 1980; Burns & Thomson

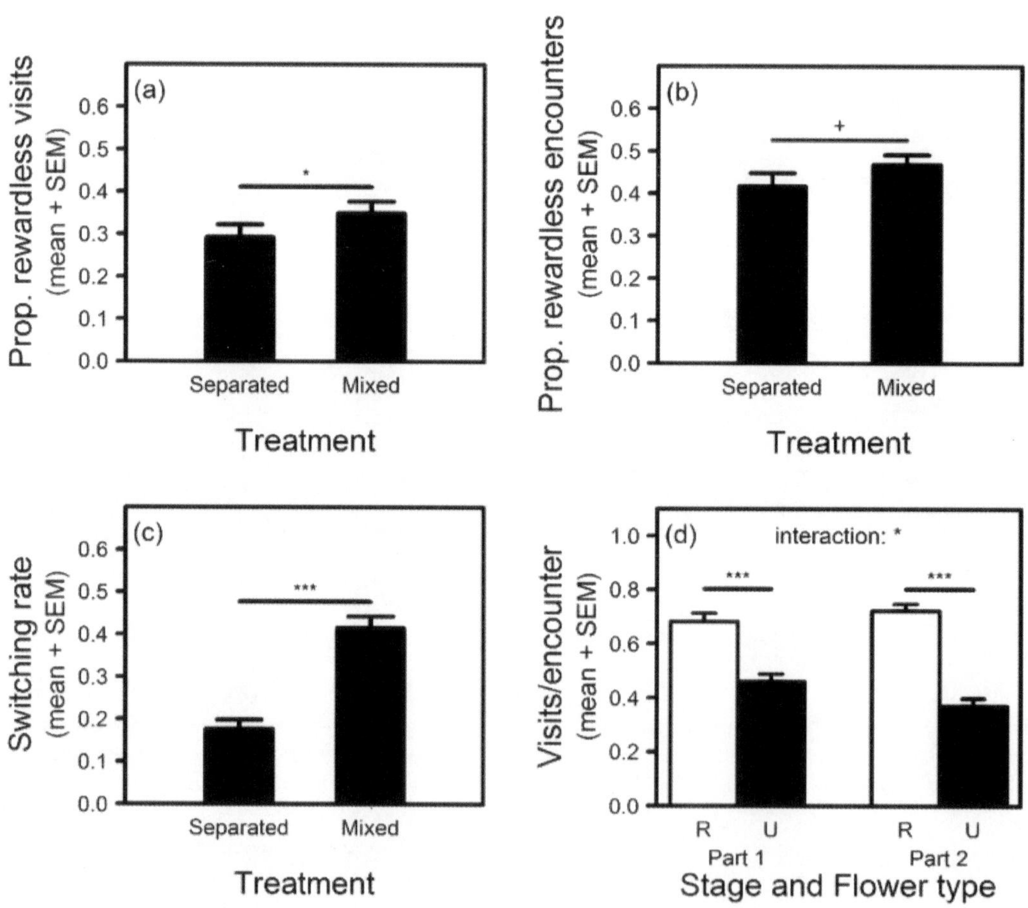

FIGURE 4. Effects of flower dispersion or forager experience on the proportion of bumblebee (*B. impatiens*) flower (a) visits and (b) encounters that were to the rewardless (purple) artificial flower type, (c) proportion of transitions that were switches between flower types, and (d) proportion of encounters to rewardless ("U", white bars) and rewarding ("R", black bars) flowers that resulted in visits. Examples of the two dispersion treatments are shown in Fig. 3. Each part of each experimental trial consisted of 15 flower visits, not including revisits. Revisits are omitted except in panel (c). Results of statistical analyses are provided in Tab. 2. + $P < 0.1$, * $P < 0.05$, ** $P < 0.01$, *** $P < 0.001$.

TABLE 2. Effects of dispersion of rewarding and rewardless flowers on the relative frequency of bumblebees' encounters of and visits to rewardless flowers, frequency of switches between flower types, and tendency to visit encountered flowers of each type. Dispersion treatments were separated or mixed, as shown in Fig. 3. Part refers to the first vs. second part of the experimental trial. Flowers that had been visited previously were excluded from analysis of all variables except switching frequency. Interactions are shown where significant; all omitted interactions have $p > 0.35$. Results are from mixed effects models, with bee identity nested within colony as random effects. Bee identity did not explain any variation in proportion of visits or encounters to rewardless flowers, and colony did not explain any variation in the proportion of encountered flowers that were visited. Proportion of visits to rewardless flowers was log-transformed to improve normality. Data are from 21 bees in the separated treatment and 20 bees in the mixed treatment, from 3 colonies.

	Factor	Den. df	*F*	*P*
Proportion of bee's visits that are to rewardless flowers	Dispersion treatment	77.4	4.3	0.042
	Part	76.8	2.1	0.152
Proportion of encountered flowers that are rewardless	Dispersion treatment	77.4	3.9	0.053
	Part	76.8	0.5	0.488
Proportion of transitions that are switches between flower types	Dispersion treatment	42.5	95.9	<0.001
	Part	46.7	0.2	0.646
Proportion of encounters that resulted in a visit	Flower type	119.3	132.3	<0.001
	Dispersion treatment	37.1	0.1	0.762
	Part	119.3	1.1	0.302
	Treatment × Part	119.3	3.5	0.065
	Flower type × Part	45.4	6.9	0.010

2006). If the species were mixed together randomly, however, this tactic would be ineffective. Waddington (1980) demonstrated this consequence of flower dispersion with honeybees, which made higher proportions of rewarding visits when rewarding flowers were clumped than when they were distributed randomly amongst unrewarding flowers with identical cues.

Consistent with predictions generated by area-restricted models of search behaviour and previous findings in honey bees, bees encountered a higher proportion of rewardless flowers when rewarding and rewardless flowers were mixed together than when they were separated into contiguous patches (again, excluding previously visited flowers), although the difference was only marginally significant ($P = 0.053$, Fig. 4b, Tab. 2).

A second reason for rewardless plants to receive more visits in our mixed treatment than in the separated treatment could have been that bees were more willing to sample a second flower species when flowers of the rewarding species were spaced farther apart, on average, forcing the bees to make longer flights between neighbouring conspecific flowers. Foragers, including bees, do often broaden their diets in response to a low density of high-quality food resources (Pyke 1984; Kunin & Iwasa 1996; Gegear & Thomson 2004; Ishii 2005). However, we found no evidence of decreased preference for the rewarding species in the separated treatment compared to the mixed treatment. Comparison of the proportion of flower encounters that resulted in flower visits (excluding revisits) revealed a significantly higher tendency to visit the rewarding than the unrewarding type in both dispersion treatments (Fig. 4d, Tab. 2). Treatment did not influence the strength of this difference in visit tendency.

Our results contrast with those of Keasar (2000), who found that bees visited rewardless flowers more often when they were arranged in distinct patches, with space in between them, than when mixed together with rewarding flowers. This difference in outcome may be because the higher energy cost of traveling 80-120 cm between patches in Keasar's separated treatment led bees to remain in the rewardless patches longer than they did in our experiment, in which the patches were contiguous. If so, then the spatial scale at which plant species are associated with each other could be critical in determining whether being intermixed with rewarding plants is favourable to a rewardless strategy. In plant communities composed of small individuals occurring in close proximity (e.g., 6 cm from nearest neighbours, as in our experiment), intermixing might be favourable for rewardless species, whereas aggregated distributions would be more favourable when individuals were spaced farther apart.

The spatial arrangement of flowers also had a significant effect on the frequency of switches between rewarding and rewardless flowers, with a higher rate of switching between flower types occurring in the mixed treatment than in the separated treatment (Fig. 4c, Tab. 2). Furthermore, the magnitude of this increase in the rate of between-species switches was considerably greater than the relatively small increase in visitation experienced by the rewardless flower type in the mixed treatment. Keasar (2000) and Internicola et al. (2007) also found that the total number of shifts between rewarding and rewardless flowers was significantly higher when the two were mixed. Higher rates of switching between species can increase interspecific pollen transfer, which is typically detrimental to both plant species (Rathcke 1983; Brown & Mitchell 2001; Morales & Traveset 2008). However, the degree to which pollinator inconstancy reduces pollination success varies across plant taxa. For instance, the packaging of pollen into pollinaria in many orchid species allows their pollen to be deposited on conspecific stigmas even after many intervening visits to other species' flowers

(Jersáková et al. 2006). As a result, rewardless orchids may be able to take advantage of increased visitation rates experienced when growing mixed together with rewarding species while experiencing little ill effect from the frequent switching between species that pollinators exhibit under those conditions. This advantage could help to explain the high prevalence of food deceptive pollination systems in the Orchidaceae.

Bees' preference for the rewarding flower type increased over the course of each trial: the difference in number of visits per flower encounter for rewarding and rewardless flowers was greatest in the second part of each trial (Fig. 4d, Tab. 2). The magnitude of this shift was similar across the two dispersion treatments. These results suggest that avoidance learning occurred in both treatments and that dispersion had little, if any, effect on learning rate. Similarly, Internicola et al. (2007) found that over time bumblebees visited fewer rewardless flowers regardless of whether the flowers occurred in mixed or homogeneous patches. In our experiment, however, the increased preference for the rewarding flower type did not translate into a significant decrease in visitation rate to the rewardless flowers over the course of an experimental trial (Tab. 2).

Our results contrast with those of previous studies in suggesting that growing mixed with a co-flowering, rewarding species can increase the quantity of visits a rewardless species experiences. However, being mixed with another species also carries a cost: pollinators switch between species more often when they are mixed, which is likely to reduce the quality of visits the rewardless species experiences. Therefore, which conditions are most favourable to a rewardless strategy will likely vary depending on the plant's susceptibility to interspecific pollen transfer as well as on the number of visits needed to achieve successful pollination.

ACKNOWLEDGEMENTS

Funding was provided by Bates College. We thank Colin A. McIntire for the artificial flower design and Xiaomeng Wang for other contributions to the methods.

REFERENCES

Brown BJ, Mitchell RJ (2001) Competition for pollination: effects of pollen of an invasive plant on seed set of a native congener. Oecologia 129:43-49.

Bruckman D, Campbell DR (2016) Pollination of a native plant changes with distance and density of invasive plants in a simulated biological invasion. American Journal of Botany 103:1458-1465.

Burns JG, Thomson JD (2006) A test of spatial memory and movement patterns of bumblebees at multiple spatial and temporal scales. Behavioural Ecology 17:48-55.

Chittka L (1992) The colour hexagon: a chromaticity diagram based on photoreceptor excitations as a generalized representation of colour opponency. Journal of Comparative Physiology A 170:533-543.

de Waal C, Anderson B, Ellis AG (2015) Relative density and dispersion pattern of two southern African Asteraceae affect fecundity through heterospecific interference and mate availability, not pollinator visitation rate. Journal of Ecology 103:513-525.

Duffy KJ, Johnson SD (2017) Effects of distance from models on the fitness of floral mimics. Plant Biology 19:438-443.

Dyer AG, Chittka L (2004) Biological significance of distinguishing between similar colours in spectrally variable illumination: bumblebees (*Bombus terrestris*) as a case study. Journal of Comparative Physiology A 190:105-114.

Gegear RJ, Thomson JD (2004) Does the flower constancy of bumble bees reflect foraging economics? Ethology 110:793-805.

Geslin B, Baude M, Mallard F, Dajoz I (2014) Effect of local spatial plant distribution and conspecific density on bumble bee foraging behaviour. Ecological Entomology 39:334-342.

Hanoteaux S, Tielbörger K, Seifan M (2013) Effects of spatial patterns on the pollination success of a less attractive species. Oikos 122:867-880.

Internicola AI, Juillet N, Smithson A, Gigord LDB (2006) Experimental Investigation of the Effect of Spatial Aggregation on Reproductive Success in a Rewardless Orchid. Oecologia 150:435-441.

Internicola AI, Page PA, Bernasconi G, Gigord LDB (2007) Competition for pollinator visitation between deceptive and rewarding artificial inflorescences: an experimental test of the effects of floral colour similarity and spatial mingling. Functional Ecology 21:864-872.

Ishii HS (2005) Analysis of bumblebee visitation sequences within single bouts: Implication of the overstrike effect on short-term memory. Behavioral Ecology and Sociobiology 57:599-610.

Jersáková J, Johnson SD, Kindlmann P (2006) Mechanisms and evolution of deceptive pollination in orchids. Biological Reviews of the Cambridge Philosophical Society 81:219-235.

Johnson SD, Peter CI, Nilsson LA, Agren J (2003) Pollination success in a deceptive orchid is enhanced by co-occurring rewarding magnet plants. Ecology 84:2919-2927.

Juillet N, Gonzalez MA, Page PA, Gigord LDB (2007) Pollination of the European food-deceptive *Traunsteinera globosa* (Orchidaceae): The importance of nectar-producing neighbouring plants. Plant Systematics and Evolution 265:123-129.

Keasar T (2000) The spatial distribution of nonrewarding artificial flowers affects pollinator attraction. Animal Behaviour 60:639-646.

Kunin W, Iwasa Y (1996) Pollinator foraging strategies in mixed floral arrays: density effects and floral constancy. Theoretical Population Biology 49:232-263.

Kuznetsova A, Brockhoff PB, Christensen RHB (2015) Package 'lmerTest'. R package version 2.

Lammi A, Kuitunen M (1995) Deceptive pollination of *Dactylorhiza Incarnate*: An experimental test of the magnet species hypothesis. Oecologia 101:500-503.

Morales CL, Traveset A (2008) Interspecific pollen transfer: magnitude, prevalence and consequences for plant fitness. Critical Reviews in Plant Sciences 27:221-238.

Pyke GH (1978) Optimal foraging: Movement patterns of bumblebees between inflorescences. Theoretical Population Biology 13:72-98.

Pyke GH (1984) Optimal foraging theory: a critical review. Annual Review of Ecology and Systematics 15:523-575.

R Core Team (2014) R: A language and environment for statistical computing. R Foundation for Statistical Computing, Vienna, Austria.

R Core Team (2017) R: A language and environment for statistical computing. R Foundation for Statistical Computing, Vienna, Austria.

Rathcke B (1983) Competition and facilitation among plants for pollination. Pollination Biology:305-325.

Renner SS (2006) Rewardless flowers in the angiosperms and the role of insect cognition in their evolution. In: Waser NM, Ollerton J (eds) Plant-pollinator interactions: from specialization to generalization University of Chicago Press, Chicago, pp 123-144.

Skorupski P, Chittka L (2010) Photoreceptor spectral sensitivity in the bumblebee, Bombus impatiens (Hymenoptera: Apidae). PLoS One 5:e12049.

Smithson A (2009) A plant's view of cheating in plant–pollinator mutualisms. Israel Journal of Plant Sciences 57:151-163.

Waddington KD (1980) Flight patterns of foraging bees relative to density of artificial flowers and distribution of nectar. Oecologia 44:199-204.

21

Movements of Floral Parts and Roles of the Tooth on the Column Wall of *Bulbophyllum praetervisum* (Orchidaceae) Flower in Pollination by Dacini Fruit Flies (Diptera: Tephritidae)

Tan, Keng Hong*[1,2] and Tan, Lin Tze[3]

[1]*Mobula Research Sdn. Bhd., 20, Jalan Tan Jit Seng, 11200 Penang, Malaysia*
[2]*Academy of Sciences Malaysia, MATRADE Tower, Jalan Sultan Haji Ahmad Shah, 50480 Kuala Lumpur, Malaysia*
[3]*Bachemer Strasse, 20, Bonn, 53179 Germany*

Abstract—*Bulbophyllum* is the largest genus of the orchid family, with about 2000 species that are mainly pollinated by flies. Flowers of many *Bulbophyllum* species under the Sections Sestochilos and Beccariana specifically attract Dacini fruit flies, *Bactrocera* and *Zeugodacus* species, as pollinators. Non-nectar producing and non-resupinate solitary flowers of *Bu. pratervisum* emit specific and pleasant floral fragrances to specifically attract and reward male fruit flies. Slippery surfaces on their lateral sepals aid in pollination. Although pollinia removal by male fruit flies has been observed frequently, deposition of pollinia to complete pollination (by two males of *Ba. albistrigata* and *Z. caudatus*) has only been observed recently. Field observations show two previously unreported movements of floral parts – a) petals and medial sepal during daily closing and reopening of flowers, and b) the spring-loaded and hinged lip during removal and deposition of pollinia by a male fruit fly. Additionally, a third (novel) movement has been observed, either when the flower closes for the night to protect the stigma, or after a fly has deposited the pollinia onto the stigma, the 'acute tooth' on each column wall folds inwards specifically to secure the newly deposited pollinia.

Keywords: Daciniphilous orchid, Bulbophyllum praetervisum, floral parts, movements, pollination, pollinators, Dacini fruit flies.

Introduction

Orchids have fascinating pollination mechanisms (Baker, H.G. 1963, Jersakova et al. 2006, Lord & Russell 2002, Tan & Nishida 2013), frequently including display (via bright colour) and deception (via fragrance/odour). The odour deception can take four different forms: a) food: b) a female mate to stimulate pseudo-copulation of a potential pollinator (Schiestl et al. 2003): c) a carcass/wound to attract carrion and flesh flies belonging to Calliphoridae and Sarcophagidae, respectively, (Ong & Tan 2011): or d) dung or/and rotting vegetation to attract black scavenger flies, *Dicranosepsis* species (Sepsidae) (first case for this family- manuscript in preparation) as pollinators.

Bulbophyllum, with approximately 2000 recognized species, is the largest genus of Orchidaceae. The flowers possess a highly mobile and functional hinged lip and are well known for releasing a fetid/unpleasant odour (Pijl & Dobson 1969; Kaiser 1993; Cingel 2001). They are mainly pollinated by flies from various families – 6 under suborder Nematocera – Cecidomyiidae, Chironomidae, Culicidae, Mycetophilidae, Psychodidae, Sciaridae; and 11 under suborder Brachycera – Anthomyzidae, Calliphoridae, Drosophilidae, Empididae, Ephydridae, Fanniidae, Lonchaeidae, Milichiidae, Muscidae, Phoridae, Sarcophagidae, Tachinidae, and Tephritidae (Christensen 1994; Larson et al. 2001; Woodcock et al. 2014 – and references therein). Recent additions of dipteran families from the suborder Brachycera are Platystomatidae (Ong & Tan 2011) and Sepsidae (unpublished data – manuscript in preparation). Some other dipteran families are known as just floral visitors, e.g. Bibionidae (Nematocera), and several under the suborder Brachycera, e.g. Bombyliidae, Conopidae, Nemestrinidae, Stratiomyidae, and Syrphidae (Kastinger & Weber 2001; Larson et al. 2001) and Sepsidae (Flügel 2013).

Many *Bulbophyllum* species under the Section Sestochilos (Breda) Benth. & Hook.f., and Beccariana Pfitz. attract carrion (Calliphoridae), flesh (Sarcophagidae), and signal (Platystomatidae) flies via deception, using fetid/carrion odour for pollination – (Ong & Tan 2011). Intriguingly, under the same Sections, there are also many daciniphilous species of *Bulbophyllum.* These produce neither nectar nor fetid odour, and attract males of tribe Dacini, mainly *Bactrocera* and *Zeugodacus* species (Tephritidae: Diptera). These daciniphilous orchids specifically release pleasant spicy

*Corresponding author: tan.kenghong@yahoo.com

or fragrant odours that are pleasant to humans. Fascinatingly, these attractants also act as floral rewards – e.g. (i) methyl eugenol (ME) to attract certain ME-responsive species (Nishida et al. 2004; Tan & Nishida 2012; Tan et al. 2002, 2006), ii) raspberry ketone (RK) for RK-responsive species (Tan & Nishida 2005), and iii) zingerone for both ME- and RK-responsive species (Tan & Nishida 2000, 2007) – exclusively for many fruit fly species, belonging to the two Dacini genera – *Bactrocera* and *Zeugodacus* (combined >500 species), as pollinators and/or visitors.

Flowers of most *Bulbophyllum* species possess a highly modified and movable lip (labellum) to assist in cross-pollination (Vermeulen 1991; Chan et al. 1994; Tan 2009, Tan and Nishida 2013). Ridley (1890) reported that of *Bu. macranthum* Lindl. was pollinated and fertilised by a 'small fly' (probably either a *Bactrocera or Zeugodacus* species). Later, this species was reported to be pollinated by tephritid fruit flies – namely *Ba. albistrigata* (Radian & Padmarsari 2012), although the actual deposition of pollinia onto the stigma was not observed. A similar pollination mechanism was observed in a closely related sibling species, *Bu. praetervisum* J.J. Verm. Flowers of *Bu. praetervisum* release mainly RK while those of *Bu. macranthum* release zingerone (except a chemotype from Philippines) to attract males of certain *Bactrocera* and *Zeugodacus* species for pollination (Nakahira et al. 2018). The floral RK also entices and rewards the attracted flies as vectors for pollination. In this mutualistic inter-relationship, both the orchid and the fruit fly gain sexual reproductive benefits (Tan and Nishida 2005). Flowers of *Bu. praetervisum* differ from those of *Bu. macranthum* by having "a triangular and acute tooth along the lower margin of each stelidium" (Vermeulen, 2002). Since there is no published report on either the movements of floral parts or the role of the floral tooth on each column lateral wall (CLW), we describe and discuss a) three movements shown by different sets of floral parts to ensure successful pollination from blooming until just before fertilisation, and b) the roles of the tooth on each CLW in *Bu. praetervisum* flowers.

Materials and Methods

Plant materials

Original plants, purchased in early 1990s from Kundasang, Sabah, East Malaysia, and as gifts from J.J. Vermeulen, were labelled and each grown on a section of either a tree branch or fern stem *ex situ* in Penang.

Estimation of torsional force of the hinged and spring-loaded lip

A solitary fresh flower was plucked, and the petals and sepals were carefully removed without ever touching the very delicate hinged and spring-loaded lip. This precaution required because the lip could be easily detached from the hinge even by the slightest lateral force. The flower was carefully and firmly attached to a movable platform using a narrow strip of parafilm. The tip of the lip, still in a naturally open position, was carefully adjusted to sit on a modified weighing pan of a torsion balance (0–50 mg with sensitivity of 0.1 mg). The torsional force of the spring-loaded lip was then estimated by measuring the weight required to lift the lip from an open to a closed position. An average of at least five weighings was obtained for each flower. This procedure was repeated for four flowers within 3 hours after anthesis.

Artificial pollination to show movement of column lateral wall (CLW) and its acute tooth

Deposition of pollinia by a male Dacini fruit fly is not commonly observed; and it has only been witnessed twice in the last two decades. Therefore, it was necessary to perform artificial pollination to study movements of CLW and its acute tooth to confirm our observation after the witnessing the two natural depositions of pollinia. Pollinia from a *Bu. pratervisum* flower were removed using the head of a stainless-steel entomological pin (38 × 0.55 mm) and kept at ambient temperature for pollinating the next flower. The pollinia were deposited onto the stigma of a freshly bloomed flower at ca 10:00 h. The movement of the tooth and the CLW was observed closely for at least 8 h and with an aid of time-lapsed photography at 15 min interval. As an experimental control, a piece of fresh carrot, delicately cut to a similar size as a package of four pollinia, was deposited onto the stigma of another *Bu. praetervisum* flower. Three replications were performed.

Field observations of floral parts movements

a) **Anthesis** – The openings of six mature floral buds were observed from 05:00 – 09:00 h.

b) **Closing of flowers** - The closing of petals and sepals was observed for six flowers between 15:00 and 23:00 h.

Results

Movements of floral parts

Movements of the lip, petals and medial sepal are apparent after a bud has fully bloomed. A bud would usually start to open between 05:30 and 06:00 h (dawn breaks ca 07:00 h), and would normally complete blooming by 08:30 h, when the petals and sepals would be fully spread out (Fig. 1). We observed three important movements involving different floral parts, including the acute tooth on each column wall (Fig. 2), during cross-pollination. They are as follows:

I. Lip movement – A flower of *Bu. praetervisum*, like its sibling species *Bu. macranthum sensu stricto*, possesses a hinged and spring-loaded lip (fresh weight 9–12 mg) which is kept in an open position. The delicate lip has a hollow core and floats for more than a day when preserved in pure ethanol, which has a relatively low surface tension under ambient conditions (Nakahira et al. 2018). The estimated minimum force required to counter the spring-loaded hinge, to overcome the inertia to change from an open to a closed position, was estimated to be between 3.0 and 5.1 mg. This is less than the weight of a male fruit fly visitor (5.9–21.4 mg, n = >200, unpublished data). When a male fruit fly is attracted to the flower via a floral synonym containing RK as a major component in three floral chemotypes (Nakahira et al. 2018), it typically probes and feeds mainly on the lateral sepals, especially on slippery surfaces, which arch distally over the floral column. While on the slippery side of the lateral sepals, particularly near the arched distal section, a fly might slip whilst feeding and fall either directly onto the column or

FIGURE 1. Anthesis of *Bulbophyllum praetervisum*. Bar: 1 cm (A) Bud begun to open (at 05:50 h). (B) 40 minutes later. (C) 1 hour later. (D) 1.5 hours later. (E) 2 hours later. (F) Fully bloom (at 08:20 h).

FIGURE 2. Acute tooth (arrow) on each column wall of *Bulbophyllum praetervisum* flower. Bar: 1 cm – Left: Flower fully bloomed in the morning. Rigth: Flower partially closed after pollinia removal – sometimes no tooth movement (at 21:00 h).

onto the lip. In the latter case, the lip immediately tips the fly backwards, abdomen first, onto the column. The fly's weight spontaneously forces the lip into a closed position. While glued to the column, the fly can repeatedly grip and release the lip, which instantly springs back to the open position when released. In the latter situation, the fly is often left to struggle vigorously – legs in the air without any support from a substratum. The movement of the lip may provide an additional outward force, allowing the fly to crawl out and free itself from being glued to the column during either removal or deposition of pollinia.

II. Movement of petals and sepals – Regardless of whether pollinia were removed or not, during late afternoon, especially when the light intensity begins to fall (at about 15:00 h) (Fig. 3A), the flower begins to close partially for the night (requiring a period of several hours). Initially, the petals move very slowly towards each other, to the point of almost

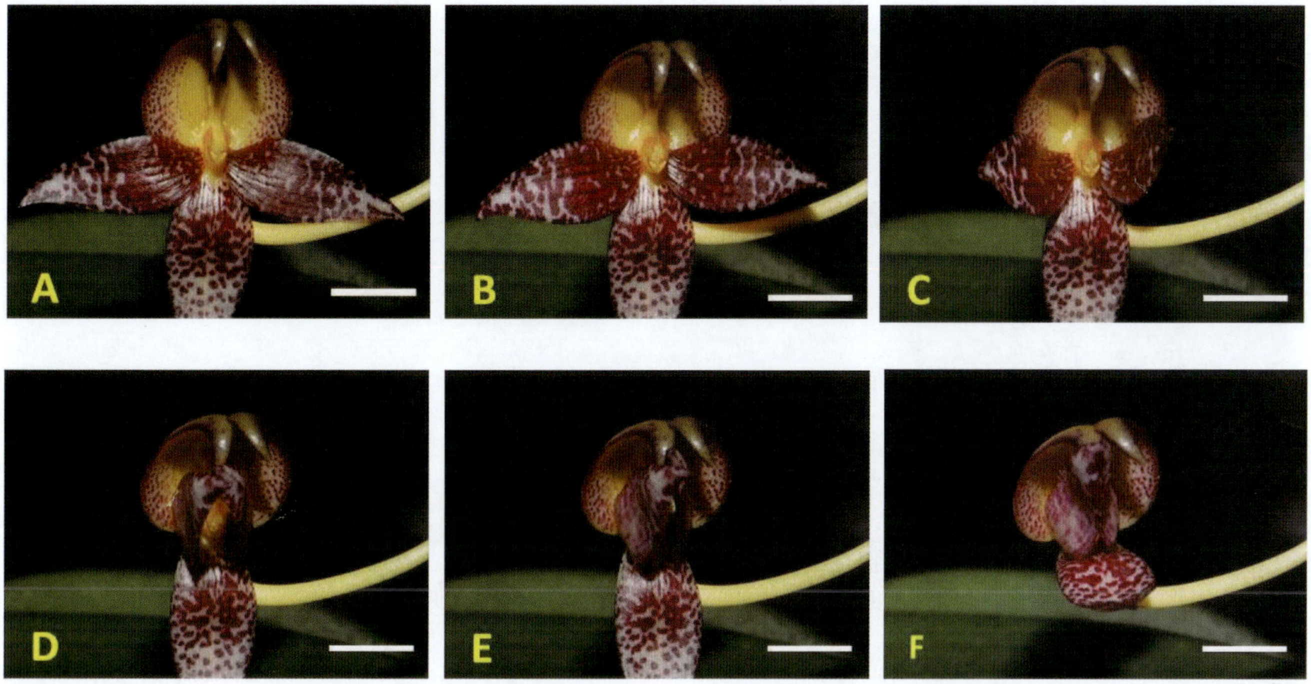

FIGURE 3. Closing of *Bulbophyllum praetervisum* flower for the night when not pollinated. (A) Process initiated with petal tips moving forward - at ca 15:00 h. (B) Tips of column wall teeth moved towards each other - at 18:00 h. (C) Tips of teeth overlap slightly and petal tips in front of the column – at 18:30 h. (D) Teeth overlap and petal tips touching over column and below lateral sepal tips – at 19:00 h. (E) Column hidden and medial sepal began folding – at 20:00 h. (F) Medial sepal moved to a horizontal position – at 23:00 h.

touching above the column (Fig. 3B-D), followed more than an hour later by the medial sepal also moving very slowly from a vertical to an almost horizontal position (Fig. 3E&F). The movement of the latter may not be obvious during the late afternoon. This partial closing of petals and sepals probably serves a) to protect the floral sexual parts from unwelcome nocturnal visitors (Fig. 3E&F), and b) to prevent floral volatiles on the petal, lateral sepal and lip surfaces from being washed away by dew or rain. It should be noted that sometimes a flower may not close on the first night after anthesis. Any reason given for this would be highly speculative, and it warrants further investigation. The movements of different floral parts of the flower during partial closing for the night were reversed the next morning. On the second day, an unpollinated flower will reopen and receive potential visitors again. Pollinia removal by males of RK-responsive species, namely, *Ba. albistrigata, Z. caudatus, Z. cucurbitae* and *Z. tau,* have been observed.

III. Movement of the tooth of the column wall – The acute tooth is slightly arched distally over the stigma cavity (Fig. 2A). It sometimes does not show any movement, even after pollinia removal, when the flower closes for the night (Fig 2B). During a normal closing for the night, after the petals started to move as described above, the folding of the teeth to the point of overlapping each other just prior to sunset (Fig. 3 B-D) may serve to prevent unwelcome nocturnal insect visitors. However, during the morning after a set of pollinia was deposited onto the stigma (Fig. 4A) (by a male fly – either *Ba. albistrigata* or *Z. caudatus,* a very rarely observed phenomenon, witnessed on only two occasions over a period of two decades), the 'tooth' began to fold inwards to face the stigma until they touched each other, and subsequently one on top of the other over the newly deposited

FIGURE 4. Natural pollination of *Bulbophyllum praetervisum* by a Dacini fruit fly. (A) Freshly deposited pollinia (arrow) on the stigma by a fruit fly. (B) Column wall teeth folded to cover the pollinia, ca 4.5 h after deposition by the fly. Bar: 1 cm

FIGURE 5. Column lateral wall (CLW) and 'tooth' movement after pollinia (arrow) were artificially deposited. Bar: 0.5 cm (A) Left tooth started to move inwards – after 2 h. (B) Left CLW folded over pollinia – after 5 h. (C) Right CLW folding over left – after 7 h. (D) CLW overlapped (arrow) over the pollinia with petals almost touching – 11 h.

pollinia (Fig 4B) – as a mother hugs her infant close to her chest. This movement can be induced by artificially pollinating a flower with pollinia from another flower. Figure 5A shows the pollinia on the stigma ca 2 h after deposition. The left tooth started to fold very slowly inwards first (Fig. 5B), followed by the right tooth ca 2 h later (Fig. 5C). The column wall movement to cover the deposited pollinia took about 11 h to complete. The completed movement approximately coincided with the closing of the petals to the point of almost touching each other over the floral column (Fig. 5D). No tooth movement was observed in the controlled flowers when they closed and eventually dropped off. Therefore, it appears that the pollinia may contain certain surface chemical/molecular factors that induce the tooth movement when a flower is pollinated.

This is the first report of the tooth movement followed by that of the CLW (in order to protect and firmly secure the deposited pollinia to the stigma). This final of the three movements shown by floral parts thereby ensures successful cross pollination by a Dacini fruit fly, leading subsequently to fertilisation of the flower and ultimately to the production of seeds within a capsule (Fig. 6). It should be pointed out that our observations showed that *Bu. praetervisum* is self-incompatible, as no seed capsule was formed after self-pollination.

Role of column lateral walls (CLW) during pollination

Observations show that the CLW serve as guiding flaps to align a fly, either falling backwards (abdomen first) from the slippery lateral sepals or being tipped backwards by the closing hinged lip (due to a weight imbalance), in the correct longitudinal position against the floral column to allow either pollinia removal or deposition. When in the correct alignment, the column walls would be on each side of the fly, especially along the fly's abdomen. In this position, the fly,

FIGURE 6. Seed capsule - 4 weeks old.

with its head and thorax above the column cavity and each wing just above the tooth, will wiggle and struggle vigorously to free itself, thereby touching the viscidium, at which point the pollinia were removed by sticking to the fly's abdomen dorsally (Fig. 7). However, if the fly bears pollinia on its abdomen, it will deposit them onto the stigma during its struggle. For the various sibling species of the *Bu. macranthum* complex, the pollinia, when removed, were invariably stuck to the fly's abdomen (Tan 2009; Ong et al 2011; Tan & Nishida 2013). This is where other daciniphilous *Bulbophyllum* species, namely *Bu. ecornutum* (previously published as *Bu. apertum*), *Bu. baileyi*, *Bu. cheiri*, *Bu. elevatopunctatum* and *Bu. vinaceum* – where pollinia are invariably stuck onto the fly's thoracic dorsum - differ from the *Bu. macranthum* species complex (Tan & Nishida 2000, 2005, 2007, 2013; Tan et al. 2002, 2006; Ong et al. 2011).

The flexible and soft tooth of the column wall may provide a certain amount of support so that a stuck fly can move from side to side to loosen the pollinia before it

FIGURE 7. Pollinia of *Bulbophyllum praetervisum* on abdomen of the melon fly. Bar: 1 cm

FIGURE 8. A *Bulbophyllum praetervisum* flower partially closed for the night. *Note:* a dead melon fly, *Zeugodacus cucurbitae*, still glued to the column. Bar: 1 cm

ultimately climbs out of the column cavity. When a fly is glued to the column, either while removing or depositing the pollinia, it normally spends about 0.5 h of intermittent struggling and resting to free itself and leave the column.

Nevertheless, sometimes either a) the tooth hindered the rapid wing movements of the fly, so that it lacked adequate lifting force to free itself; or b) the fly released the lip too early, and thereby lost the opportunity to have a substratum on which it could crawl away from the column; or a combination of both a) and b). In such situations, a fly may be glued forever and eventually die due to dehydration (witnessed on more than a dozen occasions; also reported by Ong et al. 2011) (Fig. 8). As a result, the orchid flower could neither be pollinated nor fertilised.

The second and equally important role of the tooth of the CLW is that of securing a newly deposited package of pollinia onto the stigma. Therefore, when a compatible set of pollinia has been securely deposited, cross pollination is accomplished and will yield ultimately a seed capsule.

ACKNOWLEDGEMENTS

We gratefully thank R. Nishida for his thirty years of moral support and interest on *Bulbophyllum* orchid chemical ecology. Our sincere appreciation also goes to J.J. Vermeulen for providing the original plants to grow *ex situ*.

REFERENCES

Baker HG (1963) Revolutionary mechanisms in pollination biology. Science 139: 3558.

Chan CL, Lamb A, Shim PS, Wood, JJ (1994) *Orchids of Borneo. Vol. 1 – Introduction and a selection of species.* The Sabah Society, Kota Kinabalu, Sabah, Malaysia and Bentham-Moxon Trust, England.

Christensen DE (1994) Fly pollination in the Orchidaceae., In: J. Arditti (ed) *Orchid Biology, Reviews and Perspectives,* VI. John Wiley & Sons, New York, NY, pp 415-454.

Clinger van der NA (2001) An alas of orchid pollination. America, Africa, Asia, Australia. A.A. Balkema, Rotterdam, The Netherlands.

Flügel H-J (2013) The Sepsidae from the 'Lebendiges Bienenmuseum' (Diptera) and her flower visit in Middle Europe (in German). Phillipia 16:81-93.

Jersakova J, Johnson SD, Kindlmann, P (2006) Mechanisms and evolution of deceptive pollination in orchids. Biol. Rev. 81: 219-235.

Kaiser R (1993) The scents of orchids, olfactory, and chemical investigations. Elsevier, The Netherlands.

Kastinger C, Weber A (2001) Bee-flies (*Bombylius* spp., Bombyliidae, Diptera) and the pollination of flowers. Flora Jena 196:3-25.

Larson BMH, Kevan PG, Inouye DW (2001) Flies and flowers: Taxonomic diversity of anthophiles and pollinators. Canadian Entomologist 133:439-465.

Lord E M, Russell SD (2002) The mechanisms of pollination and fertilization in plants. Annu. Rev. Cell Dev. Biol. 18: 81-105.

Nakahira M, Ono H, Wee SL, Tan KH, Nishida R (2018) Floral Synomone diversification of two *Bulbophyllum* sibling species (Orchidaceae) in attracting fruit fly pollinators (Diptera: Tephritidae). Biochemical Systematics and Ecology 81:86-95.

Nishida R, Tan KH, Wee SL, Hee AKW, Toong YC (2004) Phenylpropanoids in the fragrance of the fruit fly orchid, *Bulbophyllum cheiri,* and their relationship to the pollinator, *Bactrocera papayae.* Biochemical Systematics and Ecology 32:245-252.

Ong PT, Tan KH (2011) Fly pollination in four Malaysian species of *Bulbophyllum* (Section Sestochilus) – *B. lasianthum, B. lobbii, B. subumbellatum* and *B. virescens.* Malesian Orchid Journal 8:103-110.

Ong PT, Hee AKW, Wee SL, Tan KH (2011) The attraction of flowers of *Bulbophyllum* (Section Sestochilus) to *Bactrocera* fruit flies (Diptera: Tephritidae). Malesian Orchid Journal 8:93- 102.

Pijl van der L, Dodson CH (1969) Orchid flowers, their pollination and evolution, second edition. University of Miami Press, Florida. 214 pp.

Radian FX, Padmarsari W (2012) Pollination biology of orchids of the genus *Bulbophyllum* in Labian-Leboyan Corridor. pp 109-112. *In:* Widmann P, Tiju A, Putra MH, Syahirsyah, Wulffraat S (eds.) Connecting Diversity – People and Nature of Labian-Leboyan Corridor in the Indonesian Heart of Borneo. WWF-Indonesia. www.hobgreeneconomy.org/downloads/Connectingdiversity_peopleandnature.pdf

Ridley HN (1890) On the method of fertilization in *Bulbophyllum macranthum,* and allied orchids. Annals of Botany 4:327-336.

Schiestl FP, Peakall R, Mant JM, Ibarra F, Schulz C, Franke S, Francke W (2003) The chemistry of sexual deception in an orchid-wasp pollination system. *Science* 302:437-438.

Tan KH (2009) Fruit fly pests as pollinators of wild orchids. Orchid Digest, 73:180-187.

Tan KH, Nishida R (2000) Mutual reproductive benefits between a wild orchid, *Bulbophyllum patens,* and *Bactrocera* fruit flies via a floral synomone. Journal of Chemical Ecology 26:533-546.

Tan KH, Nishida R (2005) Synomone or Kairomone? - *Bulbophyllum apertum* (Orchidaceae) flower releases raspberry ketone to attract *Bactrocera* fruit flies. Journal of Chemical Ecology 31(3):509-519.

Tan KH, Nishida R (2007) Zingerone in the floral synomone of *Bulbophyllum baileyi* (Orchidaceae) attracts *Bactrocera* fruit flies during pollination. Biochemical Systematics and Ecology 35:334-341.

Tan KH, Nishida R (2012) Methyl eugenol – its occurrence, distribution, and role in nature, especially in relation to insect behavior and pollination. Journal of Insect Science 12:56 1-74. online: insectscience.org/12.56

Tan KH, Nishida R (2013) Pollination of bactrocerophilous *Bulbophyllum* Orchids. In: Elliott J, Kurzweil HF, O'Byrne P, Tan KW, van der Schans AS, Wong SM, Yam TM (eds.) *Proceedings of the 20th World Orchid Conference,* Singapore Botanic Gardens, Singapore, pp 273-279.

Tan KH, Nishida R, Toong YC (2002) Floral synomone of a wild orchid, *Bulbophyllum cheiri,* lures *Bactrocera* fruit flies for pollination. Journal of Chemical Ecology 28:1161-1172.

Tan KH, Tan LT, Nishida R (2006) Floral phenylpropanoid cocktail and architecture of *Bulbophyllum vinaceum* orchid in attracting fruit flies for pollination. Journal of Chemical Ecology 32:2429-2441.

Vermeulen JJ (1991) *Orchids of Borneo. vol. 2 – Bulbophyllum.* Bentham-Moxon Trust, Toihaan Publishing Co. and the Sabah Society, Kota Kinabalu, Sabah, Malaysia 342 pp.

Vermeulen JJ (2002) *Bulbophyllum praetervisum* J.J. Verm. (Orchidaceae), an overlooked species close to *B. macranthum* Lindl. Gardens' Bulletin. Singapore 54:153-154.

Woodcock TS, Larson BMH, Kevan PG, Inouye DW, Lunau K (2014) Flies and flowers II: Floral attractants and rewards. Journal of Pollination Biology 12:63-94.

FACTORS DETERMINING VISUAL DETECTION DISTANCE TO REAL FLOWERS BY BUMBLE BEES

Gidi Ne'eman[1,*] and Rina Ne'eman[2]

[1]*Department of Biology and Environment, Faculty of Science and Science Education, University of Haifa-Oranim, Oranim, Tivon 36006, Israel.*

[2]*Amirim 20115, Israel*

Abstract—Plants use visual signals to attract pollinators and direct them to their flowers. Visual capabilities of bees have been extensively studied mostly using artificial paper models. However, there is no empirical determination of the maximal detection distance (MDD) or minimal subtended visual angle (MSVA) of real flowers. Using a six armed radial maze, we tested MDD and MSVA of 12 types of natural and manipulated real flowers by bumble bee (*Bombus terrestris*) workers. Bees were initially trained to obtain sugar solution at target flowers that were presented at close range on a mobile divider at the back of one of the six arms. Bees were individually marked and tested. For bees that passed the short range test, we gradually increased the distance of the target flower, until the number of successful choices reached chance level, indicating that they could not see the target flowers. The results show that MSVA of flowers is correlated with flower diameter but not with MDD. The variation in MDD to natural flowers by bumble bee workers can be best predicted by: **MDD = flower coloured area / (contour line * green contrast)**. Contour line length determines flower dissectedness. Full circular flowers can be detected from longer distance than dissected flowers with identical diameter. We hypothesize that dissected flower shapes might be compensated by their higher attractiveness for bees. Empirical determination of real flower MDD and MSVA is important for studying bee foraging behaviour, pollinator induced evolution of flower traits and validation of neurophysiological visual models.

Keywords: Bumble bee; detection distance; flower area; flower contour; green contrast; minimal subtended visual angle

INTRODUCTION

Pollinators visit flowers for their reward (e.g. Heinrich 1975), and plants use olfactory and visual signals to advertise their reward, attract pollinators and direct them to their flowers (Chittka & Raine 2006; Dafni et al. 1997; Menzel & Shmida 1993; Schaefer et al. 2004). Sprengel (1793, in Hasse, 1996) recognized the importance of floral size and shape in "catching the eye of the insect from afar". Flower colour and shape recognition are based on visual perception and learning as shown by the classical work of Karl von Frisch (1914) and others (Hertz 1929). Kugler (1934) was the first to experimentally determine the distance from which bumble bees detect artificial flowers and change their flight direction to visit them, showing that large flowers attract bees from longer distances than small ones.

In accordance to 'honest signalling' principle (Armbruster et al. 2005; Keasar et al. 2006), large flowers are more conspicuous and detectable from longer distances (Chittka & Spaethe 2007) and they generally contain more nectar or pollen rewards than small flowers (Blarer et al. 2002; Fenster et al. 2006; Herrera 2005; Martin 2004). Pollinators typically prefer large over small flowers, visit them more frequently and pollinate them more efficiently (Conner & Rush 1966; Elle & Carney 2003; Fenster et al. 2006). Consequently, the fitness of plants with larger flowers, which can be detected from longer distances by their pollinators, should be higher than that of plants with smaller flowers. Thus pollinators could generate a selective pressure for increased flower size via pollinator mediated selection (Galen & Newport, 1987).

However, in addition to the effect of pollinator mediated selection, the evolution of flower size can also be constrained by selection imposed by plant mating systems (Goodwillie et al. 2010), interactions with flower and plant herbivores, resource costs (Galen, 1999), and a suite of other abiotic and biotic agents of selection that together may shape floral evolution (Krizek & Anderson, 2013). Variation in flower size and other floral traits, within a species, can promote reproductive isolation and ultimately speciation (Krizek & Anderson, 2013).

Giurfa et al. (1996) determined the minimal subtended visual angle (MSVA) that (for full circular paper targets) corresponds directly to maximal detection distance (MDD) from which honeybee (*Apis mellifera*) workers can detect them. By varying target diameter, Giurfa et al. (1996) showed that visual angle is the only factor needed to estimate single full circular target resolution independent of target distance and size. They found that MSVA from which honeybees could detect full circular targets was 5° for targets that provide chromatic and green receptor contrasts to their background, and 15° for targets of similar size, with chromatic but no green receptor contrast. Visual angle is a

*Corresponding author: gneeman@gmail.com

very important parameter in general vision science and it is used by researches because the actual size of a stimulus is confounded by viewing distance. Thus, targets providing green contrast, irrespective of their chromatic contrasts, could be detected from longer distances than corresponding targets with only chromatic contrast to their background (Giurfa et al. 1996; Giurfa & Vorobyev 1997; Lehrer & Bischof 1995). From these data, models and simulations were constructed to determine the distances from which honeybees could detect flowers, depending only on flower diameter and colour (Giurfa & Vorobyev 1997; Menzel et al. 1997; Vorobyev et al. 1997).

Differing from honeybees, bumblebees are able to detect targets containing both green and colour receptor contrasts at a visual angle of 2.3° and stimuli that contain only colour contrast at a visual angle of 2.7° (Dyer et al. 2008). Recently, Dyer et al. (2016) determined these values also for Australian stingless bees, 9.5° for targets that presented both chromatic- and green-receptor contrasts, and 9.3° for targets presenting only chromatic contrast. These differences between honeybees, bumblebees and stingless bees may have evolved because these bees operate in different environments, but more probably may be due to the efficient dance language that helps honeybee workers to find flowers, whilst bumblebees and stingless bees search individually and thus may require better colour acuity to find small individual flowers (Dyer et al. 2008; Bukovac et al. 2013; Dyer et al. 2016).

Many studies on bee vision have focused on perception, neurobiology, memory and learning mainly of honeybees and bumblebees (Chittka & Raine 2006; Menzel & Giurfa 2006; Srinivasan 2010; Avarguès-Weber et al. 2011; Avarguès-Weber & Giurf, 2014; Hempel de Ibarra et al. 2014; Hempel de Ibarra et al. 2015). The roles of size, shape and pattern in floral advertisement and their perception by pollinators were reviewed by Lehrer & Bischof (1995), Dafni and Kevan (1997), Dafni et al. (1997), Neal et al. (1998), Zhang et al. (2004), Kaczorowski et al. (2012), Avarguès-Weber et al. (2011), and Hempel de Ibarra et al. (2015). Critical examination of the literature shows that most of the studies used full circular coloured papers, or other artificial targets, but not real natural flowers. Streinzer et al. (2009), and Dyer et al. (2007) examined the detection of real flowers by bees, but did not determined their MDD or MSVA.

Studies using circular artificial targets, which have used diameter and colour as the most important shape parameters to determine MSVA and consequent MDD, seem to be insufficient for inferring to flower-like dissected shapes. Therefore, Ne'eman and Kevan (2001) explored MDD using blue paper flower models with dissected outlines instead of circles. They demonstrated that target coloured area, rather than the diameter of its enclosing circle, is the single most important parameter that affects MDD. Ne'eman and Kevan (2001) showed that MDD is positively correlated with squared target coloured area and negatively with the target contour length (perimeter, cm). MDD to radial dissected targets, which present longer contour lines, is shorter than to non-dissected targets of similar diameter. MDD to a full circular target with a given diameter is equal to that of a dissected target with the same area but double diameter. Therefore, MSVA cannot be used for calculating their MDD to dissected targets and flowers. Markedly lacking, however, are experiments on the distance from which bees can detect natural flowers of different sizes and shapes, and their correspondent MSVA.

Contour length has been recognized long ago as important for bee vision and behaviour. von Frisch (1914) showed that bees could discriminate between shapes that differed greatly in their contour lengths but could not discriminate between simple shapes, such as triangles, squares and circles that had similar contour lengths. In spontaneous attraction experiments bees were found to prefer shapes with higher contour lengths (Hertz 1929). Anderson (1977) studied the degree of similarity, perceived by the honey bee, between a training shape and each of a set of test shapes. He found that shape area and contour density (the ratio between the length of outline of a shape and the area it encloses) explained the observed degree of similarity perceived by the bees. Bees are capable of much more elaborate pattern detection, differentiation and learning than was suspected in these early experiments (Avarguès-Weber et al. 2011; Srinivasan 2010). In recent papers it was discovered that honeybees have the ability to group similar stimuli into categories. They learned to distinguish between images belonging to different categories: star-shaped flowers of different colours, images of flowers that were nearly circular in shape of different colours, images of plant stems of various shapes, and images of landscapes (Zhang et al. 2004). Honeybees are able to discriminate Monet paintings from Picasso ones by extracting and learning the characteristic complex visual information inherent to each painting style (Wu et al. 2013). Individual free flying honeybees can learn to use size relationship rules to choose either the larger or smaller stimulus as the correct solution in a given context, and subsequently apply the learnt rule to novel colours and shapes (Avarguès-Weber et al. 2014).

While the visual and cognitive capabilities of bees have been extensively studied using artificial models and simulations, empirical determination of actual MDD to real flowers and their correspondent MSVA, which has a crucial ecological effect on foraging bees and pollinator mediated selection on flower size is missing.

Our aim is to determine, for the first time, MDD, to natural flowers by bumble bee workers. In addition we construct an empirical regression model as a practical tool for predicting MDD to real flowers using parameters of flower shape and colour. Empirical determination of flower MDD and MSVA are important for studying bee foraging behaviour, pollinator mediated evolution of flower size shape and colour, as well as validation of psychophysical and neural models established for flower shape and colour.

MATERIALS AND METHODS

Apparatus

Following other researchers (e.g. Lehrer et al., 1995; Simonds & Plowright, 2004; Plowright et al. 2011;

FIGURE 1(A). General view of the six-armed radial maze (100 cm in diameter). A bumble bee colony (yellow box on the left) was connected with a transparent walkway and a plastic tube going under the maze floor to the entrance hole in the maze centre. At one distal corner of each arm there was a small open headed compartment (indicated by arrows) into which fresh flowers of the tested species were inserted. (B) A 'bee view' of the central emergence whole in the maze floor and the target flower (indicated by an arrow) presented on a mobile stand in the rewarding arm.

Thompson & Plowright 2014) we used a six armed radial maze (100 cm in diameter) (Fig. 1A). We used this type of maze because rejection of the null hypothesis of random choices can be achieved with lower number of attempts than in the commonly used dual choice Y shaped maze without losing statistical significance, as explained later (Bukovac et al. 2016).

A bumble bee colony was connected to the maze with a transparent Plexiglas® walkway that led to the entrance hole located in the centre of the maze floor. The walkway had three sliding shutters for controlling bee traffic so that single numbered bees could be identified and let one by one into the maze. A yellow circle was painted around the entrance, to help bees navigate back home. A yellow line connecting the entrances to the six arms marked the borders of the 'decision chamber'. When crossing this line to the rewarding arm first, the bee's choice was considered as successful. An error was recorded if the bee crossed the line to enter an unrewarding arm first. The decision line was 10 cm from the entrance hole and it served as the baseline for measuring the distance of the target flowers. Each arm was 5 cm wide at its proximal and 15 cm at its distal end; the maze was 20 cm high and covered by removable UV-transmitting Plexiglas® plates. The maze was painted dark green, and only in one arm a flower was vertically presented and 50% (weight/volume) sugar solution offered as reward. The flower was mounted on a vertical dark green cardboard fixed to a mobile stand (Fig. 1B). The flower's stem was inserted through a hole into a tube with fresh water located behind the cardboard. At the base of the stand, 5 ml of sucrose solution was presented in a small covered compartment, invisible by the bees from a distance. In a preliminary experiment four bees successfully discovered the uncovered reward from 15 cm (7/9, 10/11, 5/5, and 9/10 correct/total choices), but failed to do so with covered reward (1/8, 2/10, 1/6, and 3/10 correct/total choices); for significance level see 'data analyses'. Small marks at five cm distances on the floor at the central line of each arm were used to determine the distance of the target flower from the decision line during the experiments.

Using natural flowers in a maze could introduce a difficulty to separate between visual and olfactory attractions of the bees to the flowers. We are confident that in our radial maze, which consisted of a single enclosed space and the random changes of the arms caused equal distribution of the flowers' fragrance in all the maze's arms. However, to ensure equal distribution of the fragrance in the maze, in one distal corner of each arm there was a small open headed but hidden compartment into which fresh flowers of the tested specie were inserted (Fig 1 A).

The maze was lit with high frequency lighting (TMS 24F lamps with 4.3 kHz ballasts; (Philips, Eindhoven, The Netherlands) fitted with fluorescent Activa day light tubes.

Flower types

We used nine flower species (Tab. 1) that were freshly collected each morning (April-May 2012) in the gardens of Queen Mary University of London campus and in the nearby Victoria Park. The flowers were presented vertically, thus seen by an entering bee as if it looked at a natural flower in an 'en face' position. Inflorescences of *Bellis perennis* were presented unmanipulated, but were also manipulated to produce a 16 and 8 petal flower-like shapes; flowers of *Argyanthemum frutescens* were manipulated to produce a 10 petal flower-like shape and flowers of *Spartium junceum* were presented in an 'en face' and profile positions; in total we used 12 flower presentations differing in the following shape parameters (Tab. 1): (1) As a general indication for flower size, we calculated the diameter of the flower-enclosing circle. (2) The area of the dominant colour of the flower (petals). (3) The contour line (perimeter) of the

TABLE I. Shape parameters: flower (coloured) area, average diameter and the contour line (perimeter) of all tested flowers and the value of the green-receptor contrast between petal colour and the green background of the radial maze against which the flowers were presented to the bees.

Flower	Area (mm^2)	Diameter (mm)	Contour line (mm)	Green-receptor contrast
Allium cristophii	111	17	128	0.379
Aubrieta deltoidea	314	20	74	0.221
Argyanthemum frutescens 10	606	30	234	0.451
Bellis perennis	112	12	47	0.499
Bellis perennis 8	134	16	142	0.499
Bellis perennis 16	242	22	259	0.499
Cotoneaster microphyllus	105	12	61	0.463
Escallonia macrantha	151	15	63	0.255
Ranunculus bulbosus	174	15	55	0.468
Solanum crispum star	190	21	126	0.427
Spartium junceum (en face)	510	26	117	0.379
Spartium junceum (profile)	288	25	113	0.379

coloured area. (4) The value of the green receptor contrast between the dominant petal colour and the green background of the radial maze, against which the flowers were presented to the bees. Floral morphological parameters were measured from calibrated photographs using "CellProfiler 2.0" cell image analysis free software (Massachusetts Technological institute).

Colour Analysis

Spectral reflectance functions of the flowers and the background was measured by using a spectrometer (Ocean Optics, Dunedin, FL, S2000 with a deuterium-halogen light source). The colour parameters (relative excitation values in the bees' UV, blue, and green receptors, colour contrast, and green contrast) were calculated according to Backhaus (1991) using the colour hexagon (Chittka, 1992). The relative amount of light absorbed by each photoreceptor colour type is:

$$P = R \int_{700}^{300} I\, s(\lambda)S(\lambda)D(\lambda)d\lambda \qquad \text{[Eq. 1]}$$

$Is(\lambda)$, is the spectral reflectance function of the stimulus. $S(\lambda)$, is the spectral sensitivity function of the receptor (we used the functions of Peitsch et al. (1992) for the *B. terrestris* UV, blue, and green receptors). $D(\lambda)$ is the illuminate (standard neon light filtered through the Plexiglas cover combined with natural daylight). The sensitivity factor R in Eq. 1 is determined by:

$$R = I \int_{700}^{300} Ib\,(\lambda)S(\lambda)D(\lambda)d\lambda \qquad \text{[Eq. 2]}$$

Ib is the spectral reflection function of the background to which the receptors are adapted (Fig. 2). With this model, it is assumed that the photoreceptors display half their maximal response when stimulated by the light reflected from the adaptation background. When the maximum excitation E_{max} of the photoreceptors is normalized to 1, the photoreceptor excitation can be described by:

$$E = P/(P+1) \qquad \text{[Eq. 4]}$$

P is the stimulus strength, in units such that for $P = 1$, $E = 0.5$ [i.e., half the maximum potential; for details see Backhaus (1991) and Vorobyev et al. (1997)]. Thus, for the adaptation background, E equals 0.5 in each photoreceptor. Green contrast, then, is the degree to which any given stimulus generates an excitation value different from 0.5 in the green receptor. Because excitation can range from 0 to 1, the maximum green contrast is 0.5. Stimulus brightness is defined as the sum of all three photoreceptor excitations, so it can have any value from 0 to 3. Because the background, by definition, has a brightness of 1.5, brightness contrast can have any value up to 1.5. For calculation of hexagon colour loci from receptor excitation values, see Chittka (1992). The background colour locus lies in the centre of the colour hexagon. The distance from the centre to any of the hexagon's corners is unity. Perceptial colour differences of stimuli in the colour hexagon is described by a sigmoidal relationship with colour distance (Dyer & Chittka 2004) perceived as differently coloured.

Pictures and relative reflectance curves of the flowers are presented in Fig. 2. The location of the colours of all studied flowers are presented in a bumble bee colour-hexagon (Fig. 3), where the dark green radial maze background is located at the centre, and the distance of each flower from the centre indicates its colour contrast with the background (Chittka 1992).

Bicoloured patterns are detected by bees worse than single coloured and patterns with higher L-contrast in their outer than in their inner pattern are better detected than that with a reverse L-contrast distribution (Hempel de Ibarra et al. 2001). However, because of their small size we could not measure the central parts of the bicoloured flowers and measured only the colour of the dominant peripheral petals. It is generally accepted that 'nectar guides' can be seen only

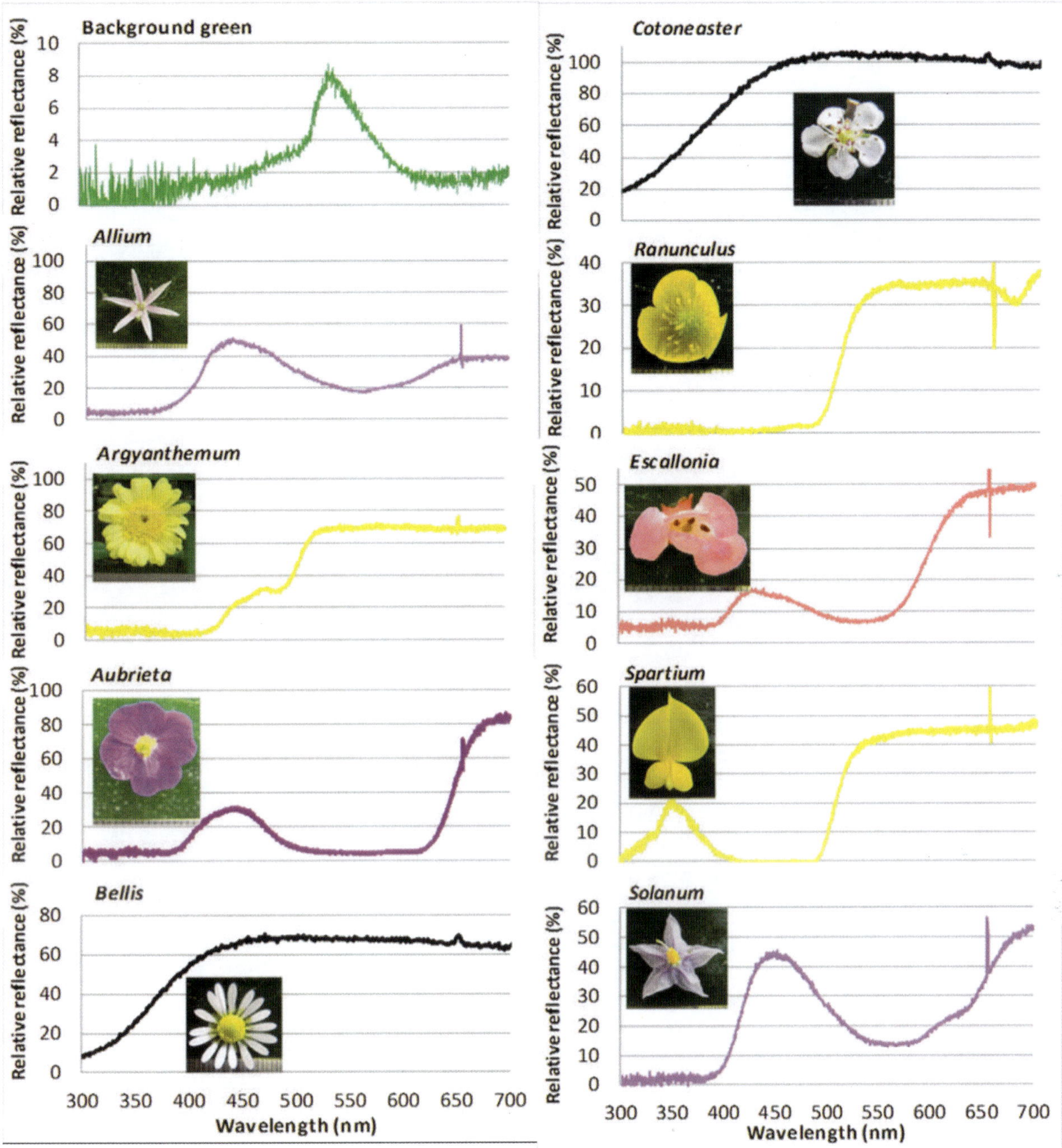

FIGURE 2. The relative reflectance curves of all studied flower species: *Allium cristophii, Argyanthemum frutescens, Aubrieta deltoidea, Bellis perennis, Cotoneaster microphyllus, Escallonia macrantha, Ranunculus bulbosus, Solanum crispum* and *Spartium junceum*. For bicoloured flowers (with yellow centres) the reflectance of the petals' major colour was measured.

from short distances and thus should have no or minimal effect on MDD.

Bees, training and experimental procedure

A bumble bee (*Bombus terrestris*) colony was purchased from Syngenta Bioline Bees, The Netherlands. The colonies were kept in dark wooden boxes and the bees were reared in a dark, colour-devoid environment, until training began.

Foraging bees were first allowed to freely enter the maze for two days to acclimatise. During acclimatization, the bees were pre-trained to feed at the feeding stand with a yellow paper target (5 cm in diameter) placed close to the decision line. Bees that were observed to successfully find the reward were marked with numbered tags for identification. Marked bees were then trained to forage at the feeding stand with a flower target at a distance of 5 cm from the decision line. During a single training and testing session, we worked with up to seven bees, but only one bee was permitted to enter the maze at a time. After each foraging bout, the target was randomly moved to another arm, using a random number chart. The training period enabled bees to associate the sucrose reward with the flower target. The randomisation of the target placement and rewarding arm ensured that bees were prevented from linking the reward to any particular arm (spatial learning). After a minimum of three training foraging bouts, the bees were tested. Only bees that successfully passed the test from 15 cm, were tested with gradually increasing distances to determine MDD to any tested flower type. All the flowers were tested with sister bees of one colony.

After entering the maze, each bee was observed until it arrived at the feeding stand. Bees that entered the rewarding arm first, were considered to have a successful trial. Bees that after entering crossed first a decision line of an empty arm were considered to have an error. All bees were left to find the target and feed before returning to the colony. After every foraging bout the target was randomly moved to another arm and the distance increased with 10 cm steps, until all individual bees' performance decreased to chance levels. At this stage the target stand was moved 5 cm back toward the decision line and retested. MDD was determined as the average between the longest distance from which a bee could see the target and the shortest one from which it failed. After the MDD was determined for each target, the colony was once again permitted to freely enter the radial maze, so that more bees could be recruited through training and marking. Each target was tested with five to nine bees.

MSVA was calculated from MDD and the average flower diameter (D): MSVA = 2arctan (D / 2MDD).

Data analyses

The binomial test was used to determine, whether or not, the target could be detected by a given bee from any tested distance. The probability of random choice (1/6) is 0.17, but to be on the safe side, we used the probability threshold of 0.2. In each session (testing a flower from a given distance), each bee was tested during up to 12 trials. In order to be successful in a session (choices not to be random at a 95% confidence level) a bee had to be successful for at least in 4/5, 4/6, 4/7, 4/8, 5/9, 5/10, 5/11, and 6/12 successful/total choices.

To establish a model for the best prediction of MDD we used linear regression with the following parameters as independent variables: diameter of enclosing circle, coloured area, contour line (perimeter), green receptor contrast and colour distance between flower and the background. We used Kolmogorov-Smirnov test for testing normal distribution of the data. SPSS 19 (IBM Corp., 2010) was used for all statistical analyses.

RESULTS

Flower types

Morphological shape parameters and green receptor contrast of all flower species used in the experiment are presented in Tab. 1, the relative reflectance curves are presented in Fig. 2, and the locations of the studied flower species in the bumble bee colour hexagon in Fig. 3.

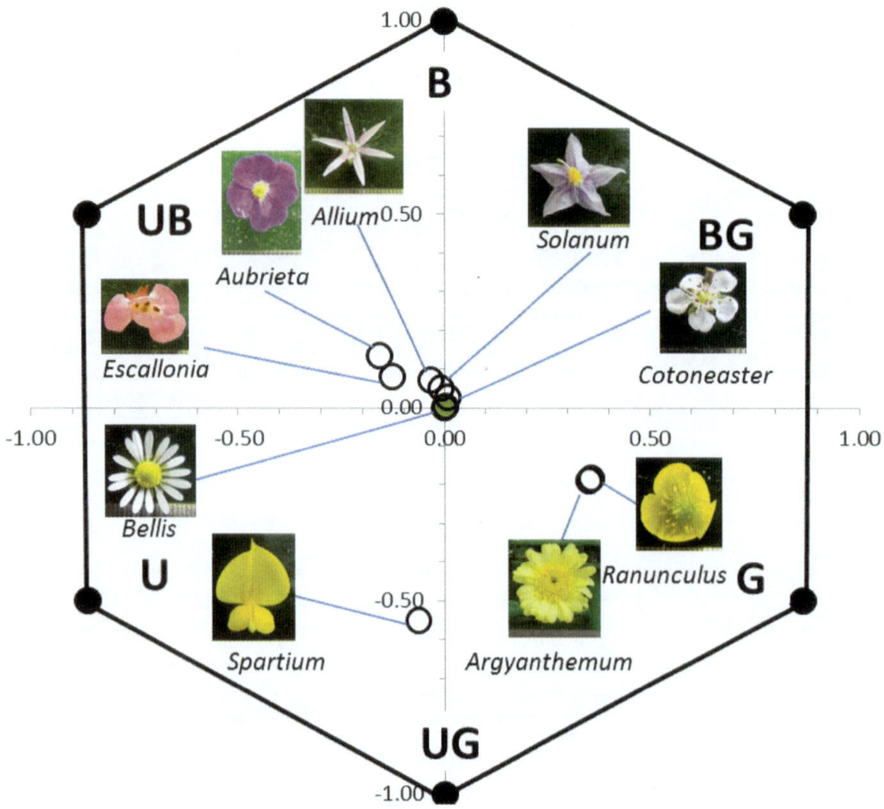

FIGURE 3. The location of the colours of studied flowers (Allium cristophii, Argyanthemum frutescens, Aubrieta deltoidea, Bellis perennis, Cotoneaster microphyllus, Escallonia macrantha, Ranunculus bulbosus, Solanum crispum and Spartium junceum) in the bumble bee colour-hexagon. The dark green radial maze background is located at the centre and the distance of each flower from the centre indicates its colour contrast with the background (Chittka 1992).

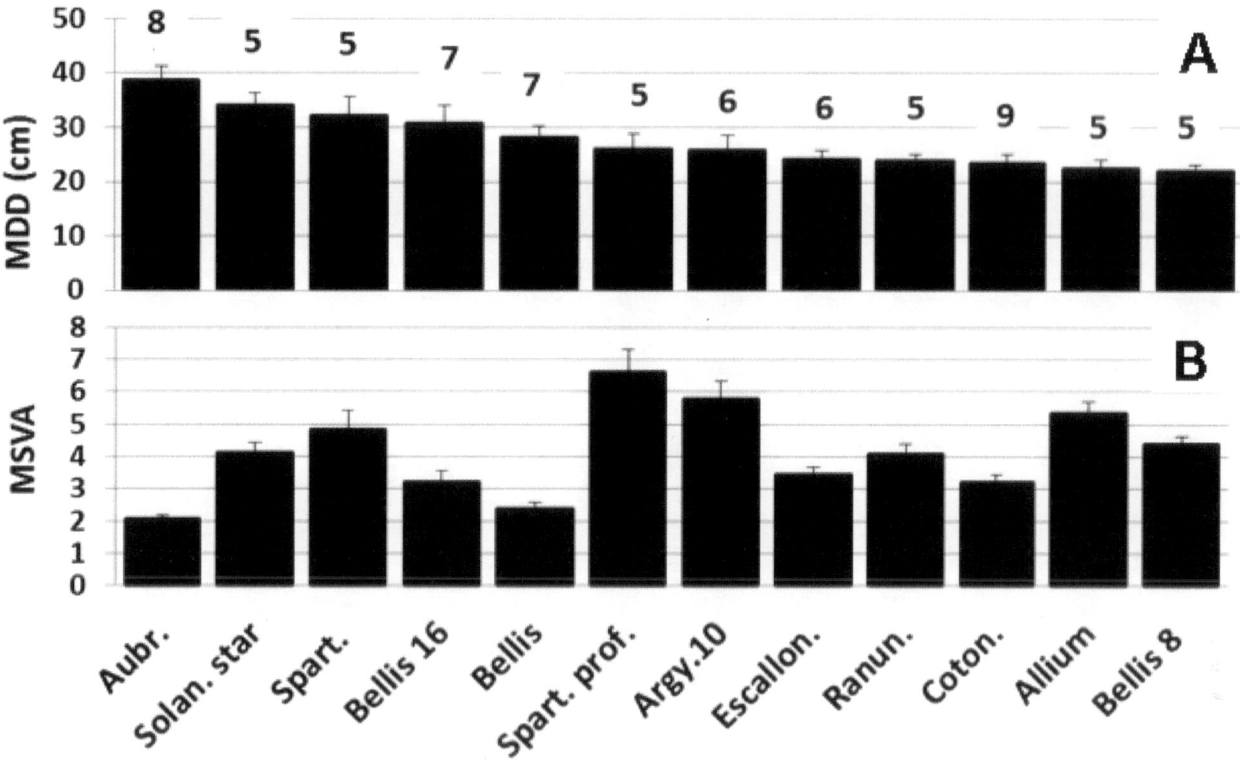

FIGURE 4. The average (±SE) maximal detection distance (MDD) (cm) (A), and the minimal subtending visual angle (MSVA) (B) of the studied flower types: *Aubrieta deltoidea* (Arbur.), manipulated *Solanum crispum* (Solan. star), *Spartium junceumI* (en-face) (Spart.), manipulated *Bellis perennis* (Bellis 16), reduced diameter *Bellis perennis* (Bellis), *Spartium junceumI* (profile) (Spart. prof.), manipulated *Argyanthemum frutescens* (Argy.10), *Escallonia macrantha* (Escallon.), *Ranunculus bulbosus* (Ranun.), *Cotoneaster microphyllus* (Coton.), *Allium cristophii* (Allium), manipulated *Bellis perennis* (Bellis 8). The number of tested bees for each flower is indicated above the bars in A. Manipulated flowers have reduced number of petals as indicated by the number following their name in the horizontal axis. The visual parameters of all flower types are presented in Tab. 1.

Maximal detection distance and minimal subtending visual angle

For each tested flower type, the average percentage of correct choices as a function of the distance made by *Bombus terrestris* workers; the correspondent binomial probability (*P*-value) that the number of correct choices was higher than 20% of the total number of trials (average = 7, range 5-9) and the number of bees tested for each flower type are presented in Appendix I. The distance from which *P* value exceeded 0.05 was determined as the MDD for each flower type and was used for calculation of its MSVA. The MDD of each individual tested bee to each of the tested flowers and their main traits is presented in Appendix II.

The average diameter (± STD) of the 12 tested flower was 1.9 ± 0.6 cm. The MDD from which flowers were detected by bumble bee workers ranged from 22.0 cm to 38.8 cm with an average (± STD) of 27.6 ± 5.2 cm (Fig. 4A), and the differences among flower types were significant (ANOVA, $F_{11,63} = 6.458$, $P < 0.001$).

We used linear regressions to find the best significant model for predicting MDD as a function of flower shape and colour traits as well as various combinations of them (Tab. 2). The results prove that and MDD cannot be predicted by: (1) Flower diameter (cm); (2) contour density (contour line length/area) (cm^{-1}); (3) flower area (cm^2); (4) flower area divided by flower contour line (cm); (5) Green receptor contrast between the flower and background; (6) Euclidean colour distance between the flower and the background in bumble bee colour hexagon), none of which was significant (Tab. 2).

Multiple linear regression using uncorrelated predicting variables (flower area, flower contour line, green receptor contrast and chromatic distance) was also not significant ($F_{4,11} = 0.564$, $P = 0.697$), as were Pearson correlation coefficients between MDD and each of the variable presented in Tab. 2.

The only significant regression mode for prediction of MDD to the flowers (Fig. 5) was:

$$\mathbf{MDD = A/(C*GC)} \qquad \text{[Eq. 4]}$$

A = floral coloured area (cm^2), **C** = floral contour line (cm), and **GC** = degree of floral green-receptor contrast to the background.

The average MSVA (± STD) for the tested flowers was 4.12 ± 1.36 cm (Fig. 4B), but it was not correlated with MDD (one tailed Pearson correlation $r = -0.409$, $P = 0.094$).

Independent variable	$F_{1,11}$	P	R^2
Diameter (cm)	3.245	0.102	0.1153
Contour density (contour line /area) (cm^{-1})	1.403	0.264	0.1238
Area (cm^2)	1.399	0.264	0.1228
Area/contour line (cm)	3.245	0.102	0.2450
Green receptor contrast	0.708	0.420	0.0661
Euclidean colour distance (hexagon)	1.797	0.210	0.1523
Area /(contour line *green contrast) (cm^2)	6.176	0.032	0.3822

TABLE 2. Results of regression tests (*F*, *P* and R^2) of MDD as dependent variable and various flower shape and colour parameters as independent variables ($N = 12$). Green-receptor contrast was calculated between flower and the green maze background.

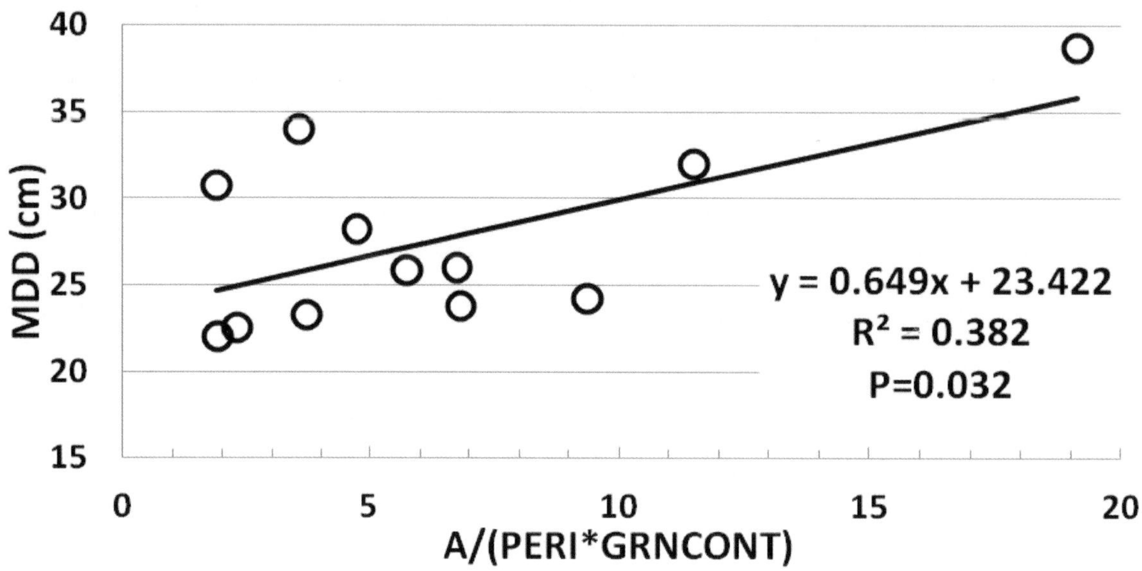

FIGURE. 5. Maximal detection distance (MDD) (cm) as a function of floral area (A cm^2) divided by floral contour line (C) and the degree of floral green contrast (GC). Regression line formula, its significance (*P*) and the proportion of explained variability are presented (R^2).

MSVA of the flowers was correlated with average flower diameter (one tailed Pearson $r = 0.605$, $P = 0.0185$), and it was negative correlated with relative flower area (the proportion of actual flower area out of an area of a circle with a diameter equal to the flower average diameter) (one tailed Pearson $r = -0.545$, $P = 0.034$). MSVA can be predicted from its significant regression line with the average diameter of the flower: MSVA = 0.142*Diameter + 1.371 ($F_{1,10} = 5.766$, $R^2 = 0.366$, $P = 0.037$).

DISCUSSION

Here we present empirical data for determination of MDD to various intact or manipulated real flowers by *Bombus terrestris* workers and their MSVA. To the best of our knowledge, this is the first empirical measurement of MDD to natural flowers, or their correspondent MSVA. The average diameter of the tested flower was almost 2 cm, the average MDD by bumble bee workers was only 28 cm. That is in agreement with the predictions of a previous theoretical model that used photographs of flowers that bees can visually detect flowers only from close range (Vorobyev et al., 1997). The average MSVA was 4° and it was not correlated with MDD. The lack of correlation is apparently because of differences in shape and green contrast of flowers that shared similar diameters. Therefore, and because MDD is the parameter that dictate the behaviour of bees in the field, and thus affects their foraging efficiency as well as the selection they impose on floral traits, we propose that when discussing detectability distances to real flowers MDD is the most relevant and important variable, as is MSVA for neurophysiological models.

MDD to flowers under natural conditions is probably even shorter than those obtained here under laboratory conditions. This is mainly because bees naturally approach flowers from a variety of angles from which the perceived area of the flowers is smaller than the *en face* view tested here which, in most cases, presents flowers at their maximal area (Spaethe 2001). In addition, natural backgrounds are not uniformly green, as our experimental maze, but are commonly multi coloured and heterogeneous, which can make visual detection harder.

The diameter and colour properties of circular paper targets has been commonly used as the only variable for calculating MSVA (α_{min}) by bees (Chittka & Raine 2006;

Giurfa et al. 1996; Giurfa & Vorobyev 1997; Lehrer & Bischof 1995; Menzel et al. 1997; Menzel & Giurfa 2006; Vorobyev et al. 1997) and its corresponding MDD. However, compound non-radial and deeply dissected flowers have no single clear diameter. Therefore, Ne'eman and Kevan (2001) examined the MDD to artificial human-blue paper targets with various dissected flower shapes by honeybee workers. They demonstrated that the diameter could not explain the variation in MDD to flower shapes differing in their diameter and degree of dissectedness. Instead they found that the product of coloured flower area squared (cm4) over its contour length (perimeter) provided the best significant predictor for the MDD to human blue paper models by honeybees (Ne'eman & Kevan, 2001).

Here, working with bumble bees and natural flowers that vary simultaneously in their size, shape and colour, we found that this index was insufficient for predicting MDD (Tab. 2). This may be due to the effects of the green receptor contrast on the visual system of bees in detection of far targets and in determination of vision fields. For the detection of a full circular target with a green-receptor contrast, honeybee workers employ a system having a MSVA of $\alpha_{min} = 5°$, and a target image must excite a minimum of seven ommatidia to elicit detection. The MSVA of similar targets that lack green-receptor contrast is $\alpha_{min} = 15°$, and the minimal number of ommatidia that must be excited is 59 (Chittka & Raine 2006; Giurfa & Vorobyev 1997; Hempel de Ibarra et al. 2002). For bumble bees, targets lacking green-receptor contrast have MSVA of $\alpha_{min} = 2.7°$, and the minimal number of ommatidia that must be excited is seven, and for the detection of a target with a green-receptor contrast, bumble bee workers employ a system having a MSVA of $\alpha_{min} = 2.3°$ and target image must excite a minimum of three ommatidia to elicit detection, and this varies with bee size (Spaethe and Chittka, 2003). Bumble bees can see the same flower from larger distance then honeybees, especially flowers with relatively little green receptor contrast (Spaethe & Chittka 2003; Dyer et al. 2008; Kapustjansky et al. 2010).

Green colour receptor contrast has been shown to be important for detection of distant targets, while chromatic contrast for near targets (Guirfa et al. 1996). Green receptor contrast also improves the detection of three spatially separated coloured discs, by honeybees and bumble bees, compared with discs that presented chromatic but no green-receptor contrast (Wertlen et al. 2008).

Here, we determined MDD to natural flowers, by bumble bee workers, and propose an empirical regression model for predicting MDD from parameters of flower shape and colour. To do so, we used 12 natural flower types that varied much in their morphology (Tab. 1) and colour (Fig. 2), and determined MDD to each of them by bumble bee workers. Based on previous studies, by applying linear regression, we tested the ability of: flower diameter, contour density, coloured area, area/contour line, green contrast, euclidean colour distance (cm) to predict MDD, none of which yielded significant results. The only significant regression model for predicting MDD by bumble bee workers to natural and manipulated flowers is presented in (Eq. 4) and Fig. 5.

For a bee to detect a visual stimulus a minimal number of ommatidia must be excited and this depends on stimulus size (Dyer et al., 2008; Giurfa & Vorobyev, 1997). It is thus reasonable to assume that flower area is the most important factor affecting MDD (Eq. 4). Ne'eman and Kevan (2001) found that only target area was not sufficient for their explaining model and only the area squared provided a significant linear model, which is not in concert with current models of bee vision. Bukovac et al. (2016) demonstrated that even for simple colour disc detection in a standard Y-maze, the current models of visual detection seem inadequate but such behaviour fits modelling of what flower colours have actually evolved. Their finding might indicate that flower detection is more complex than previously thought.

When two flowers have equal areas but one has a longer contour line it will inevitably be also more dissected than a flower with a shorter contour line. Although sharing the same diameter or area a dissected flower will probably fully stimulate less ommatidia and consequently will have also a longer MDD by bees relatively to a less dissected flower. This explains the location of the contour length in the denominator of our regression model (Eq. 4).

While chromatic contrast is used for short range vision of details, the green contrast signal is used as far-distance flower detection, and therefore it would always be detected before chromatic contrast during an approach flight (Spaethe & Chittka 2003). The green contrast between the flowers, used as targets in this study, and their background ranged from 0.255 to 0.499 (Tab. 1), where maximum green contrast is 0.5 (Spaethe et al. 2001). The green contrast in all our tested flowers is within the range of the colours tested by Spaethe et al. (2001) that have used the colour hexagon model as we did. High green contrast is essential for distance detection, and its values are always smaller than 0.5; these facts explains the location of the value of the green contrast in the denominator of our proposed linear regression model for predicting MDD (Eq. 4). This indicates that green contrast may have a quantitative effect, which contradicts other studies that proposed a threshold effect (Giurfa et al., 1996), which to the best of our knowledge was never explicitly determined. A threshold effect should not lead to an increase in the distance from which an object can be detected. Working with artificial flowers, Spaethe et al. (2001) found a significantly negative correlation between mean search time and green contrast value of small flowers (< 8 mm). Searching time of *Ophrys heldreichii* flowers and its coloured paper models by male solitary bees from close range (< 30 cm) was negatively correlated with green receptor contrast between flower perianth and background Streinzer (2009). These two cases may also indicate a quantitative rather than threshold effect of green contrast on bee behaviour. Recently it was discovered that sufficient input to the green photoreceptor channel is needed for honeybees to learn and use size relationship rules to choose the larger or smaller stimulus as the correct solution in a given context, and subsequently apply it to novel colours and shapes (Avarguès-Weber et al. 2014).

Our model (Eq. 4) is significant, but it explains only 38.2% of the variation in the MDD data, implying that there are additional variables, not included in the model, which affecte MDD. One such main factor, in addition to measurement errors, is the variation in body size of bumble bees. In bumble bees, eye optics is correlated with body size, and an increase of 33% in bumble bee body size is accompanied by 100% increase in precision of single object resolution, since larger bees have larger eyes with better spatial resolution (Spaethe & Chittka 2003). Thus, bee body size is a major candidate for explaining additional part of the variation in MDD to flowers by bumble bee workers.

Our results clearly show that dissected flower shape reduces detection distance by bees, which could negatively affect its visitation rate by pollinators and probably also the plant fitness. If so, why are star shaped and other dissected radial flowers are so common? Floral nectar guides are colour markings on flower corolla that guide pollinators to the location of nectar reward in flowers. Therefore, pollinators spend less time in finding and extracting the nectar, they increase visitation rate and increase pollination probability of flowers that have nectar guides (Dafni & Kevan 1997; Dafni et al. 1997; Dafni & Guirfa 1999; Waser & Price 1985). In addition to the response of experienced bees to nectar guides, flower-naive bumble bees are attracted to the margins of flowers and follow them using the radial flower shape itself to guide them toward the reward (Lunau et al. 2006). Leonard and Papaj (2011) proved that star shaped flowers and flowers with radial nectar guides are preferred by bumble bees, and this preference continues even after reward stopped. When detected, dissected shape presents radial signals that are more attractive to bees and direct them to the reward located in the centre of the flower. Conventional nectar guides can be detected only after a bee has landed on the flowers, while star shaped corolla can be seen from longer distance than nectar guides on a flower of similar size. Dafni and Kevan (1997) proposed that dissected margins of flowers may guide scanning pollinators to flower rewards. An analysis of the Israeli flora indicated that small flowers (< 5 cm^2) have disproportionately longer contour lines and more highly dissected shapes. Thus, small flowers may compensate for their reduced detectability by having highly dissected edges vis-a-vis large flowers (Dafni & Kevan, 1997). To this point we hypothesize that the reduced detectability of radial dissected flowers might be compensated by their higher attractiveness for bees and that nectar guide should be less frequent in dissected than in none dissected radial flowers.

Conclusion

To conclude, we present an empirical linear regression model that can be used for predicting MDD to natural flowers by bumble bee workers using flower coloured area, the length of its contour line and the value of green receptor contrast between the flower reflectance and the green background (Eq. 4). This model requires refinements using more flower types differing in their shapes and colours as well as more near to natural testing background colour and illumination conditions. Further development can also be scaling up the model from individual flowers to inflorescences. Developing procedures to determine MDD in the field is most important but a difficult future task. Such improvements are essential for better understanding of bees' foraging behaviour under natural conditions, where various plant species co-flower.

Empirical determination of MDD and MSVA to real flowers is most important and should be used as a novel parameter for bee foraging behaviour, pollinator induced evolution of flower size shape and colour, as well as validation of psychophysical and neural models established for flower shape and colour.

Acknowledgements

We are pleased to thank Lars Chittka, Queen Mary University of London, for hosting us in his lab, his generous help, advice and comments on an early version of this paper. We thank also editor AG Dyer and two unknown reviewers for suggestions on the manuscript.

Appendices

Additional supporting information may be found in the online version of this article:

Appendix I. The average percentage of correct choices as a function of the distance by *Bombus terrestris* workers, and the number of bees tested for each flower type in each distance for all studied flower types.

Appendix II. Maximal detection distance (MDD) of all tested individual bees (each line presents a single bee) to all tested flowers; flower (coloured) area, average diameter and the contour line (Perimeter) of all tested flowers and the value of the green-receptor contrast between petal colour and the green background of the radial maze against which the flowers were presented to the bees. See also Tab. 2.

References

Anderson AM (1977) Shape perception in the honey bee. Animal Behaviour 25:67-79.

Armbruster WS, Antonsen L, Pélabon C (2005) Phenotypic selection on *Dalechampia* blossoms: honest signalling affects pollination success. Ecology 86:3323-3333.

Avarguès-Weber A, Deisig N, Giurfa M (2011) Visual cognition in social insects. Annual Review of Entomology 56:423-443.

Avarguès-Weber A, Giurfa M (2014) Cognitive components of color vision in honey bees: how conditioning variables modulate color learning and discrimination. Journal of Comparative Physiology A 200:449-461.

Avarguès-Weber A, d'Amaro D, Metzler M, Dyer AG (2014) Conceptualization of relative size by honeybees. Frontiers in Behavioral Neurosciences 8:80.

Backhaus W (1991) Color opponent coding in the visual system of the honeybee. Vision research 31:1381–1397.

Blarer A, Keasar T, Shmida A (2002) Possible mechanisms for the formation of flower size preferences by foraging bumblebees. Ethology 108:341-351.

Bukovac Z, Dorin A, Dyer A (2013) A-Bees See: a simulation to assess social bee visual attention during complex search tasks. In: Lio P, Miglino O, Nicosia G, Nolfi S, Pavone M (eds) Advances in artificial life, ECAL 2013. Proceedings of the twelfth European conference on the synthesis and simulation of living systems, Taormina, September 2013. Complex Adaptive Systems, MIT Press, Cambridge, London, pp 276–283

Bukovac Z, Dorin A, Finke V, Shrestha M, Garcia J, Avarguès-Weber A, Burd AM, Schramme J, Dyer A (2016) Assessing the ecological significance of bee visual detection and colour discrimination on the evolution of flower colours. Evolutionary Ecology 30:1-20.

Chittka L (1992) The colour hexagon: a chromaticity diagram based on photoreceptor excitations as a generalized representation of colour opponency. Journal of Comparative Physiology A 170:533-543.

Chittka L, Raine NE (2006) Recognition of flowers by pollinators. Current Opinion in Plant Biology 9:428-435.

Chittka L, Spaethe J (2007) Visual search and the importance of time in complex decision making by bees. Arthropod-Plant Interactions 1:37-44.

Conner JK, Rush S (1996) Effects of flower size and number on pollinator visitation to wild radish, *Raphanus raphanistrum*. Oecologia 105:509-516.

Dafni A, Giurfa M (1999) The functional ecology of floral guides in relation to insects behaviour and vision. In: Waser S (ed) Evolutionary theory and processes: modern perspectives. Kluwer Academic Publishers, Dordrecht, pp. 363-383.

Dafni A, Kevan PG (1997) Flower size and shape: implications in pollination. Israel Journal of Plant Sciences 45:201-211.

Dafni A, Lehrer M, Kevan P (1997) Spatial flower parameters and insect spatial vision. Biological Review 72:239-282.

Dyer AG, Chittka L (2004) Biological significance of distinguishing between similar colours in spectrally variable illumination: bumblebees (*Bombus terrestris*) as a case study. Journal of Comparative Physiology A 190:105-114.

Dyer AG, Whitney HM, Arnold SE, Glover BJ, Chittka L (2007) Mutations perturbing petal cell shape and anthocyanin synthesis influence bumblebee perception of *Antirrhinum majus* flower colour. Arthropod-Plant Interactions 1:45-55.

Dyer AG, Spaethe J, Prack S (2008) Comparative psychophysics of bumblebee and honeybee colour discrimination and object detection. Journal of Comparative Physiology A 194:617-627.

Dyer AG, Streinzer M, Garcia J (2016). Flower detection and acuity of the Australian native stingless bee *Tetragonula carbonaria* Sm. Journal of Comparative Physiology A 202:629-639.

Elle E, Carney R (2003) Reproductive assurance varies with flower size in *Collinsia parviflora* (Scrophulariaceae). American Journal of Botany 90:888-896.

Fenster CB, Cheely G, Dudash MR, Reynolds RJ (2006) Nectar reward and advertisement in hummingbird-pollinated *Silene virginica* (Caryophyllaceae). American Journal of Botany 93:1800-1807.

Frisch Kv (1914) Der Farbensinn und Formensinn der Biene. Gustav Fischer, Jena.

Galen C (1999) Why do flowers vary? The functional ecology of variation in flower size and form within natural plant populations. Bioscience 49:631-640.

Galen C, Newport M (1987) Bumble bee behavior and selection on flower size in the sky pilot, *Polemonium viscosum*. Oecologia 74:20-23.

Giurfa M, Vorobyev M (1997) The detection and recognition of color stimuli by honeybees: performance and mechanisms. Israel Journal of Plant Sciences 45:129-140.

Giurfa M, Vorobyev M, Kevan P, Menzel R (1996) Detection of coloured stimuli by honeybees: minimum visual angles and receptor specific contrasts. Journal of Comparative Physiology A 178:699-709.

Goodwillie C, Sargent RD, Eckert CG, Elle E, Geber MA, Johnston MO, Kalisz S, Moeller DA, Ree RH, Vallejo-Marin M (2010) Correlated evolution of mating system and floral display traits in flowering plants and its implications for the distribution of mating system variation. New Phytologist. 185, 311-321.

Heinrich B (1975) Energetics of pollination. Annual Review of Ecology and Systematics 15:139-170.

Hempel de Ibarra NH, Giurfa M, Vorobyev M (2001). Detection of coloured patterns by honeybees through chromatic and achromatic cues. Journal of Comparative Physiology A 187: 215-224.

Hempel de Ibarra N, Giurfa M, Vorobyev M (2002) Discrimination of coloured patterns by honeybees through chromatic and achromatic cues. Journal of Comparative Physiology A 188:503-512.

Hempel de Ibarra NH, Vorobyev M, Menzel R (2014) Mechanisms, functions and ecology of colour vision in the honeybee. Journal of Comparative Physiology A 200:411-433.

Hempel de Ibarra NH, Langridge KV, Vorobyev M (2015) More than colour attraction: behavioural functions of flower patterns. Current Opinion in Insect Science 12:64-70.

Herrera J (2005) Flower size variation in *Rosmarinus officinalis*: individuals, populations and habitats. Annals of Botany 95:431-437.

Hertz M (1929) Die Organisation des optischen Feldes bei der Biene. Journal of Comparative Physiology A 8:693-748.

IBM Corporation (2010) IBM SPSS Statistics for Windows, Version 19.0. Armonk, NY IBM Corp.

Kaczorowski RL, Seliger AR, Gaskett AC, Wigsten SK, Raguso RA (2012) Corolla shape vs. size in flower choice by a nocturnal hawkmoth pollinator. Functional Ecology 26, 577-587.

Kapustjansky A, Chittka L, Spaethe J (2010) Bees use three-dimensional information to improve target detection. Naturwissenschaften 97:229-233.

Keasar T, Pollak G, Arnon R, Cohen D, Shmida A (2006) Honesty of signalling and pollinator attraction: the case of flag-like bracts. Israel Journal of Plant Sciences 54:119-128.

Krizek BA, Anderson JT (2013) Control of flower size. Journal of Experimental Botany 64:1427-1437.

Kugler H (1934) Blütenokologische Untersuchungen mit Hummeln VII Die Anlockung von "Neulingen" durch Blüten. Planta 23:692-714.

Lehrer M, Bischof S (1995) Detection of model flowers by honeybees: the role of chromatic and achromatic contrast. Naturwissenschaften 82:145-147.

Lehrer M, Horridge GA, Zhang SW, Gadagkar R (1995) Shape vision in bees: Innate preference for flower-like patterns. Philosophical Transactions of the Royal Society B Biological Sciences 347:123–137.

Leonard AS, Papaj DR (2011) 'X' marks the spot: the possible benefits of nectar guides to bees and plants. Functional Ecology 25:1293-1301.

Lunau K, Fieselmann G, Heuschen B, van de Loo A (2006) Visual targeting of components of floral colour patterns in flower-naive bumblebees (*Bombus terrestris*; Apidae). Naturwissenschaften 93:325-328.

Martin NH (2004) Flower size preferences of the honeybee (*Apis mellifera*) foraging on *Mimulus guttatus* (Scrophulariaceae). Evolutionary Ecological Research 6:777-782.

Menzel R, Giurfa M (2006) Dimensions of cognition in an insect, the honeybee. Behavioural and Cognitive Neuroscience Reviews 5:24-40.

Menzel R, Gumbert A, Kunze J, Shmida A, Vorobyev M (1997) Pollinators' strategies in finding flowers. Israel Journal of Plant Sciences 45:141-156.

Menzel R, Shmida A (1993) The ecology of flower colours and the natural colour vision of insect pollinators: the Israeli flora as a study case. Biological Review 68:81-120.

Neal PR, Dafni A, Giurfa M (1998) Floral symmetry and its role in plant-pollinator systems: terminology, distribution, and hypotheses. Annual Review of Ecology and Systematics 29:345-373.

Ne'eman G, Kevan PG (2001) The effect of shape parameters on maximal detection distance of model target by honeybee workers. Journal of Comparative Physiology A 187:653-660.

Peitsch D, Fietz A, Hertel H, de Souza J, Ventura DF, Menzel R (1992) The spectral input systems of hymenopteran insects and their receptor-based colour vision. Journal of Comparative Physiology A 170:23–40.

Plowright CMS, Evans SA, Leung JC, Collin CA (2011) The preference for symmetry in flower-naïve and not-so-naïve bumblebees. Learning and Motivation 42:76 83.

Schaefer HM, Schaefer V, Levey DJ (2004) How plant–animal interactions signal new insights in communication. Trends in Ecology and Evolution 19:577-584.

Simonds V, Plowright CMS (2004) How do bumblebees first find flowers? Unlearned approach responses and habituation. Animal Behaviour 67:379-386.

Spaethe J, Chittka L (2003) Interindividual variation of eye optics and single object resolution in bumblebees. Journal of Experimental Biology 206:3447-3453.

Spaethe J, Tautz J, Chittka L (2001) Visual constraints in foraging bumblebees: flower size and color affect search time and flight behavior. Proceedings of The National Academy of Sciences Biology 98:3898-3903.

Sprengel CK (1793/1996) Discovery of the secret of nature in the structure and fertilization of flowers (translated by Hasse P). In: Lloyd DG, Barrett SC (eds) Floral biology: studies on floral evolution in animal-pollinated plants. Springer, New York, pp 3-43.

Srinivasan MV (2010) Honey bees as a model for vision, perception, and cognition. Annual Review of Entomology 55:267-284.

Streinzer M, Paulus HF, Spaethe J (2009) Floral colour signal increases short-range detectability of a sexually deceptive orchid to its bee pollinator. Journal of Experimental Biology 212:1365-1370.

Thompson EL, Plowright CM (2014) How images may or may not represent flowers: picture–object correspondence in bumblebees (*Bombus impatiens*)? Animal Cognition 17:1031-1043.

Vorobyev M, Gumbert A, Kunze J, Giurfa M, Menzel R (1997) Flowers through insect eyes. Israel Journal of Plant Sciences 45:93-101.

Waser NM, Price MV (1985) The effect of nectar guides on pollinator preference: experimental studies with a montane herb. Oecologia 67:121-126.

Wertlen AM, Niggebrugge C, Vorobyev M, Hempel de Ibarra N (2008) Detection of patches of coloured discs by bees. Journal of Experimental Biology 211:2101-2104.

Wu W, Moreno AM, Tangen JM, Reinhard J (2013) Honeybees can discriminate between Monet and Picasso paintings. Journal of Comparative Physiology A 199:45-55.

Zhang S, Srinivasan MV, Zhu H, Wong J (2004) Grouping of visual objects by honeybees. Journal of Experimental Biology 207:3289-3298.

23

DO HONEYBEES ACT AS POLLEN THIEVES OR POLLINATORS OF *DATURA WRIGHTII*?

Andrew C. McCall*[,1], Sarah Richman[2,3], Eric Thomson[1], Monica Edgerton[1], Skyler Jordan[1], Judith L. Bronstein[2]

[1]*Department of Biology, Denison University, Granville OH 43023 USA*
[2]*Department of Ecology and Evolutionary Biology, University of Arizona, Tucson AZ 85721 USA*
[3]*Present address: Department of Biology, University of Nevada, Reno; Reno NV 89557 USA*

Abstract—Datura wrightii (Solanaceae), a common shrub in the southwestern United States, bears massive, white, night-blooming flowers that attract and reward hawkmoth pollinators. However, *Apis mellifera* (honeybee) foragers are often observed on its flowers, especially at dusk and dawn hours. Their foraging activities are focused on the anthers, suggesting they could be pollen thieves. We used a series of observations and manipulative experiments to determine if honeybees are detrimental or beneficial to *D. wrightii.* We found that honeybees were the most frequent visitors to *D. wrightii* flowers at both dusk and dawn, and that they removed and carried large amounts of *D. wrightii* pollen. Flowers were capable of being pollinated at dusk and dawn and a single visit by a honeybee was sufficient to pollinate the flowers and produce fruit. There was no evidence that restricting visitation to diurnal hours yielded fruit set that was different from when we restricted visitation (likely by hawkmoths) to evening hours. These results suggest that honeybees are capable of effective pollination of *D. wrightii.* Although honeybees might interfere with pollen transmission mediated by their highly specialized hawkmoth pollinators, they may also increase plant fitness by pollinating *D. wrightii* when hawkmoths are not present.

Keywords: Pollen theft, pollen thieves, Datura wrightii, Apis mellifera

INTRODUCTION

Floral tissue can be damaged by animals in a number of ways. For example, it may be consumed (florivory; McCall & Irwin 2006), or may be damaged by foragers that then feed upon nectar without concomitant pollination (nectar robbery; *sensu* Inouye 1980; Irwin et al. 2010). In pollen theft, as defined by Hargreaves et al. (2009) and Solís-Montero et al. (2015), thieves are visitors that can consume pollen without causing significant cross-pollination. Although florivory and nectar robbery have received recent attention in the literature (see Irwin et al. 2010 and González-Browne et al. 2016 for reviews), there are fewer studies that consider the effect of pollen thieves on plant reproductive success (Hargreaves et al. 2009).

Pollen theft is a product of interactions between the potential thief and the host plant's floral structure and development. For example, if there are temporal differences between the timing of pollen presentation and stigma receptivity (dichogamy), then a potential thief could consume pollen and have pollen present on its body without effecting pollination (Hargreaves et al. 2009; Hargreaves et al. 2012). Vaughton (1996) found that pollen-collecting *A. mellifera* (honeybees) visited male-phase *Grevillea barklyana* (Proteaceae) but seldom visited female-phase flowers. This habitual theft reduced plant fitness, as flowers caged to allow only honeybee visits set fewer fruit than bagged control flowers. Pronounced spatial separation of anthers and stigmas (herkogamy) in flowers could also contribute to theft because a visitor may not be able to transfer pollen to stigmas if the distance between the reproductive organs is too great (Hargreaves et al. 2009; Solís-Montero et al. 2015).

Pollen theft may be more prevalent in plants with specialized flowers than in plants with more generalized flowers because unique floral structures could preclude efficient pollination by generalist floral visitors (Aslan et al. 2016). Visitors such as honeybees may act as pollen thieves because they are frequent visitors to many flowers but are often inefficient in transferring pollen to stigmas or do not demonstrate the specific behavior needed for pollen release. For example, only 18% of honeybee visits to the buzz-pollinated *Solanum rostratum* resulted in contact to both an anther and the stigma, even though honeybees made up 95.8% of total insect visits (Solís-Montero et al. 2015). In *Syncolostemon densiflorus* (Lamiaceae), a plant with a bird-pollinated flower, Wester and Johnson (2017) found that honeybee foragers were frequent floral visitors but they did not pollinate effectively, most likely because foragers did not contact the stigmas.

The perennial herb *Datura wrightii* (Solanaceae) would appear to be particularly susceptible to pollen theft because honeybees frequent collect only pollen (Fig. 1A, A. McCall, pers. obs.), and flower exhibit pronounced herkogamy,

*Corresponding author: mccalla@denison.edu

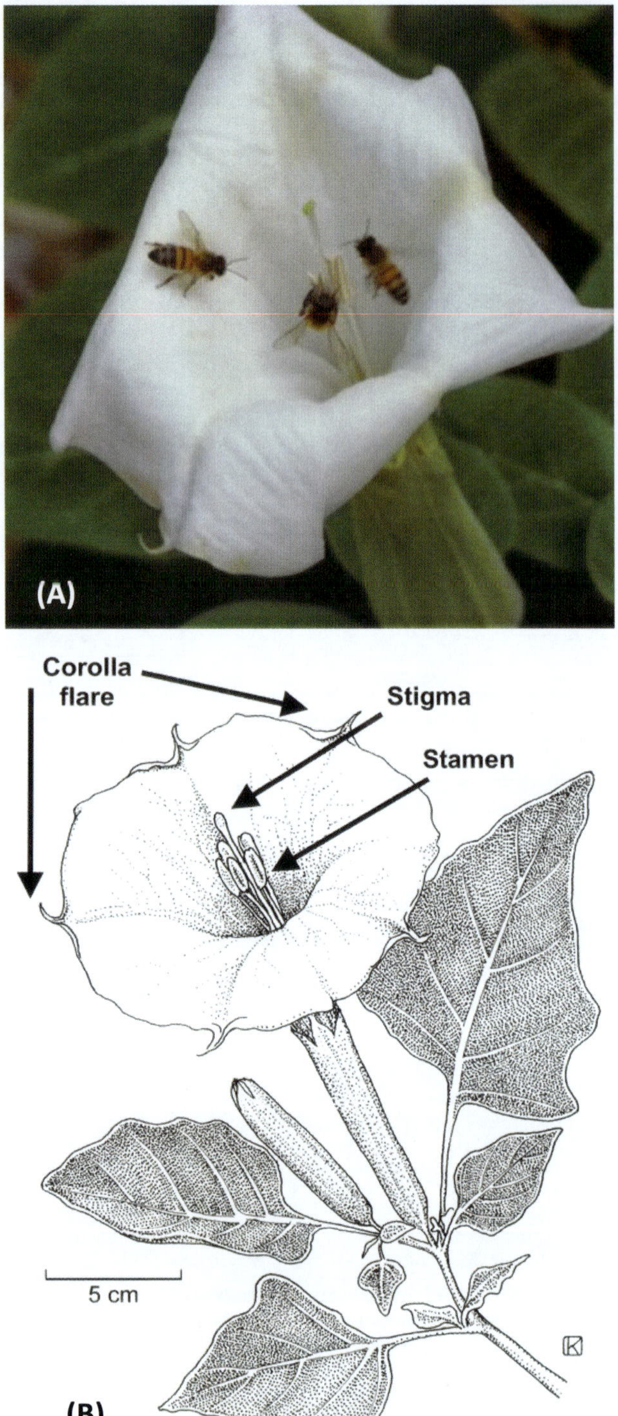

FIGURE 1. (A) Photo of honeybees entering the corolla headspace and collecting pollen from anthers. Photo used with permission from Deborah Wilson. (B) Line drawing of *D. wrightii* flowers and habit. Stamens, the exerted stigma and the corolla flare are labeled. Drawing reprinted with permission from the Jepson Herbarium, UC Berkeley.

sometimes precluding contact of the stigma by the foragers (Fig. 1B, Bronstein et al. 2009). The main pollinators in this system are hawkmoths, in southern Arizona most notably *Manduca sexta* (Sphingidae), which also lays its eggs on the plants whose flowers it visits (Bronstein et al. 2009). However, flowers are also visited at dusk and dawn by honeybees (*Apis mellifera* (Apidae)), introduced species in Arizona that appear to collect large amounts of pollen without contacting stigmas (A. McCall, pers. obs.). This would suggest that honeybees might remove pollen without providing efficient pollen transfer in the process. Furthermore, because hawkmoths are most active at night, when honeybees do not fly, pollen collection by honeybees at dusk could reduce the amount of pollen available for more effective pollinators. This is possible because individual *D. wrightii* flowers are typically only open in a single interval, from dusk until dawn in a 24-hour period. Alternatively, honeybee visitation during dawn hours could add pollen deposition and removal beyond that caused by hawkmoth activity (Barker and Bronstein 2016).

We examined whether honeybees might act as conditional pollen thieves of *D. wrightii*. Our first set of questions focus on the specific attributes of honeybees or *D. wrightii* that might promote conditional pollen theft: 1) Do honeybee foragers visit *D. wrightii* flowers before hawkmoths visit, and do they contact anthers and stigmas during these visits? 2) Do honeybee foragers remove pollen from *D. wrightii* anthers? and 3) Do honeybee foragers carry pollen on their bodies? We then consider honeybee attributes that might contribute to their ability to serve as pollinators: 1) Are *D. wrightii* flowers able to set fruit when hand-pollinated during dawn hours, just before the single-night flowers close and senesce, mimicking the actions of dawn-foraging bees? 2) Is a single visit from a honeybee forager sufficient to set fruit? Finally, 3) Are flowers exposed to honeybees able to set fruit at the same rate as those flowers exposed to either nocturnal, presumably pollinating moths or control flowers open to all visitors? If honeybees cause lower rates of fruit set than moths, then there would be evidence that they are conditional pollen thieves. Alternatively, if fruit set is similar when flowers are exposed to honeybees or exposed to moths, then there would be evidence that honeybees are legitimate pollinators within this hawkmoth pollination system.

MATERIALS AND METHODS

Study system

Datura wrightii (Solanaceae) is a perennial herb found from Western Texas to California and Mexico (Kearney & Peebles 1960). Flowering occurs from mid-April to early November. Flowers are large (maximum length is ~25 cm), white, and tubular. The flowers open at dusk and remain open for a single night, although flowers can remain turgid in the morning hours if the temperatures are cool (A. McCall, personal observation). Flowers frequently exhibit pronounced herkogamy, with anthers positioned over the stigmas (see Fig. 1A and B, Bronstein et al. 2009), which may preclude self-pollination by pollinators. Flowers are self-compatible and frequently set fruit without pollinator activity (Bronstein et al. 2009). *Manduca sexta* individuals forage exclusively nocturnally and are highly effective pollinators at this site (Bronstein et al. 2009); the same is likely true of the rarer congener *M. quinquemaculata* (Alarcón et al. 2010).

Previous observations revealed that honeybee foragers frequently visit *D. wrightii* flowering buds, especially anthers, even before floral anthesis and nightfall (A. McCall, pers.

obs.). These foragers often land on the large floral buds and will sometimes force their way into the bud to collect pollen. These activities can occur 2-3 hours before *M. sexta* visits the fully-opened flowers, allowing honeybees to remove pollen before *M. sexta* arrives (A. McCall, pers. obs.). Although we observed several species of native bees visiting *D. wrightii* flowers, some of which are small enough to collect pollen from anthers without contacting stigmas (Thorp 2000), we chose to focus on honeybees because they are usually the most abundant floral visitor at dusk in the *D. wrightii* populations we examined. We also observed frequent honeybee visitation at dawn, before flowers wilted due to heat.

Sites

Most of this study was conducted at the University of Arizona Campus Agricultural Center in Tucson, Arizona (the "Farm population", N 32.280, W 110.938). *Datura wrightii* seeds were collected in 2007 from a wild population at the Santa Rita Experimental Range (SRER, N 31.788, W 110.828). In 2008 these seeds were planted at the Agricultural Center. They were grown under field conditions with occasional watering. In late 2008 we collected fruit from these plants and massed their seed to produce the seed source for our experimental plot. In 2009, we planted 96 seeds in each of two adjacent tilled and cleared blocks. Blocks were occasionally watered and weeded but were otherwise unmanipulated. The second site was a natural population of *D. wrightii* located along Box Canyon Road (the "Box Canyon" population, N 31.775, W 110.715) in the Santa Rita Mountains, adjacent to the Santa Rita Experimental Range.

Pollinator surveys

In order to quantify floral visitation to *D. wrightii,* we observed visitor behavior at the Box Canyon population. Beginning on 11 July 2011 and over a period of one week, we observed floral visitors on as many flowering plants as possible. Our observations took place as the flowers were opening, at approximately 1900h. A 'visit' was defined as taking place when an insect broke the plane of the corolla flare into the headspace of the flower (see Fig. 1B). We observed an insect visitor until it left the corolla or until five minutes had elapsed. Because there were often numerous insects approaching many flowers on a plant, we focused on a single flower and a single individual insect for each replicate. We described the insect's behaviors using the following mutually-exclusive categories: gathering pollen on anthers, contacting the stigma, contacting both the stigma and anthers, or not contacting any primary reproductive structures. A total of 228 individual visitors of ten different species were observed.

Pollen removal experiments

Honeybee foragers might be more prevalent during dawn versus dusk hours because they are primarily diurnal insects, possibly leading to more pollen removed in the morning. To determine if honeybee foragers removed different pollen amounts at different times of the day, we conducted an experiment at the Farm population on 18 Sept, 27 Sept and 4 Oct 2012. On each day, we identified all plants with nearly-open floral buds and haphazardly chose 60 of them to be bagged with bridal veil. We then randomly assigned buds to one of three treatments: 1) Controls with no visitation, 2) Honeybee visitation at dusk only, and 3) Honeybee visitation at dawn only. For each of the treatment flowers, we collected and weighed anthers as a proxy for pollen mass. In order to move the anthers from flowers to collection envelopes without spilling pollen, it was necessary to include a small portion of the filaments for all treatments.

We removed the bags from 20 flowers at approximately 1900h, prior to hawkmoth activity. For the dawn treatment, we removed bags from 20 flowers at 0500h and allowed bees to visit for the same length of time as in the dusk treatment (1.5 hr), then removed anthers. As a control, we collected five anthers from each of 20 bagged, unvisited flowers. Collected anthers were placed in glassine envelopes (1 envelope/flower). Anther samples were dried for 1 week at 60° C and then weighed.

We constructed a linear mixed model with weight of anthers per flower as the dependent variable, treatment (control, dusk, or dawn) as the fixed independent variable and day and day by treatment interactions as the random independent effects. The model was constructed using the 'lmer' function in the R package 'lme 4' (v. 1.1-6; Bates et al. 2015). The effect of the treatment was determined by constructing another linear mixed model identical to the first, except without the treatment effect. The two models were then tested against each other using log likelihood ratio tests implemented by the 'Anova' function in the R package 'car' (Fox & Weisberg 2011). If the overall treatment effect was significant, we performed pairwise Wald tests to determine which treatment levels were significantly different from each other.

Examination of honeybees and pollen

To determine if honeybees carry pollen in their corbicula, at 1800h on 13 July 2011 we collected 11 honeybee foragers on *D. wrightii* flowers from the Farm population. We removed pollen from the corbicula of each bee and examined it under a dissecting microscope. We compared these samples to reference slides of *D. wrightii* pollen to determine whether each pollen load included *D. wrightii* pollen. We did not identify pollen grains other than *D. wrightii.* There was a small possibility that pollen grains from the congener *D. discolor* were also present, but the two types of pollen are quite distinct and can easily be differentiated under a microscope (G. Smith, personal communication).

Dusk versus Dawn pollination

Honeybees could visit flowers at dusk or dawn, but stigmas may not be competent to receive pollen in the morning, as flowers are usually only open one night. To test if *D. wrightii* flowers have the capability to set fruit when pollinated at dusk or dawn, we conducted an experiment on all plants in the Farm population. From 17-29 July 2013 we haphazardly chose an open flower on a given plant as the pollen donor for both evening and morning-pollinated flowers on that plant. We did this to mimic within-plant visitation on plants bearing more than one flower, which is often observed on larger plants when honeybees are foraging at dusk. On each evening of the experiment, we bagged floral buds that were about to open. As they opened, each flower was randomly assigned to either a 'dusk' or 'dawn' pollination treatment. For dusk-pollinated flowers, we removed the bag and brushed two anthers from the donor flower across the stigma five times at anthesis, after which the bag was closed. Dawn-pollinated flowers were allowed to open inside their bags in the evening and remained bagged until approximately 0500 the following morning, just as honeybee activity began and after *M. sexta* activity ceased. We then took two of the remaining anthers from the open donor flower and brushed them across the flower's stigma five times. The flower was then rebagged to prevent honeybee foragers from adding additional pollen to the flower. It is important to note that, for the dawn-pollinated plants, we used pollen from the donor flower that had been collected the previous evening. We did this to ensure that we would have sufficient pollen available to deposit similar amounts of pollen at both time periods.

Both dusk and dawn pollinated flowers were bagged for two days until fully wilted. In total, 15 flowers were pollinated at dawn and 23 at dusk. Fruit set was recorded on 31 Aug and 3 Sept 2013. We analyzed the data using a generalized linear mixed model (GLMM) with fruit presence or absence as the response variable, pollination time treatment as a fixed variable, plant as a random variable, and a binomial error structure. The model was implemented using the *glmer* function under the *lme4* package in R. An identical model but without the treatment effect was constructed, and the two models were tested against each other using log likelihood ratio tests.

Single bee visits

To determine if honeybees can act as pollinators, we conducted an experiment to measure the effect of a single visit. On 2 July 2012 and ending 2 Aug 2012, we randomly assigned floral buds in the Farm population to experimental treatments. We visited the site every other day at midday and used those plants that were flowering on each day to supply the buds for treatments. In treatment 1, we bagged flowers, then rolled down the top of the bag at dusk (approximately 1915h) to allow all pollinators access to the flower until evening (2030h), when the bags were closed. In treatment 2, buds were allowed to reach anthesis inside the bags. Bags were then removed at dusk to allow a single honeybee to enter and contact the reproductive organs of each flower. After a single visit, the bags were immediately closed. Finally, the control treatment consisted of buds that were allowed to open in the bag but were never exposed to pollinators. All bags remained on the flowers until midday on the following day, when the flowers began to wilt. Our procedure yielded a total of 69 closed control flowers, 24 open-pollinated flowers, and 41 flowers visited by a single bee. Fruit set was recorded on 6 and 10 Aug 2012 and again on 12 Nov 2012.

Data were analyzed using a GLMM with fruit presence or absence as the response variable and Julian date of anthesis, plant (random effect), and treatment as the independent variables. We used a binomial error structure in the model. We constructed a full model with the treatment factor and a reduced model without the treatment factor and these models were tested against each other using log likelihood ratio tests.

If the overall treatment effect was significant, we performed pairwise Wald tests to determine which treatment levels were significantly different from each other.

Pollinator exclusion

We investigated whether floral visitors active at different times differed in their ability to pollinate *D. wrightii*. We first exposed individual flowers to visitors during diurnal, nocturnal, and crepuscular periods at the Farm population from 5 July to 14 Aug 2010. As plants came into flower, each flower was randomly assigned to one of five treatments: a bridal veil bag opened from anthesis until dawn (07:30) ($N = 25$), a bag left on to permit self-pollination only ($n = 30$), a bag open only from 19:00 to 20:30 (dusk, $N= 29$), a bag open from 20:30 until 04:30 (night, $N= 28$), and a bag open from 04:30 until 07:30 (dawn, $N= 25$). Not all treatments were applied on the same day because plants often had fewer than five flowers open on a single night. After two days, bags were removed from all flowers and the flowers, now fully wilted, were tagged. Fruit set was recorded on 15 Aug 2010.

We analyzed the data using a GLMM with presence of fruit as the dependent variable and time treatment (fixed), plant (random), and Julian date of manipulation (continuous covariate) as independent variables. We specified a binomial error structure and implemented the analysis using the *glmer* function under the *lme4* package in R. An identical model but without the treatment effect was constructed in the same way, and we tested the two models against each other using log likelihood ratio tests in R.

Unless otherwise noted, all analyses were performed using R (R Core Team, 2014, v. 3.01, "Good Sport").

RESULTS

Pollinator surveys

We identified the following visitor species to *Datura wrightii* flowers: honeybees (*Apis mellifera*), two beetle species (*Cyclocephala melanocephala* and an unidentified species (Scarabaeidae)), three hawkmoth species (*Hyles lineata, Manduca sexta* and an unidentified species (Sphingidae)), and several unidentified species of ants, wasps, and flies. The four most common visitors, making up 91.2% of all visits, were honeybees (143 visits), *Manduca sexta* (14 visits), *Cyclocephala melanocephela* (10 visits), and the unidentified scarab beetle (33 visits).

The proportion of visits in which the visitor contacted the primary reproductive tissue varied among the visitors, with honeybees and *M. sexta* contacting these tissues in the highest proportion of visits (Table 1).

Pollen removal and pollen on honeybees

We conducted this experiment to find if honeybees removed pollen and if the amount of pollen removed depended on the time of day. Whether plants were exposed to honeybee visits at dusk only, at dawn only, or were unvisited affected the mass of pollen in the anthers ($\chi^2 = 74.06$, 2 df, $P < 0.0001$, Fig. 2). Pollen mass in dawn-visited anthers was significantly less than in closed control anthers and pollen mass in dusk-visited anthers was less than in control flowers.

TABLE 1. The number of individual insects contacting reproductive organs among the four most common floral visitors to *D. wrightii*. Categories are mutually exclusive. Also shown is the percent of visits resulting in contact with any reproductive organs for each visitor.

Visitor	Anther	Stigma	Stigma and anther	No reproductive organs	% visits contacting any reproductive organ	Total visits
A. mellifera	29	1	32	82	42.67	143
C. melanocephala	0	0	0	10	0	10
M. sexta	0	0	11	3	78.57	14
Unidentified scarab	1	2	1	29	12.12	33

We also found that anthers visited at dawn had significantly lower pollen masses than anthers visited at dusk, suggesting that dawn visits removed more pollen than dusk visit did. All honeybee foragers were carrying *D. wrightii* pollen.

Dusk and dawn pollination

In this experiment, we tested if there was a difference in pollination efficacy between dusk and dawn hours. The model with the time of pollination effect explained significantly more of the deviance than the model without the time of pollination effect (With treatment AIC = 50.3, Without treatment AIC = 54.0, χ^2= 5.73, 1 df, P= 0.017). In particular, there was a significant difference in the proportion of flowers setting fruit between the dawn and dusk treatments (Dusk pollination = 78.3%, Dawn pollination = 21.7%, z= 2.31, P= 0.021).

Single bee visits

We tested if flowers receiving a single honeybee forager visit set fruit at a different rate than flowers without a visit (i.e., that only had the opportunity for self-pollination) or flowers open to all pollinators. There was a significant effect of pollination treatment on fruit set (With treatment AIC = 136.02, Without treatment AIC = 139.36, χ^2= 7.34, 2 df, P = 0.025). Among the treatments, there were significant differences in the proportion of flowers setting fruit between the closed control and bee treatments (z = 2.329 P = 0.02) and between the control and the open pollination treatments (z = 2.080, P = 0.037), but no evidence of a difference between the bee and open pollination treatments (z = 0.241, P= 0.809, Fig. 3).

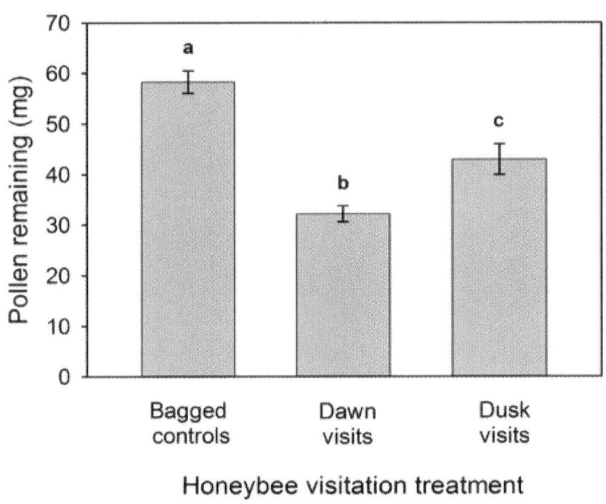

FIGURE 2. Mass of pollen remaining in unvisited control flowers, flowers open to pollinators at dawn only, and flowers open to pollinators at dusk only. Lines represent means and +/- 1 SE. Letters indicate percentages that are significantly different from each other at the P= 0.05 level.

Pollinator exclusion

To determine if pollinator activity at different times of day affected fruit set, we exposed open flowers during dusk, nighttime, and dawn hours to potential pollinators. There were no significant differences between the model with the experimental treatment effect and model without the experimental treatment effect (χ^2 = 0.17, 4 df, P = 0.99, percent flowers setting fruit in closed controls = 13.3%, open controls = 4.0%, open at dusk = 17.9%, open at night = 21.4%, open at dawn = 20.0%, Fig. 4).

DISCUSSION

Pollen theft is an understudied plant-insect interaction, particularly from the perspective of its effect on plant fitness (Hargreaves et al. 2009). The purpose of this study was to investigate if honeybee foragers are conditional pollen thieves or pollinators of *Datura wrightii*. It appears that honeybees are not conditional thieves because they are able to effect pollination. Honeybees were frequent dusk visitors to *D. wrightii*. We show here that foragers collect pollen during their visits, and that large amounts of *D. wrightii* pollen can be found on their corbicula. A single honeybee visit significantly increased the proportion of flowers setting fruit compared to unvisited flowers. The percentage of flowers setting fruit during diurnal or dusk/dawn periods, when honeybees are common visitors, did not differ significantly from the percentage of flowers setting fruit during nocturnal periods, when hawkmoths, their specialist pollinators, are most active.

In the last experiment mentioned, it is important to note that the quality or quantity of the seeds from fruit may have differed depending on the bagging treatment. Seed quality may depend on whether seeds come from autogamous selfing,

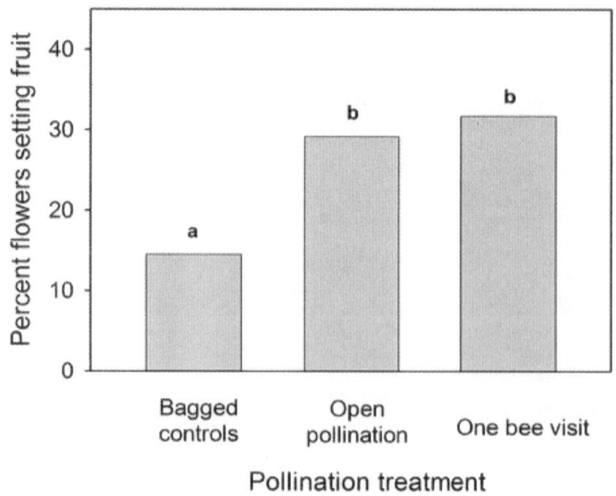

FIGURE 3. Percent of flowers setting fruit in closed and bagged flowers, flowers open to all visitors, and flowers visited by a single honeybee forager. Letters indicate percentages that are significantly different from each other at the $P = 0.05$ level.

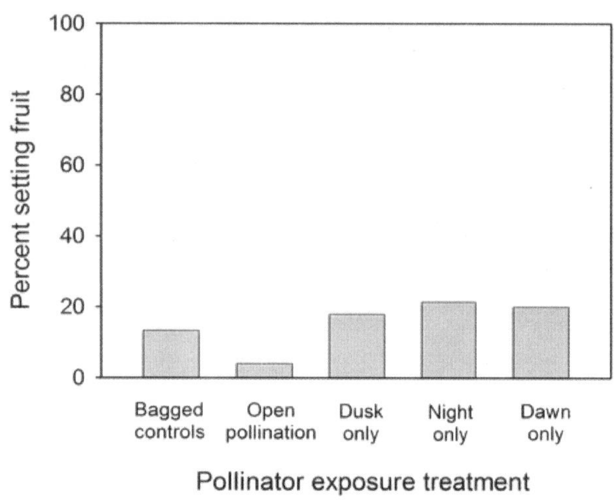

FIGURE 4. Percent of flowers setting fruit under different bagging treatments. There were no significant differences in the percentages.

geitonogamous selfing, or outcrossing. For example, progeny from autogamous pollination resulted in progeny with lower resistance to herbivory than progeny from open pollination in the *D. wrightii* congener *Datura stramonium* (Bello-Bedoy and Núñez-Farfán 2010). In *D. wrightii*, autonomously selfed fruit do not make as many seeds as fruit pollinated by hawkmoths (Bronstein et al. 2009). Thus, although we had no evidence that the percent of flowers setting fruit in the experiment differed, it is possible that different pollinators produced fruit with different numbers of seeds or with seeds of differing quality, both of which may have important consequences for the overall fitness of the maternal plant. Investigating these possibilities lay beyond the scope of the present study, but are important to consider in future work.

Honeybees are pollen thieves in other systems. For example, Hargreaves et al. (2010) found that the addition of *A. mellifera scutellata* hives reduced the anther pollen loads and the amount of pollen receipt on stigmas of *Aloe maculata* (Asphodelaceae) in eastern South Africa. Moreover, honeybee additions reduced the pollen loads found on sunbirds, the primary pollinator in that system. Vaughton (1996) similarly found that *Grevillea barklyana* (Proteaceae) plants caged with *A. mellifera* set 50% fewer fruit than did plants caged with both bees and birds. In *D. wrightii*, reductions of pollen mass on anthers caused by honeybee visits prior to hawkmoth visitation could negatively affect the ability of *M. sexta* to remove or deposit pollen during the night. In terms of effects on behavior, hawkmoths might not avoid flowers visited by honeybees because honeybees cannot reach *D. wrightii* nectar unless they resort to nectar larceny while hawkmoths don't actively seek pollen.

There are other examples of organisms that generally decrease plant fitness but can have neutral or positive effects in the absence of pollinators that move conspecific pollen to stigmas. Florivores reduce plant fitness either directly through consumption of gametes, or indirectly by affecting pollinator activity (McCall & Irwin 2006; González-Browne 2016) but there are also cases where florivores have a negligible effect on fitness components (Pohl 2006; McCall 2007), Another relationship is nectar robbery, wherein an organism acquires nectar without pollination. Nectar robbing often results in decreased female plant fitness (González-Browne et al. 2016) but neutral (Rojas-Nossa et al. 2016; Heiling et al. 2018) or positive effects on plant fitness are also possible (see Irwin et al. 2010 and Singh et al. 2014).

Honeybees can act as pollen thieves or pollinators both in and out of their native range, having significant impacts on plant populations across large regions. Honeybees often pollinate plants outside of the bees' native range, sometimes having a net positive effect on fitness (Aslan et al. 2016; Russo 2016). In general, per-visit pollinator efficiency rates for honeybees may be lower than those for some native pollinators (Hung et al. 2018), but the sheer number of visits by honeybees could result in a net positive effect on plant fitness. Alternatively, frequent visitation could have negative effects on fruit set. For example, Young (1988) found that increased visits by beetles increased fruit set up to a point, but decreased fruit set thereafter in *Dieffenbachia longispatha* (Araceae). In another example, Willmot and Búrquez (1996) found that up to three hawkmoth visits to *Merremia palmeri* (Convolvulaceae) led to an increase in fruit set, but thereafter more visits decreased fruit set. These negative effects of increased pollinator visitation could be caused by pollen dislodgement, pollen tube crowding, or damage to primary reproductive tissues (Young and Young 1992).

The positive effects of honeybee visits on plant fitness may be especially important when there are low numbers of native pollinators. For example, the typical bird pollinators of the Hawaiian tree species *Metrosideros polymorpha* have become rare, but honeybee foragers have become effective pollinators in the system (Junker et al. 2010). Habitat fragmentation can negatively affect native pollinator numbers, but honeybees can compensate for these losses in some species in Argentina (Aizen & Feinsinger 1994). In general, *D. wrightii*'s native hawkmoth pollinators are apparently not in decline, but their numbers vary from year to year and there may be years of very low *M. sexta* activity (Alarcón et al. 2010). In these low years *A. mellifera* foragers could compensate for reduced hawkmoth services through the number of their visits and their morning pollination activity which doesn't typically occur with hawkmoths.

Other plants have similar bimodal or dual pollination systems, some of which include hawkmoth pollinators. In the genus *Tritonioposis* there are instances of species that are pollinated by both sunbirds and nectivorous insects (Manning and Goldblatt 2005). In *Clarkia breweri*, both nocturnal hawkmoths and diurnal bees are frequent floral visitors and carry *C. breweri* pollen, suggesting that the plant could benefit from both types of insect (Miller et al. 2014). In *Polemonium brandegei*, researchers found that both hummingbirds and hawkmoths visited flowers in high numbers and might select for the variation in herkogamy seen across native plant populations (Kulbaba and Worley 2008).

Honeybees are frequent pollinators of both agricultural and native plants and often have similar per-visit pollinator efficiencies as other pollinators. Their pollination activities may or may not result in a net fitness gain in natural populations, depending on the identity of the total pollinator fauna, the degree to which a plant may be specialized for certain types of flowers, and the abundance of native pollinators, among other factors (Hung et al. 2018). It is possible that their role in maintaining native plant populations may increase with continued habitat fragmentation, urbanization, and losses in insect diversity, especially if native pollinators are negatively impacted by continued human activity. To understand how these changes will affect our natural areas, we will need more studies that include monitoring of both honeybee and native pollinators, experiments establishing pollinator efficiencies of both honeybees and coexisting pollinators, and studies, like this one, that manipulate both honeybee and primary pollinator visitation rates in a field setting.

ACKNOWLEDGEMENTS

We would like to thank Denison University and the Battelle Foundation for funding E.T., S.J., and M.E. for three summers of research. A.C.M. thanks the Denison University Biology department and the Department of Ecology and Evolutionary Biology at the University of Arizona for their space, support, and time during the writing of this manuscript. Two anonymous reviewers helped to improve this manuscript. We also thank Dr. Lyn Loveless and Emily McCall for fieldwork help.

REFERENCES

Aizen MA, Feinsinger P (1994) Forest fragmentation, pollination, and plant reproduction in a Chaco Dry Forest, Argentina. Ecology 75:330–351.

Alarcón R, Riffell JA, Davidowitz G, Hildebrand JG, Bronstein JL (2010) Sex-dependent variation in the floral preferences of the hawkmoth *Manduca sexta*. Animal Behaviour 80:289–296.

Barker JL, Bronstein JL (2016) Temporal structure in cooperative interactions: what does the timing of exploitation tell us about its cost? PLOS Biology 14:e1002371.

Bates D, Mächler M, Bolker B, Walker S (2015) Fitting linear mixed-effects models using lme4. Journal of Statistical Software 67 [online] URL: http://www.jstatsoft.org/v67/i01/ (accessed 29 June 2018).

Bello-Bedoy R, Núñez-Farfán J (2010) Cost of inbreeding to resistance to herbivores in *Datura stramonium*. Annals of Botany 105:747-753.

Bronstein JL, Huxman T, Horvath B, Farabee M, Davidowitz G (2009) Reproductive biology of *Datura wrightii*: the benefits of a herbivorous pollinator. Annals of Botany 103:1435–1443.

Fox J, Weisberg S (2011). An {R} Companion to Applied Regression, Second Edition. Sage, Thousand Oaks, CA.

González-Browne C, Murúa MM, Navarro L, Medel R (2016) Does plant origin influence the fitness impact of flower damage? A meta-analysis. PLOS ONE 11:e0146437.

Hargreaves AL, Harder LD, Johnson SD (2009) Consumptive emasculation: the ecological and evolutionary consequences of pollen theft. Biological Reviews 84:259–276.

Hargreaves AL, Harder LD, Johnson SD (2010) Native pollen thieves reduce the reproductive success of a hermaphroditic plant, *Aloe maculata*. Ecology 91:1693–1703.

Hargreaves AL, Harder LD, Johnson SD (2012) Floral traits mediate the vulnerability of aloes to pollen theft and inefficient pollination by bees. Annals of Botany 109:761-772.

Heiling JM, Ledbetter TA, Richman SK, Ellison HK, Bronstein JL, Irwin RE (2018) Why are some plant - nectar robber interactions commensalisms? Oikos [online] URL: http://doi.wiley.com/10.1111/oik.05440 (accessed 23 June 2018).

Hung K-LJ, Kingston JM, Albrecht M, Holway DA, Kohn JR (2018) The worldwide importance of honey bees as pollinators in natural habitats. Proc. R. Soc. B 285:20172140.

Inouye DW (1980) The terminology of floral larceny. Ecology 61:1251–1253.

Irwin RE, Bronstein JL, Manson JS, Richardson L (2010) Nectar robbing: ecological and evolutionary perspectives. Annual Review of Ecology, Evolution, and Systematics 41:271–292.

Junker RR, Bleil R, Daehler CC, Blüthgen N (2010) Intra-floral resource partitioning between endemic and invasive flower visitors: consequences for pollinator effectiveness. Ecological Entomology 35:760–767.

Kearney TH, Peebles RH, Howell HT, McClintock E (1960) Arizona Flora, Second edition, University of California Press, Berkeley.

Kulbaba MW, Worley AC (2008) Floral design in *Polemonium brangegei* (Polemoniaceae): genetic and phenotypic variation under hawkmoth and hummingbird pollination. International Journal of Plant Sciences 169:509-522.

Manning JC, Goldblatt P (2005) Radiation of pollination systems in the Cape genus *Tritoniopsis* (Iridaceae: Crocoideae) and the development of bimodal pollination strategies. International Journal of Plant Sciences 166:459-474.

McCall AC (2007) Leaf damage and gender but not flower damage affect female fitness in *Nemophila menziesii* (Hydrophyllaceae). American Journal of Botany 94:445–450.

McCall AC, Irwin RE (2006) Florivory: the intersection of pollination and herbivory. Ecology Letters 9:1351–1365.

Miller TJ, Raguso RA, May KA (2014) Novel adaptation to hawkmoth pollinators in *Clarkia* reduces efficiency, not attraction of diurnal visitors. Annals of Botany 113:317-329.

Pohl N, Carvallo G, Botto-Mahan C, Medel R (2006) Nonadditive effects of flower damage and hummingbird pollination on the fecundity of *Mimulus luteus*. Oecologia 149:648–655.

Rojas-Nossa SV, Sánchez JM, Navarro L (2016) Effects of nectar robbing on male and female reproductive success of a pollinator-dependent plant. Annals of Botany 117:291–297.

Russo L (2016) Positive and negative impacts of non-native bee species around the world. Insects 7:69.

Singh VK, Barman C, Tandon R (2014) Nectar robbing positively influences the reproductive success of *Tecomella undulata* (Bignoniaceae). PLoS ONE 9 [online] URL: https://www.ncbi.nlm.nih.gov/pmc/articles/PMC4103821/ (accessed 23 June 2018).

Solís-Montero L, Vergara CH, Vallejo-Marín M (2015) High incidence of pollen theft in natural populations of a buzz-pollinated plant. Arthropod-Plant Interactions 9:599–611.

Thorp RW (2000) The collection of pollen by bees. Plant Systematics and Evolution 222:211-223.

Vaughton G (1996) Pollination disruption by European honeybees in the Australian bird-pollinated shrub *Grevillea barklyana* (Protaceae). Plant Systematics and Evolution 200:89–100.

Wester P, Johnson SD (2017) Importance of birds versus insects as pollinators of the African shrub *Syncolostemon densiflorus* (Lamiaceae). Botanical Journal of the Linnean Society 185:225–239.

Willmott AP, Búrquez A (1996) The pollination of *Merremia palmeri* (Convolvulaceae): can hawk moths be trusted? American Journal of Botany 83:1050-1056.

Young H (1988) Differential importance of beetle species pollinating *Diffenbachia lonispatha* (Araceae). Ecology 69:832-844.

Young H, Young TP (1992) Alternative outcomes of natural and experimental high pollen loads. Ecology 73:639-647.

FALSE-COLOUR PHOTOGRAPHY: A NOVEL DIGITAL APPROACH TO VISUALIZE THE BEE VIEW OF FLOWERS

Christian Verhoeven[1], Zong-Xin Ren[2], and Klaus Lunau[1]

[1]*Institute of Sensory Ecology, Department Biology, Heinrich-Heine-University Düsseldorf, Universitätsstr. 1, 40225 Düsseldorf, Germany*
[2]*Key Laboratory for Plant Diversity and Biogeography of East Asia, Kunming Institute of Botany, Chinese Academy of Sciences, Kunming, Yunnan 650201, People's Republic of China*

Abstract—The colour vision system of bees and humans differs mainly in that, contrary to humans, bees are sensitive to ultraviolet light and insensitive to red light. The synopsis of a colour picture and a UV picture is inappropriate to illustrate the bee view of flowers, since the colour picture does not exclude red light. In this study false-colour pictures in bee view are assembled from digital photos taken through a UV, a blue, and a green filter matching the spectral sensitivity of the bees' photoreceptors. False-colour pictures demonstrate small-sized colour patterns in flowers, e.g. based on pollen grains, anthers, filamental hairs, and other tiny structures that are inaccessible to spectrophotometry. Moreover, false-colour pictures are suited to demonstrate flowers and floral parts that are conspicuous or inconspicuous to bees. False-colour pictures also direct the attention to other ranges of wavelength besides ultraviolet demonstrating for example blue and yellow bulls' eyes in addition to UV bulls' eyes which previously have been overlooked. False-colour photography is a robust method that can be used under field conditions, with various equipment and with simple colour editing.

Keywords: bee, colour vision, flower colour, colour pattern, bee view, false colour

INTRODUCTION

Flowers serve as filters signalling attractants for pollinators as well as deterrents for flower antagonists (Junker & Blüthgen 2010). Due to differences in the number of photoreceptors types, the spectral range of sensitivity of the colour photoreceptors, and differences in the colour preference of various flower visitors, flowers can be attractive to pollinators and unattractive to illegitimate visitors, i.e. pollen thieves and nectar robbers, at the same time (Lunau & Maier 1995; Lunau et al. 2011). Since many flower visitors, such as hummingbirds (Herrera et al. 2008), bats (Winter et al. 2003), bees (Peitsch et al. 1992), hoverflies (Lunau 2014), beetles (Martínez-Harms et al. 2012), butterflies and moths (Kelber et al. 2002; Arikawa 2003) are sensitive to ultraviolet light, these flower visitors necessarily view different aspects of the spectral reflectance of flowers as compared to UV-blind humans. Many of these flower visitors possess a different number of colour photoreceptor types compared to humans, e.g. one in bats, two in beetles, three in bees, four in hummingbirds and flies, and from three up to fifteen in butterflies (Briscoe 2008; Chen et al. 2016). This results in colour vision systems hardly conceivable to that in humans. However, bees have trichromatic colour vision just like humans, but are sensitive to ultraviolet, blue and green light rather than blue, green and red light as humans (Peitsch et al. 1992; Fig. 1). Neuronal processing of colour photoreceptor inputs in bees and humans is similar, both bees and humans have colour opponent neurons (Backhaus 1991) and colour constancy (Neumeyer 1981). To understand flower colours and floral colours pattern as bees might see them Kevan

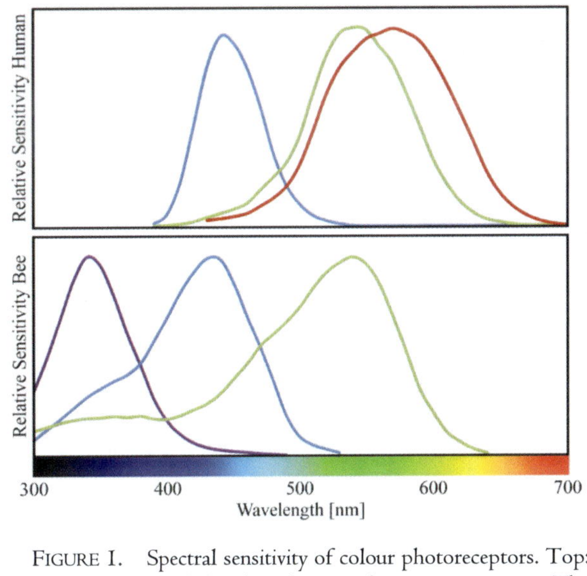

FIGURE 1. Spectral sensitivity of colour photoreceptors. Top: Relative sensitivity of the three human photoreceptor types. The curves peak at 420 nm, 534 nm and 564 nm (Bowmaker et al. 1980). In the UV range of wavelength between 300 nm and 400 nm the photoreceptors are not sensitive (modified from Stockman et al. 1993). Bottom: Relative sensitivity of the three photoreceptor types of the honeybee (*Apis mellifera*). The curves peak at 344 nm, 436 nm and 544 nm. In the red range of wavelength between 630 nm and 700 nm (red light) the photoreceptors are not sensitive (modified from Peitsch et al. 1992).

*Corresponding author: Klaus.Lunau@hhu.de

(1978, 1979, 1983) claimed the need for a bee-subjective colorimetric analysis of flower colours.

Three methods are frequently used to study floral colours and colour patterns as perceived by bees. The measurement of the spectral reflectance by a spectrophotometer allows for the comparison of the spectral reflectance of a floral tissue with that of a black standard and of a white standard. A database provides access to spectral reflectance measurements for many plant species (Arnold et al. 2010). Reflection spectroscopy is limited by the size and surface properties of the measuring spot, which does not allow for measuring the reflectance of small-sized, curved or rough-textured structures like pollen grains, stamens, and filamental hairs. The standard method of reflection spectroscopy is a measurement angle of 45° without integrating sphere (Chittka & Kevan 2005; but see Stavenga & van der Kooi 2016). This method largely ignores the impact of the surface of petals being flat, conical or other, since most white standards possess a smooth if not glossy surface. Moreover, the transparency of the petals is ignored, since the flower, e.g. a petal, is placed on different backgrounds being black, white or another petal for measurements, which disregards light shining through the petal from the backside and the impact of the colour chosen as background.

Another method is the synopsis of a colour picture and a UV-picture showing the flower as appearing to humans and the ultraviolet reflection pattern on different pictures (Daumer 1958; Silberglied 1979). This method possesses the disadvantage that the readers might be successful in notionally superimposing the colour picture and the UV-picture but fail to eliminate the red content of the colour picture at the same time. Red, however, is a component of yellow, magenta, purple and white colours. Moreover, the interpretation of UV-pictures showing UV-absorbing parts of flowers is sometimes misleading, e.g. if it is concluded that red and UV-absorbing parts are invisible to bees (Chittka & Waser 1997). A television camera equipped with UV-transmitting lens and UV-transmitting filter allowed direct observation and recording on video tape (Eisner et al. 1969; Silberglied 1976; Gronquist et al. 2001). Also, a modified image converter and image amplifier combined with UV-transmitting filter and UV-transmitting lens enabled to screen the UV-reflectance properties of many flowers (Rosen & Barthlott 1991) in the field. However, classical UV-photography using a UV-sensitive camera mounted on a tripod, UV-transmitting filter and UV-transmitting lens and a UV-sensitive film were still preferred to produce UV-photos with high quality (Biedinger & Barthlott 1993; Burr & Barthlott 1993; Burr et al. 1995; Lunau 2000). With the turn of the millennium, digital cameras were used mostly after having their internal UV- and IR-blocking filter removed and sometimes with the support of a UV-lamp for illumination (Koski & Ashman 2013).

Several attempts have been made to develop methods to create colour pictures in bee view. As early as in 1958 Daumer tried to evaluate floral colours considering the bees' colour perception. He took photographs with a yellow, a blue and a UV filter and calculated the "bee-colour" (Daumer 1958), but of course he did not have the option to merge the photographs to a bee view-picture. Vorobyev et al. (1997) developed a method based on video recordings of flowers made by a UV-sensitive camera and specially selected set of filters to reconstruct images of flowers as bees might see them by calculating the quantum catches of the honeybee photoreceptors for each ommatidium dependent on the distance between bee and flower. Multispectral images, a series of black-and-white photos taken through different bandpass filters (Kevan 1972; Chittka and Kevan 2005) from alpine flowers of the Rocky Mountains, Colorado, are available from the Floral Image Database (http://liu.edu/flower). Benitez-Vieyra et al. (2007), Williams & Dyer (2007), Brito et al. (2015) and Hempel de Ibarra et al. (2015) presented false colour photos of flowers in bee view, some of which require sophisticated calculations or calibrations. Photographic methods have also been used for simultaneous quantitative measurements of spatial and spectral components of floral colour patterns as seen by bees (Vorobyev et al. 1997; Benitez-Vieyra et al. 2007; Williams & Dyer 2007; Hempel de Ibarra & Vorobyev 2009; Garcia et al. 2014; Hempel de Ibarra et al. 2015; White et al. 2015).

Recent studies presenting multispectral images from flowers in bee view (Garcia et al. 2014; Brito et al. 2015; Hempel de Ibarra et al. 2015; White et al. 2015; Vasas et al. 2017) are mostly focusing on one or few target species and /or require sophisticated calculations or calibrations. In this study we introduce a simple and cheap method for general use based on a slightly modified and supplemented standard camera equipment to create UV-pictures and false-colour pictures in bee view and to highlight the rationale of this method for field studies. The estimated prices for a low-cost and a standard equipment as used in this study are shown in Supplement S1. This equipment is much less expensive than custom-built multispectral cameras, e.g. Spectrocam Multispectral Wheel Cameras (Ocean Optics). The pictures are intended to demonstrate the bee-subjective colour patterns of flowers in bee view with details that cannot be measured using a normal spectrophotometer. Moreover, it is demonstrated how false-colour pictures in bee view are used to better understand the impact of the bees' sensitivity for ultraviolet and insensitivity for red light. For this purpose, false-colour pictures of floral colour changes and small-sized floral structures are shown as false-colour pictures. In addition, yet undescribed forms of colour patterns are described using false colour pictures such as blue bull's eyes rather than ultraviolet bull's eyes (Silberglied 1979; Koski & Ashman 2014). The robustness of this method is tested by the comparison of false-colour pictures taken with different cameras, lenses, filters, different picture editing programs and in field conditions.

MATERIALS AND METHODS

False-colour photography of flowers and floral colour patterns in bee view is an extension to the well-established UV-photography. The editing of digital pictures allows to split a colour picture into the colour channels and to reassemble a colour picture from the colour channels and thus to use a UV-channel instead of blue, a blue channel instead of green and a green channel instead of red in order to shift the range of wavelength for about 100 nm. To match the bee photoreceptors' sensibility as close as possible specific filters might be used. Here the principal method of false-colour photography is described and the quality of results obtained with different equipment and with different processing is compared.

Equipment

To capture photos in bee view we used a modified Panasonic GH-1 camera. The low-pass filter was removed in order to increase the sensitivity for ultraviolet light. We used an Ultra-Achromatic-Takumar 1:4.5/85 lens made of fused quartz that transmits UV light. Since the modified camera is sensitive to ultraviolet and infrared light, we used a UV-/IR-Cut filter to capture a normal reference picture. This filter transmits light between 400 nm and 700 nm only (Fig. 2). To simulate the spectral sensitivity of typical bee photoreceptor types we used special filters that only transmit light of certain ranges of wavelengths (Fig. 3). Other equipment used included a Nikon D40 camera with low-pass filter, a 18-55 mm Nikkor lens and a 105 mm UV-Nikkor lens. In addition, only a UV picture and a colour photo split into the colour channels were used for picture processing (see below). A robust tripod and calm conditions are necessary for taking the different photos since changing the filter in between can be time-consuming. A white Teflon disc reflecting equal amounts of light in a range of wavelength from 300 nm to 700 nm was used for manual white balance before taking photos. Alternatively, a 20% Spectralon Grey Standard was used (see below).

Photography

Since most flowers are small we set the aperture values between f/8 and f/16 to ensure good image resolution. The exposure time and ISO value of the camera were adjusted to the daylight. All pictures were focused manually because the viewfinder is completely dark after attaching the UV transmitting filter. Four photos were taken, one photo with UV-/IR-Cut filter, one photo with the UV filter, one photo with the blue, and one photo with the green filter. All pictures were saved in the camera's RAW format. This allows to adjust the white balance in post-production. As a reference we used a Spectralon Grey Standard that reflects all wavelengths of the needed 300 nm to 700 nm spectrum with 20%. The grey

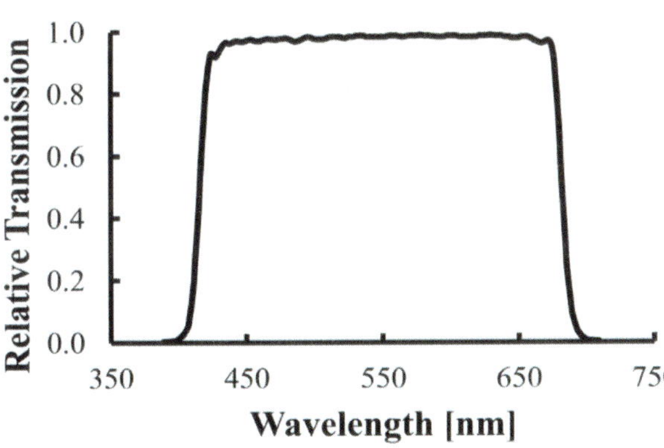

FIGURE 2. Relative transmission of the UV-/IR-Cut filter (Baader). Only light between 400 nm and 700 nm can transmit the filter (changed according to Baader Planetarium 2017).

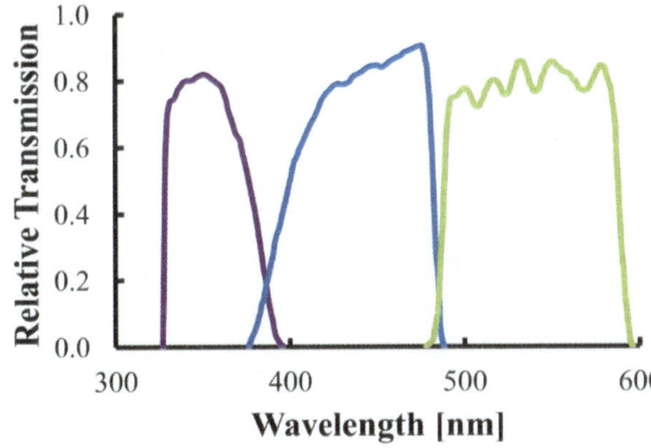

FIGURE 3. Relative transmission of the three colour filters (Baader). Violet: UV-filter transmits UV light only. Blue: Blue filter that only transmits blue light. Green: Green filter that only transmits light of the green part of the spectrum (changed according to Schmitt K., personal communication, 21. September 2015).

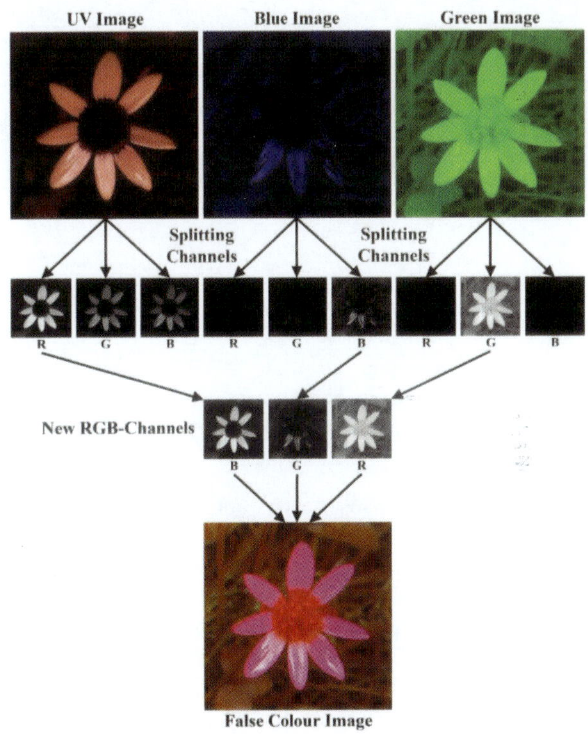

FIGURE 4. The splitting and merging process in detail.

reference was captured for every channel (normal, blue, green, UV) with the flowers or in separate pictures and always under the identical light and with identical camera settings.

Picture editing

The pictures taken with the UV, blue, and green filter were split into the colour channels in order to convert the merged picture into the RGB system which is visible for humans. From the picture with UV filter the red channel, from that with blue filter the blue channel, and from that with green filter the green channel was used for merging (Fig. 4). In windy conditions the frequent change of filters might not be useful. The alternative option is to take only two photos, one with the UV-/IR-Cut filter and one with the UV-filter. The photo taken with the UV-/IR-Cut filter then is split to

FIGURE 5. Colour picture of the ram's horn *Proboscidea louisianica* (Martyniaceae) showing the human view and false-colour picture showing the bee view. The merging of the false-colour picture was done using the UV, blue and green channel and by omitting the red channel.

the colour channels and the merging of the false-colour photo is done with the split colour photo and the split UV photo. To merge the channels to a bee view picture all captured pictures were pre-adjusted using Adobe Lightroom CC (version 2015 or newer). The merging was done in Photoshop or ImageJ. Moreover, preset white balance values can be used to reduce time between the two photos.

If only two photos were taken using manual white balance with a white and UV-reflecting Teflon disc, one photo was taken with UV-/IR-Cut filter and one with the UV-filter. Then the picture taken with the UV-/IR-Cut filter was split into the colour channels and the blue and green channels were used for merging. The red channel was discarded. Also, the photo taken with the UV-filter was split into the colour channels and the red channel was used for merging. In addition, we resigned the use of the RAW format or used a modified Nikon D40 camera and a plastic lens Nikon AF-S DX NIKKOR 18-55 mm. The merging was done in Photoshop or ImageJ.

For producing false-colour pictures the UV channel was converted to the new blue channel, the blue channel to the green, and green channel to the new red channel (Fig. 5). To analyse the pictures the following colour translation chart is useful (Tab. 1). For a picture that shows colours in bee view we need a UV, a blue and a green colour channel. These three colour channels need to be converted into the RGB system, which is visible for humans. Since the red channel is not needed it can be discarded. The bees' S-type of photoreceptors is sensitive for UV light. The humans' S-type of photoreceptors is sensitive for blue light. As a consequence, we convert the UV channel to the new blue channel, the blue channel is shown as green and green as red.

Grey Standard

All pictures were saved in the camera's RAW format. This allows to adjust the white balance in post-production. As a reference we used a Spectralon Grey Standard Labsphere SRS-20-010 that reflects all wavelengths of the needed 300 nm to 700 nm spectrum with 20%. The grey reference was captured for every photo, resp. channel (normal, blue, green, UV) with the flowers or in separate pictures and always under the identical light and with identical camera settings. To merge the channels to a bee view picture all captured pictures were pre-adjusted using Adobe Lightroom CC (version 2015 or newer).

Preparing the pictures using Adobe Lightroom

By using the histogram in Adobe Lightroom we adjusted the pictures so that the containing grey standard has the same brightness in all pictures. The histogram shows how many pixels in a picture (y-axis) have a certain brightness (x-axis). This is shown for the blue, the green and the red channel. Since we only use the blue channel of the picture captured with the blue filter, the brightness of the picture was adjusted so that the peak of the blue channel is exactly in the middle of the histogram (Fig. 6). We did the same with the peak of the green channel in the picture that was captured using the green filter. Because red channel of our cameras seems to be more UV sensitive than the other channels we adjusted the UV picture so that the peak of the red channel is in the middle of the histogram. In pictures where we captured the grey standard separately we adjusted the picture containing the grey standard and copied the values to the picture containing just the flower.

Real colour	Colour we see	Colour the bees see	Colour we see in the false-colour picture
UV	Black	Bee-Ultraviolet	Blue
Blue +UV	Blue	Bee-UV-Blue	Cyan
Green +UV	Green	Bee-Purple	Purple
Red +UV		Bee-Ultraviolet	Blue
Blue	Blue	Bee-Blue	Green
Blue +Green	Cyan	Bee-Blue-Green	Yellow
Blue +Red	Purple	Bee-Blue	Green
Green	Green	Bee-Green	
Green +Red	Yellow	Bee-Green	
Red		Bee-Black	Black
White -UV	White	Bee-Blue-Green	Yellow
White +UV	White	Bee-White	White
Black -UV	Black	Bee-Black	Black
Black +UV	Black	Bee-Ultraviolet	Blue
Grey -UV	Grey	Bee-Blue-Green	Yellow
Grey +UV	Grey	Bee-Grey	Grey

TABLE 1. Colour Translation Chart

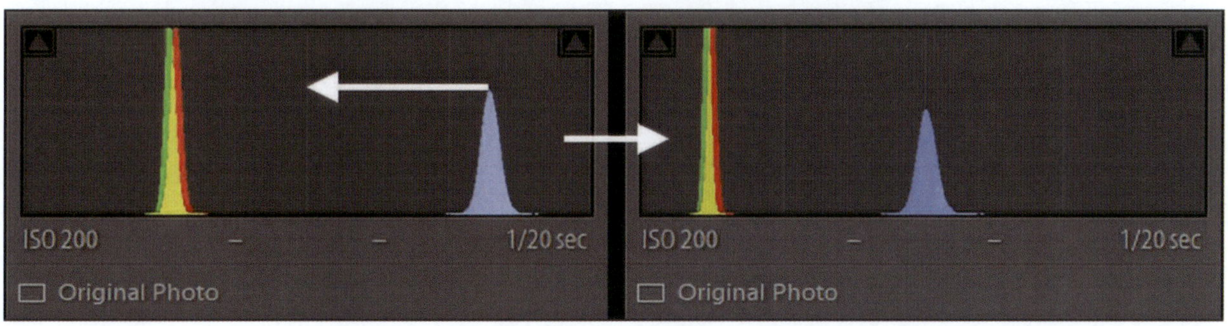

FIGURE 6. Histogram of a picture taken with the blue filter. On the left: Picture without adjustments. It is clearly visible that the blue channel of the picture is overexposed. On the right: The same picture after adjusting the brightness. The peak of the blue channel is now located in the middle and has the correct brightness.

For the normal picture we used the white balance dropper tool with the grey standard as a reference point to set the white balance.

Merging the photos in Adobe Photoshop

The prepared green, blue and UV pictures were opened in Photoshop as layers. At this point every picture still consists of three colour channels (RGB). By using the command "Image > Apply Image" we discarded the channels not needed. Figure 7 shows the settings that we used to discard the green and the red channel of the blue image/layer. This command overwrites the blue image with its own blue colour channel.

We used the same dialogue to overwrite the green image with its own green colour channel and to overwrite the UV image with its own red channel.

Using the command "Layer > Layer Style > Blending Options" we converted the layers into their new colour channels. Under "Advanced Blending" we unchecked the colour channels not needed. For example: The blue image

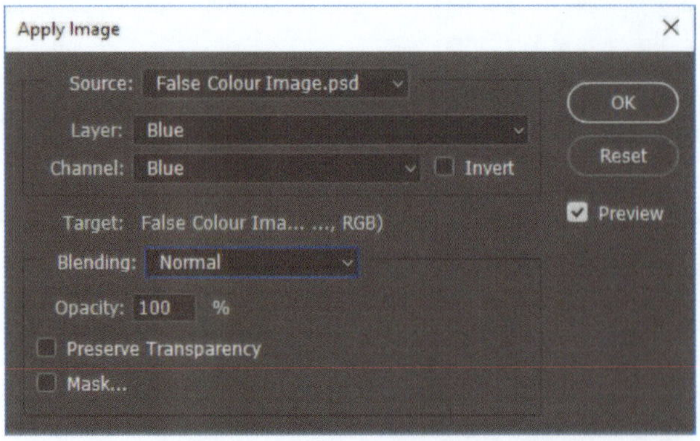

FIGURE 7: The "Apply Image" dialogue in Photoshop. The layer with the blue image gets overwritten with its own blue colour channel. The not needed green and red channels get discarded.

becomes the new green channel so we unchecked "R" and "B" (Fig. 8).

In some cases, the images are a little bit shifted to one another. To compensate this, we re-aligned the images by clicking "Select > All Layers" and "Edit > Auto-Align Layers". We used the projection method "Reposition" or "Auto". If this method did not work we aligned the layers manually using the "Move Tool".

If the grey standard was photographed next to the flower it should be grey in the false-colour image as well. If it shows a colour tint the new created colour channels are not perfectly equally bright. These small colour shifts were corrected by using a "Levels Adjustment Layer". By creating a new empty layer in the uppermost position and using the shortcut "Ctrl + Alt + Shift + E" (on Windows) we merged the whole false-colour picture to a single layer. With the "Marquee Tool" we selected the surface of grey standard and blurred the selected area with the command "Filter > Blur > Average". This ensures that the influence of the grainy structure of the standard and noise from the camera's sensor are as small as possible.

We created a "Levels Adjustment Layer" by clicking "Layer > New Adjustment Layer > Levels" in the uppermost position. By using the Tool "Sample in image to set grey point" on the blurred area this layer corrects all the colours in the image using the grey standard as a reference (Fig. 9).

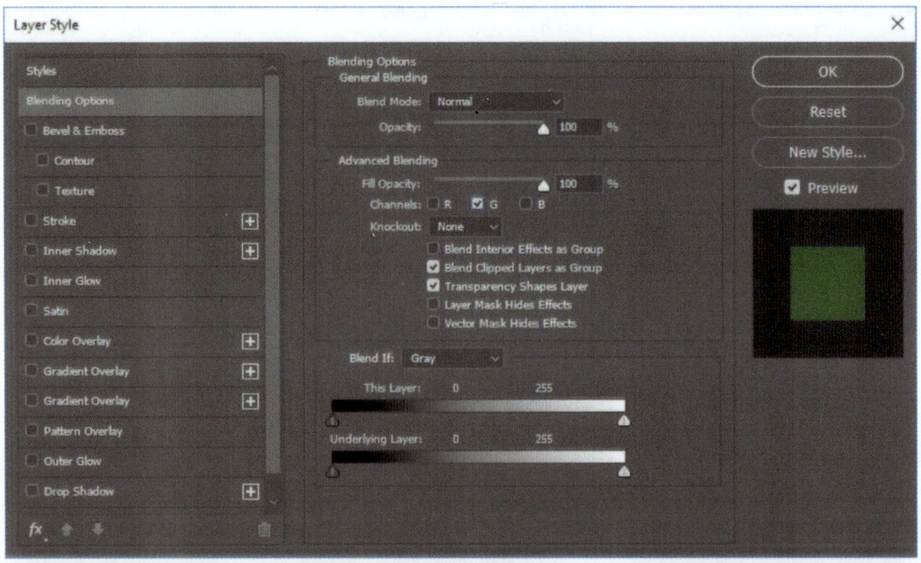

FIGURE 8. The "Blending Options" dialogue in Photoshop. Under "Advanced Blending" we can decide to which colour channel the image/layer gets converted to. This figure shows the settings for the blue image/layer that gets converted to the new green channel.

FIGURE 9. Levels Adjustment Layer. If the grey standard was photographed in separate pictures and not next to the flower we also converted these photos into a false-colour picture and followed all described steps. Then we copied the "Levels Adjustment Layer" to the false-colour picture of the flower. The pictures were saved in the Photoshop PSD format and as JPGs (8 Bit, sRGB).

Merging the photos in ImageJ

To verify the merging process and to show an alternative to Photoshop we also merged some images using ImageJ and the plugin "MICA Toolbox" (Troscianko & Stevens 2015).

Because we used special colour filters we had to create a custom *.txt file that tells the program which colour channels from which image are getting used (Fig. 10). How this works shall not be explained in detail in this paper because there is a comprehensive user guide available on the programmer's website.

To merge the images, we opened ImageJ used the command "Plugins > Multispectral Imaging > Generate Multispectral Image". In the upcoming dialogue we adjusted some settings. Under "Settings" we chose our custom *.txt file. All other settings are visible in Fig. 11.

After starting the process by clicking "OK" ImageJ automatically asks for the different pictures and guides the user through different steps that we followed. First, we opened the green image containing the grey standard and selected the standards surface area. Then the program asks for the green image without the standard. The steps repeat with the blue and the UV image. In between we manually aligned the images since we selected "Manual Align" (Fig. 11).

When all images were opened and aligned we used the command "Plugins > Multispectral Imaging > Tools > Make Presentation Image". In the following dialogue box, we decided that the green channel of the green image becomes the new red channel, the blue channel of the blue image becomes the new green channel and the red channel of the UV image becomes the new blue channel. We also chose "Square Root" under "Transform" and checked the "Convert to RGB colour" box (Fig. 12).

By clicking "OK" the program merges the images to a false-colour image. We saved the images as JPEG files.

Interpretation of false-colour pictures

In order to test the value of false-colour photography we studied floral colour changes and floral colour patterns and compared false-colour pictures and UV-pictures as a basis for the interpretation of the flower colours as seen by bees. A series of false-colour pictures of sub-alpine and alpine flowers were taken to test the reliability of this method. Whether the results of false-colour photography are similar with deviant equipment, deviant picture processing and different filters were also tested by using the selected flowers.

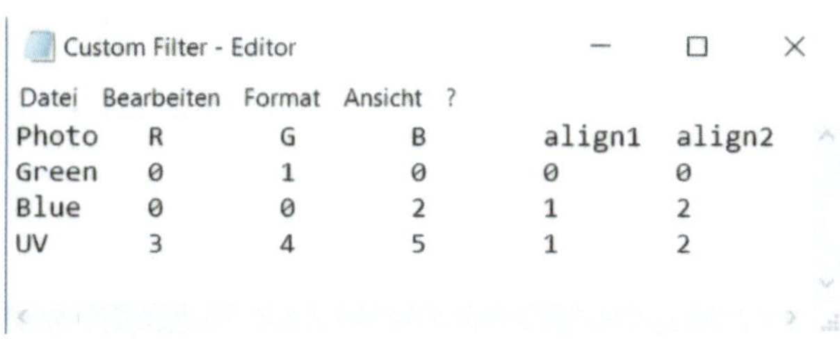

Photo	R	G	B	align1	align2
Green	0	1	0	0	0
Blue	0	0	2	1	2
UV	3	4	5	1	2

FIGURE 10: The custom *.txt file created to tell the program which colour channel to use.

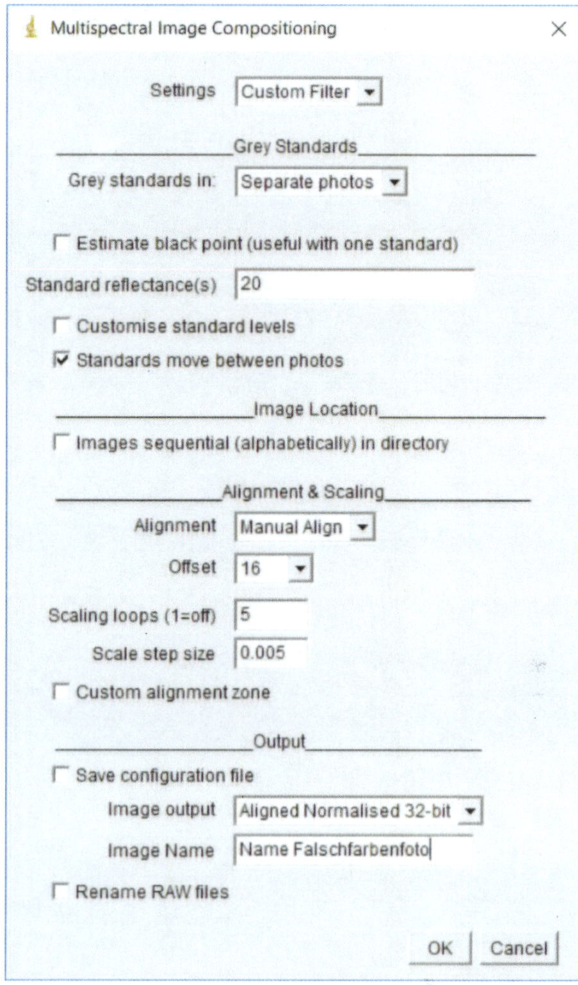

FIGURE 11. The settings that were used to create a false-colour image with ImageJ. We told the program that the grey standard was captured in separate photos and that is moves between them. We also set the reflectance to 20%. Under 'Alignment & Scaling' we chose manual align. We set the output to 'Aligned Normalised 32-bit'.

RESULTS

Colour change

Floral colour changes in the Shrubby Milkwort *Polygala chamaebuxus* (Polygalaceae) affects the conspicuousness of the post-change colour in the false-colour picture but is not detectable in the ultraviolet range of wavelength (Fig. 13 A, B, C). The pre-change as well as the post-change colour of the Common Lungwort *Pulmonaria officinale* (Boraginaceae) are reflecting ultraviolet light and are conspicuous for bees according to the false-colour picture (Fig. 13 D, E, F). The Field Forget-me-not *Myosotis arvensis* (Boraginaceae) displays a yet undescribed threefold colour change. The purplish early flowing phase and middle bluish flowering phase display a colour change visible to the human eyes. But the flower petals also change from UV-absorbing to UV-reflecting. This colour change is visible to bees' eyes but invisible to the human eyes. The colour change from the bluish and UV-absorbing middle flowering phase to the equally bluish late UV-reflecting flowering phase is visible to bees' eyes as is the colour change from human-yellow to human-white of the floral guide (Fig. 13 G, H, I).

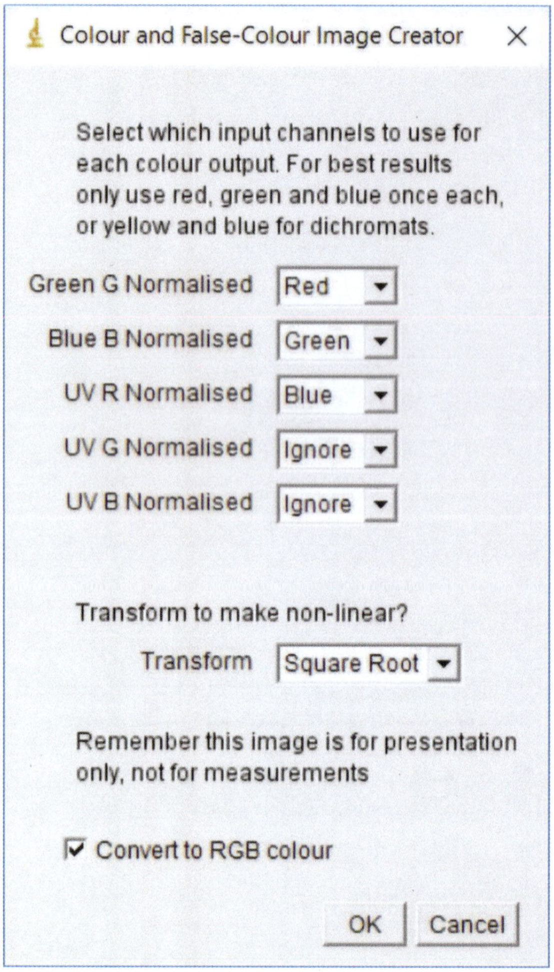

FIGURE 12. Dialogue box to select that the green channel from the picture taken with the green filter becomes the new red channel, blue channel from the picture taken with the blue filter becomes green and the red channel from the picture taken with the UV filter becomes blue.

Bull's eyes

Bull's eye refers to colour patterns that highlight the rewarding center of a flower. Some of these bull's eyes are visible to bees but invisible to humans (e. g. UV bull's eyes). Some colour patterns can be visible to humans too, but there are also non-UV colour patterns that are invisible to humans. The Yellow Archangel *Lamium galeobdolon* (Lamiacae) displays a typical all-yellow flower colour and an additional ultraviolet bull's eye resulting in a conspicuous colour pattern in bee view (Fig. 14 A, B, C). The flower displays two shades of yellow which are congruent with the UV-pattern (Fig. 14 A, B, C). The Marshmarigold *Caltha palustris* (Ranunculaceae) is uniformly yellow, but nevertheless displays a bee-visible colour pattern due to a pattern of reflection in the ultraviolet light (Fig. 14 D, E, F). The Yellow Daffodil *Narcissus pseudonarcissus* (Amaryllidaceae) shows no ultraviolet, but a blue bull's eye resulting in a colour pattern visible to bees (Fig. 15). The Key of Heaven *Primula veris* (Fig. 16 A, B, C) and the Oxlip *Primula elatior* (Fig. 16 D, E, F; Primulaceae) have all-yellow flowers, reflect uniform in the ultraviolet range of wavelength, but nevertheless display a colour pattern visible to bees. The Great Mullein *Verbascum phlomoides* (Scrophulariaceae) has all-yellow flowers and displays an ultraviolet bull's eye. Due to an additional blue bull's eye the floral colour pattern is rather complex in bee view (Fig. 17).

Tiny structures

False-colour pictures are able to display small-scale floral colour patterns visible to bees as shown for stamens and florets of the Cornflower *Centaurea cyanus* (Asteraceae) (Fig. 18 A, B, C) pollen and white staminal protuberances of the Flower-of-an-hour *Hibiscus trionum* (Malvaceae) (Fig. 18 D, E, F), a colour gradient of small-sized floral guides of the Round-Leaved Saxifrage *Saxifraga rotundifolia* (Saxifragaceae) (Fig. 18 G, H, I) and the androeceum of the Wood Anemone *Anemone nemorosa* (Ranunculaceae) (Fig. 18 J, K, L).

Didactics

The conspicuousness or inconspicuousness of human-red flowers for bees' eyes is dependent of the ultraviolet reflection as demonstrated for the UV-reflecting flower of the Winged Pea *Lotus tetragonolobus* (Fig. 19 A, B, C; Fabaceae) and the Red Poppy *Papaver rhoeas* (Fig. 19 G, H, I; Papaveraceae) as well as the UV-absorbing flower of Forrests Rhododendron *Rhododendron forrestii* (Fig. 19 D, E, F; Ericaceae). The underside of the petals in the Lousewort *Pedicularis gruina* (Orobanchaceae) display a rare white and UV-reflecting colour (Fig. 19 J, K, L), whereas the white flowers Common Snowdrop *Galanthus nivalis* (Amaryllidaceae) stand out against the snow due to absorption of ultraviolet light (Fig. 19 M, N, O). The Bittersweet Nightshade *Solanum dulcamara* (Solanaceae) displays yellow anthers and green floral guides on the petals with possess a similar colour for bees (Fig. 19 P, Q, R).

Colour pattern of sub-alpine and alpine flowers on the mountains of eastern Himalayas

A series of photos taken under field conditions in the surroundings of the Lijiang Forest Ecosystem Research Station operated by the Kunming Institute of Botany, Chinese Academy of Sciences (China) using automatically controlled exposure to light for UV-photos and for colour photos, a single white balance value for all colour photos and another for all UV-photos and later manual adjustment of the brightness of the channels reveals satisfactory results (Fig. 20, 21). This equipment was easily carried on an international flight and proved to stay functional at high altitude with low temperature, low oxygen and high UV-radiation. The Edelweiss flower of *Leontopodium* sp. (Asteraceae) possesses hairy petals inaccessible to spectrophotometry (Fig. 20 A, B, C). The white part of the fused dorsal petals of the Lousewort *Pedicularis cephalantha* (Orobanchaceae) displays a colour that is rarely found in bee-pollinated plants, i.e. a white and UV-reflecting colour (Fig. 20 D, E, F). The flower of *Tibetia yunnanensis* (Fabaceae) displays a white, yet UV-absorbing floral guide similar to pollen-mimicking floral guides described above (Fig. 20 G, H, I). The flower of *Primula poissonii* (Primulaceae) possesses no UV-pattern, yet a bee-visible floral guide (Fig. 20 J, K, L). The flower of *Salvia hylocharis* (Lamiaceae) displays an unusual type of floral guide absorbing all visible wavelengths as well as ultraviolet

FIGURE 13. Colour pictures (left), UV-pictures (middle) and false-colour pictures (right) of floral colour change in *Polygala chamaebuxus* L. (A, B, C), *Pulmonaria officinalis* L. (D, E, F), and *Myosotis arvensis* Hill (G, H, I).

FIGURE 14. Colour pictures (left), UV-pictures (middle) and false-colour pictures (right) of bull's eyes in *Lamium galeobdolon* (L.) (A, B, C) and *Caltha palustris* L. (D, E, F).

FIGURE 15: Colour picture, UV-picture and false-colour picture of *Narcissus pseudonarcissuss* L.. Additional pictures show the blue channel and the green channel.

FIGURE 16. Colour pictures (left), UV-pictures (middle) and false-colour pictures (right) of *Primula elatior* (L.) (A, B, C) and *Primula veris* L. (D, E, F).

FIGURE 17. Colour picture, UV-picture and false-colour picture of *Verbascum phlomoides* L.. Additional pictures show the blue channel and the green channel.

FIGURE 18: Colour pictures (left), UV-pictures (middle) and false-colour pictures (right) of *Centaurea cyanus* L. (A, B, C), *Hibiscus trionum* L. (D, E, F), *Saxifraga rotundifolia* L. (G, H, I) and *Anemone nemorosa* L. (J, K, L).

light (Fig. 20 M, N, O). The pollinators of *Paris polyphylla* (Melanthiaceae) are still unknown (Ren et al. 2014); although the flower displays no UV-pattern it has a bee-visible colour pattern (Fig. 20 P, Q, R). The Saxifrage *Saxifraga signata* (Saxifragaceae) possesses flowers with tiny yellow protuberances that uncommonly reflect ultraviolet light and might mimic the colour and gloss of nectar in addition to orange and UV-absorbing floral guides (Fig. 21 A, B, C). The yellow flower of *Potentilla griffithii* (Rosaceae) displays basal orange floral guides on each petal that are also visible to bees which, however, cannot be read from the UV-picture (Fig. 21 D, E, F). The yellow flower of the Lousewort *Pedicularis longiflora* (Orobanchaceae) displays two red floral guides which are invisible on the UV-picture, but clearly visible on the false-colour picture (Fig. 21 G, H, I). The flowers of the Lousewort *Pedicularis tricolor* (Fig. 21 J, K, L; Orobanchaceae) and of *Trollius yunnanensis* (Fig. 21 M, N, O; Ranunculaceae) display different shades of yellow and no UV-pattern which results in clearly bee visible colour pattern.

Robustness to changes in equipment and processing

We compared different false-colour images using the three colour filters with false-colour images that were taken with just the UV-/IR-Cut filter and the UV filter (Fig. 22 A, B). The principal details about the colour change of *Myosotis arvensis* (Fig. 13, see above) are also found using deviant techniques. For the merging of the false-colour photos Adobe Photoshop (Fig. 22 C) and ImageJ (Fig. 22 D) were compared.

Using different cameras and lenses also revealed similar results; even the use of a cheap plastic lens with less transmittance of ultraviolet light than the quartz lenses seem adequate for creating false-colour pictures (Fig. 23).

To see if the method works without setting the white balance we created a false-colour picture without any brightness adjustments. The camera was set to auto (Fig. 24 A). We also used styrofoam as a reference standard to match the brightness levels in post (Fig. 24 C) to see if styrofoam can be a cheap alternative to grey standards. For reference we also took a false-colour image by using a spectralon grey standard and the described method (Fig. 24 D).

The result shows that the automatically exposed false-colour picture without adjustments has a green colour cast (Fig. 24 A). The camera overexposed the blue channel because the flower and the background don't reflect a lot of blue light. Because of the blue filter the camera sees a dark picture and tries to compensate by overexposing the picture. This leads to a green colour cast in the false-colour image (Fig. 24 A). This problem can be solved by not using the blue and green filter and instead splitting the normal RGB image that were also taken by using the automatic exposure (Fig. 24 B). The resulting image looks similar to the false-colour picture taken with the grey reference standard (Fig. 24 D). This method however can still lead to colour casts since camera sees a dark picture through the UV filter and compensates by overexposing.

FIGURE 19. Colour pictures (left), UV-pictures (middle) and false-colour pictures (right) of *Lotus tetragonolobus* L. (A, B, C), *Rhododendron forrestii* B. Balfour ex Diels (D, E, F), *Papaver rhoeas* L. (G, H, I), *Pedicularis gruina* Franch ex Maxim. (J, K, L), *Galanthus nivalis* L. (M, N, O) *Solanum dulcamara* L. (P, Q, R).

Using the styrofoam as a reference standard dramatically improves the results using the colour filters (Fig. 24 C). The false-colour image looks very similar to the false-colour picture taken with the grey reference standard (Fig. 24 D).

DISCUSSION

In this study a low-cost and easy-to-use multispectral imaging equipment is proposed and the advantages of false colour imaging over UV-photography and spectrophotometry are explained with some examples. Our equipment was proved to be effective to detect new floral color patterns and to be efficiently working in a high elevation mountain region in the Himalayas.

The equipment is configured for false colour photography in bee view. Bees use colour cues of flowers to find nectar and pollen sources (von Frisch 1915; Dyer at al. 2014; Orban & Plowright 2014). Experiments with honeybees and bumblebees have shown that flower visits consist of a series of behavioural reactions mostly based on visual cues. Initially bees are colour blind in their search for flowers in that they process only input to the green type of photoreceptors (Giurfa et al. 1996). Only if the visual angle viewing the target flower surpasses a threshold the bees use colour vision to find flowers (Dyer et al. 2008). Naïve and inexperienced bees' innate colour preferences guide them towards promising flowers (Lunau & Maier 1995; Lunau et al. 1996). Bees exhibit a preference for flowers displaying colour contrast against the background (Lunau et al. 1996), flower colours of high bee-subjective saturation (Lunau et al. 1996), and prefer distinct colour hues, bee-subjective blue in the first place (Giurfa et al. 1995; Gumbert 2000; Milet-Pinheiro et al. 2015). When bees have approached a distinct flower, floral colour patterns are important for the flower to guide the bees towards a suitable site for the first physical contact with the flower by means of the antennae, landing (Lunau et al. 1996) and proboscis extension (Wilmsen et al. 2017). Floral colour patterns may include small-sized structures such as floral guides, veins, staminodes, stamens, anthers, filamental hairs and others (Lunau & Wester 2017; Lunau et al. 2017). False colour images show flower colour patterns as seen by bees although the bee colours are visualized as human visible colours, UV as blue, blue as green and green as red (Kevan 1978, 1983; Kevan and Backhaus 1998). False colour images of flowers in bee view have been introduced several times (Kevan 1972; Vorobyev et al. 1997; Chittka and Kevan 2005; Benitez-Vieyra et al. 2007; Williams & Dyer 2007; Brito et al. 2015; Hempel de Ibarra et al. 2015). In this study, the

FIGURE 20. Colour pictures (left), UV-pictures (middle) and false-colour pictures (right) of flowers taken under field conditions in Lijiang (Yunnan, China). *Leontopodium* sp. (A, B, C), *Pedicularis cephalantha* Franch. ex Maxim. (D, E, F), *Tibetia yunnanensis* (Franch.) Tsui (G, H, I), *Primula poissonii* Franch. (J, K, L), *Salvia hylocharis* Diels (M, N, O), *Paris polyphylla* var. *yunnanensis* (Franch.) Hand.-Mazz (P, Q, R).

FIGURE 21: Colour pictures (left), UV-pictures (middle) and false-colour pictures (right) of flowers taken under field conditions in Lijiang and Shangri-La (Yunnan, China). *Saxifraga signata* Engl. et Irmsch. (A, B, C), *Potentilla griffithii* var. *velutina* Card. (D, E, F), *Pedicularis longiflora* Rudolph (G, H, I), *Pedicularis tricolor* Hand.-Mazz. (J, K, L), *Trollius yunnanensis* (Franch.) Ulbr. (M, N, O).

FIGURE 22. False-colour picture of *Myosotis arvensis* L. merged from photos taken with UV, blue and green filter (A), merged from photos taken with UV filter and the UV-/IR-Cut filter split into colour channels (B), Adobe Lightroom and Photoshop (C), and ImageJ and MICA Toolbox (D).

advantages of false colour imaging over UV-photography and spectrophotometry are clearly demonstrated by multiple examples, especially by the undetected colour pattern of sub-alpine and alpine flora on the high mountain of eastern Himalayas in China.

Interpreting flowers as seen by bees requires the consideration of ultraviolet and bee-visible ranges of wavelength (Kevan 1978, 1979, 1983), although ultraviolet reflections in general are not more important than blue or green ones (Kevan et al. 2001). Rarely the bees' lower sensitivity in the red range of wavelength has brought into focus (Cure & Wittmann 1990; Chittka & Waser 1997; Lunau et al. 2011; Bergamo et al. 2016). The fact that bees are red blind or less sensitive to red light has caused lot of confusion about how bees might perceive red colours (Hinton 1976; Johnson & Midgley 2001; Rodríguez-Gironés & Santamaría 2004; Martínez-Harms et al. 2010). It was found that some bees have developed sensitivity in the red range of wavelength (Cure & Wittmann 1990; Chittka & Waser 1997) or the red colour of flowers is used to exclude bee as potentially nectar-robbing flower visitors (Lunau et al. 2011; Bergamo et al. 2016), since bees see red colours without additional reflection in the UV as black. Sensory exclusion of bees was also proposed for white, UV-reflecting bird-pollinated flowers, since UV-reflecting white colours are not spontaneously preferred by bees (Lunau et al. 1996; 2011). In the Lousewort *Pedicularis gruina* a similar phenomenon is demonstrated for the buds of the violet flowers which display the white, UV-reflecting underside of the petals. The Snowdrop *Galanthus nivalis* represents the common type of white bee-pollinated flowers (Kevan et al. 1996). In the Red Poppy *Papaver rhoeas* (Papaveraceae) the red flowers in the Eastern Mediterranean are pollinated by red sensitive glaphyrid beetles (Martínez-Harms et al. 2012), whereas the flowers in the Western part of Europe have evolved reflection in the red range of wavelength are pollinated by bees (Burkhardt 1997).

Floral colour change is known as a semi-honest indicator of floral nectar reward (Lunau & Wester 2017). The fact that the flowering phase offering reward is attributed with a more attractive colour (Lunau 1996) is nicely demonstrated with the false-colour picture of *Myosotis arvensis*. In the Lungwort both the pre-change as well as the post-change flower colour seem to be attractive to bees (Oberrath & Böhning-Gaese 1999) which is supported by false-colour pictures of a related *Pulmonaria* species. However, false-colour pictures might reveal yet unknown details of floral colour changes (Weiss & Lamont 1997; Ohashi et al. 2015). The bee view might be enhanced if the optical resolution is also considered (Rodríguez-Gironés & Ruiz 2016; Lunau & Verhoeven 2017).

One typical pattern is the so-called ultraviolet bull's eye (Silberglied 1979; Schlangen et al. 2009; Koski & Ashman 2014) of all-yellow flowers displaying a UV-reflecting periphery and a UV-absorbing center part either by pollen, anthers of pollen-mimicking floral guides (Lunau 2000, 2007; Lunau et al. 2017). In this study an all-yellow flower displaying a blue bull's eye as well as all-yellow flowers displaying a bee-visible colour pattern with different shades of yellow were found. It is known that the centre part of floral colour pattern directs the bees' approaches towards the first physical contact and landing site of flowers due to higher colour saturation and colour contrast against the corolla (Lunau et al. 1996; Papiorek et al. 2013; Wilmsen et al. 2017). In many cases the ultraviolet picture does not indicate the higher salience of the flowers' centre part rather than the false-colour picture. Even the absence of UV-reflection pattern is of significance for flower visitors, e.g. for yellow bird-pollinated flowers (Papiorek et al. 2013), and for yellow

FIGURE 23. Colour pictures (A-D) and false-colour pictures (E-H) of the Pilewort *Ficaria verna* L. (Ranunculaceae) taken with different camera equipment which was Panasonic GH-I camera and Pentax 85 mm Ultra Achromatic Takumar lens (A, E), Panasonic GH-I camera and 105 mm UV-Nikkor lens (B, F), Nikon D40 camera and 18-55 mm Nikkor lens (C,G), and Nikon D40 camera and 105 mm UV-Nikkor lens (D, H).

FIGURE 24: *Bellis perennis* A: False-colour picture taken with the camera set to auto and using the colour filters. The picture was assembled without any adjustments. B: False-colour picture taken with the camera set to auto. The picture was assembled without any adjustments and by splitting the RGB image. C: Before merging the pictures to the false-colour picture the brightness of the pictures were adjusted so that the styrofoam has the exact same brightness in every colour channel (white balance). D: False-colour image assembled by using the spectralon grey reference standard.

Crocus species (Lunau et al. 2016). A common type of colour pattern in all-yellow flowers is the combination of an orange floral guide with a yellow periphery without any UV-pattern like in the California Poppy *Eschscholzia californica* (Papaveraceae) and Common Toadflax *Linaria vulgaris* (Plantaginaceae). It is still an open question whether and how this colour pattern is able to direct flower visitors towards the centre part of the flower (Lunau 1992). If the spectral reflectance in the ultraviolet and blue ranges of wavelengths are negligible the orange colour might appear darker to the bees than the yellow colour and thus appear more bee-subjective saturated, since darker colours appear at least to humans more saturated than lighter colours of similar chroma (Lübbe 2010).

The false-colour images will be helpful for plant-pollinator interaction studies in these underexplored regions, such as Himalayas. We demonstrated the diversity of floral colours on the high elevation mountains in Lijiang and Shangri-La, eastern Himalayas. These flowers of the Chinese flora reveal a large diversity of bee-visible colour pattern which has not yet been described to our knowledge despite high intensity of studies in visual aspects of pollination biology in China (Shi et al. 2009; Eaton et al. 2012; Huang et al. 2015; Ma et al. 2015; Zhang et al. 2017; Tao et al. 2018; Yan et al. 2018). Unusual flower colour patterns were detected in the bumblebee buzz-pollinated genus *Pedicularis* and indicated a highly diverse signal system in this genus. On the mountains of southwestern China, *Pedicularis* species exhibit a spectacular level of endemism and floral diversity (> 300 species; Yang et al. 1998). The different *Pedicularis* species are pollinated by the same bumblebee species, but the floral mechanisms that ensure pollen deposits and pickups on different parts of the pollinator's body ("*Pedicularis*-type" of mechanical isolation) were well characterized in a few co-existing species in China (Wang & Li 2005; Tong & Huang 2016, 2018; Liang et al. 2018). However, the visual and olfactory signal aspects of this model system have not been explored yet.

The flora of China is extremely diverse with more than 30,000 species of angiosperms (Wu et al. 2013). The mountains of southwest China are regarded as one of global hotspots of biodiversity with the highest diversity of species and endemism of all temperate floras (Myers et al. 2000). The floral diversity of high elevation mountains of this region is strongly shaped by generalist bee species, such as Chinese native honeybee *Apis cerana* and bumblebees (Ren et al. 2014;

Zhao et al. 2016). This region harbours the half of global species of bumblebees (Williams et al. 2009). The limited number of false-colour photos presented in this study indicate that the diverse bee pollination system in Himalayas requires more researches on their visual signals in the future. Our low-cost and easy-to-use multispectral imaging equipment will be helpful for the future when we explore these understudied areas.

In conclusion, this study shows that false colour images of flowers in bee view provide similar information as classical UV-photos and additional information about the floral colour pattern that is not available from UV-photos. The false colour images of flowers are easy to interpret due to the consistent bathochromic shift for about 100nm during the reassembling to the false colour image resulting that all kinds of bee-visible colour patterns are seen as patterns rather than uniform colours. This also holds for very tiny structure like pollen grains, anthers, filamental hairs, and veins which is advantageous compared to spectrophotometry.

APPENDICES

Additional supporting information may be found in the online version of this article:

APPENDIX I. Estimated costs for a low-budget version and an optimal equipment for false colour photography in bee view.

REFERENCES

Arikawa K (2003) Spectral organization of the eye of a butterfly, *Papilio*. Journal of Comparative Physiology A 189:791-800.

Arnold SEJ, Faruq S, Savolainen V, McOwan PW, Chittka L (2010) FReD: The Floral Reflectance Database — A web portal for analyses of flower colour. PLoS ONE 5(12): e14287.

Backhaus W (1991) Color opponent coding in the visual system of the honeybee. Vision Research 31:1381-1397.

Benitez-Vieyra S, Hempel de Ibarra N, Wertlen AM, Cocucci AA (2007) How to look like a mallow: evidence of floral mimicry between Turneraceae and Malvaceae. Proceedings of the Royal Society B Biological Sciences 274:2239-2248.

Bergamo PJ, Rech AR, Brito VLG, Sazima M (2016) Flower colour and visitation rates of *Costus arabicus* support the "bee avoidance" hypothesis for red-reflecting hummingbird-pollinated flowers. Functional Ecology 30:710-720.

Biedinger N, Barthlott W (1993) Untersuchungen zur Ultraviolettreflexion von Angiospermenblüten. I Monocotyledonae. Tropische und Subtropische Pflanzenwelt 86:1-122.

Bowmaker JK, Dartnall HJ (1980) Visual pigments of rods and cones in a human retina. The Journal of Physiology 298:501-511.

Briscoe AD (2008) Reconstructing the ancestral butterfly eye: focus on the opsins. Journal of Experimental Biology 211:1805-1813.

Brito VLG, Weynans K, Sazima M, Lunau K (2015) Trees as huge flowers and flowers as oversized floral guides: the role of floral color change and retention of old flowers in *Tibouchina pulchra*. Frontiers in Plant Science 6:362.

Burkhardt D (1997) Die Welt mit anderen Augen—Wie Insekten und Vögel die Welt und ihre Farben sehen. Biologie in unserer Zeit 19:37-46.

Burr B, Barthlott W (1993) Untersuchungen zur Ultraviolettreflexion von Angiospermenblüten II. Magnoliidae, Ranunculidae, Hamamelididae, Caryophyllidae, Rosidae. Tropische und Subtropische Pflanzenwelt 87:1-193.

Burr B, Rosen D, Barthlott W (1995) Untersuchungen zur Ultraviolettreflexion von Angiospermenblüten III. Dilleniidae und Asteridae s.l.. Tropische und Subtropische Pflanzenwelt 93:1-185.

Chen P-J, Awata H, Matsushita A, Yang E-C, Arikawa K (2016) Extreme spectral richness in the eye of the Common Bluebottle Butterfly, *Graphium sarpedon*. Frontiers in Ecology and Evolution 4:18.

Chittka L, Kevan PG (2005) Flower colour as advertisement. In: Dafni A, Kevan PG, Husband BC (eds) Practical Pollination Biology. Enviroquest Ltd., Cambridge, ON, Canada, pp 157-196.

Chittka L, Menzel R (1992) The evolutionary adaptation of flower colors and the insect pollinators' color vision systems. Journal of Comparative Physiology A 171:171-181.

Chittka L, Shmida A, Troje N, Menzel R (1994) Ultraviolet as a component of flower reflections, and the colour perception of Hymenoptera. Vision Research 34:1489-1508.

Chittka L, Waser NM (1997). Why red flowers are not invisible for bees. Israel Journal of Plant Sciences 45:169-183.

Clark C (1979) Ultraviolet absorption by flowers of the Eschscholzioideae (Papaveraceae). Madrono 26:22-25.

Cure JR, Wittmann D (1990) *Callonychium petuniae*, a new panurgine bee species (Apoidea, Andrenidae) oligolectic on *Petunia* (Solanaceae). Studies on Neotropical Fauna & Environment 25:153-156.

Daumer, K (1958) Blumenfarben, wie sie die Bienen sehen. Zeitschrift für vergleichende Physiologie 41:49-110.

Dyer AG, Boyd-Gerny S, Mcloughlin S, Rosa MGP, Simonov V, Wong BBM (2012) Parallel evolution of angiosperm colour signals: common evolutionary pressures linked to hymenopteran vision. Proceedings of the Royal Society B: Biological Sciences 279:3606-3615.

Dyer AG, Spaethe J, Prack S (2008) Comparative psychophysics of bumblebee and honeybee colour discrimination and object detection. Journal of Comparative Physiology A 194:617-627.

Eaton DAR, Fenster CB, Hereford J, Huang SQ, Ree RH (2012) Floral diversity and community structure in *Pedicularis* (Orobanchaceae). Ecology 93:182-194.

Eisner T, Silberglied RE, Aneshansley D, Carrel JE, Howland HC (1969) Ultraviolet video-viewing: the television camera as an insect eye. Science 166:1172-1174.

Garcia JE, Greentree AD, Shrestha M, Dorin A, Dyer AG (2014) Flower colours through the lens: Quantitative measurement with visible and ultraviolet digital photography. PLoS ONE 9(5):e96646.

Giurfa M, Núñez J, Chittka L, Menzel R (1995) Colour preferences of flower-naive honeybees. Journal of Comparative Physiology A 177:247-259.

Giurfa M, Vorobyev M, Kevan P, Menzel R (1996) Detection of coloured stimuli by honeybees: minimum visual angles and receptor specific contrasts. Journal of Comparative Physiology A 178:699-709.

Gronquist M, Bezzerides A, Attygalle A, Meinwald J, Eisner M, Eisner T (2001) Attractive and defensive functions of the ultraviolet pigments of a flower (*Hypericum calycinum*). Proceedings of the National Academy of Sciences USA 98:13745-13750.

Guldberg LD, Atsatt PR (1975) Frequency of reflection and absorption of ultraviolet light in flowering plants. The American Midland Naturalist 93:35-43.

Gumbert A (2000) Color choices by bumble bees (*Bombus terrestris*): innate preferences and generalization after learning. Behavioral Ecology and Sociobiology 48:36-43.

He Q, Shen Y, Wang M, Huang M, Yang R, Zhu S, Wang L, Wu R (2011) Natural variation in petal color in *Lycoris longituba* revealed by anthocyanin components. PLoS ONE 6(8): e22098.

Hempel de Ibarra N, Langridge KV, Vorobyev M (2015) More than colour attraction: behavioural functions of flower patterns. Current Opinion in Insect Science 12:64e70.

Hempel de Ibarra N, Vorobyev M (2009) Flower patterns are adapted for detection by bees. Journal of Comparative Physiology A 195:319-323.

Herrera G, Zagal JC, Diaz M, Fernández MJ Vielma A, Cure M, Martinez J, Bozinovic F, Palacios AG (2008) Spectral sensitivities of photoreceptors and their role in colour discrimination in the green-backed firecrown hummingbird (*Sephanoides sephaniodes*). Journal of Comparative Physiology A 194:785-794.

Hinton HE (1976) Possible significance of the red patches of the female crab-spider, *Misumena vatia*. Journal of Zoology 180:35-39.

Huang Y, Li N, Ren ZX, Chen G, Wu ZK, Ma YP (2015) Reproductive biology of *Primula beesiana* (Primulaceae), an alpine species endemic to Southwest China. Plant Ecology and Evolution 148:289-296.

Johnson SD, Midgley JJ. (2001) Pollination by monkey beetles (Scarabaeidae: Hopliini): do colour and dark centres of flowers influence alighting behavior? Environmental Entomology 30:861-868.

Junker RR, Blüthgen N (2010) Floral scents repel facultative flower visitors, but attract obligate ones. Annals of Botany 105:777-782.

Kevan PG, Chittka L, Dyer AG (2001) Limits to the salience of ultraviolet: Lessons from color vision in bees and birds. Journal of Experimental Biology 204:2571-2580.

Kelber A, Balkenius A, Warrant EJ (2002) Scotopic colour vision in nocturnal hawkmoths. Nature 419:922-925.

Kevan PG (1972) Floral colors in the high arctic with reference to insect-flower relations and pollination. Canadian Journal of Botany 50:2289-2316.

Kevan PG (1978) Floral coloration, its colorimetric and significance in anthecology. In: Richards AJ (ed.) The pollination of flowers by Insects. Linnean Society Symposium Series Vol. 6:51-78.

Kevan PG (1979) Vegetation and floral colours revealed by ultraviolet light: interpretational difficulties for functional significance. American Journal of Botany 66:749-751.

Kevan P (1983) Floral colours through the insect eye: what they are and what they mean. In: Jones CE, Little RJ (eds.) Handbook of Experimental Pollination Biology. Van Nostrand Reinhold, New York, pp 3-30.

Kevan PG, Backhaus WG (1998) Color vision: ecology and evolution in making the best of the photic environment. In: Backhaus W, Kliegl R, Werner J (eds.) Color vision: perspectives from different disciplines. Walter de Gruyter, Berlin, pp 163-183.

Kevan PG, Chittka L, Dyer AG (2001) Limits to the salience of ultraviolet: lessons from colour vision in bees and birds. Journal of Experimental Biology 204:2571-2580.

Kevan P G, Giurfa M, Chittka L (1996) Why are there so many and so few white flowers? Trends in Plant Sciences 1:280-284.

Koski MH, Ashman TL (2013) Quantitative variation, heritability, and trait correlations for ultraviolet floral traits in *Argentina anserina* (Rosaceae): implications for floral evolution. International Journal of Plant Sciences 174:1109-1120.

Koski MH, Ashman TL (2014) Dissecting pollinator responses to a ubiquitous ultraviolet floral pattern in the wild. Functional Ecology 28:868-877.

Koski MH, Ashman TL (2015a). Floral pigmentation patterns provide an example of Gloger's rule in plants. Nature Plants 1:14007.

Koski MH, Ashman TL (2015b) An altitudinal cline in UV floral pattern corresponds with a behavioral change of a generalist pollinator assemblage. Ecology 96:3343-3353.

Liang H, Ren ZX, Tao ZB, Bernhardt B, Li DZ, Wang H. (2018) Impact of pre- and post-pollination barriers on pollen transfer and reproductive isolation among three sympatric *Pedicularis* (Orobanchaceae) species. Plant Biology; DOI:10.1111/plb.12833

Lübbe E (2010) Colours in the Mind - Colour Systems in Reality - A formula for colour saturation. [Book on Demand].

Lunau K (1992) Innate recognition of flowers by bumblebees – orientation of antennae to visual stamen signals. Canadian Journal of Zoology 70:2139-2144.

Lunau K (1995) Notes on the colour of pollen. Plant Systematics and Evolution 198:235-252.

Lunau K (1996) Unidirectionality of floral colour changes. Plant Systematics and Evolution 200:125-140.

Lunau K (2000) The ecology and evolution of visual pollen signals. Plant Systematics and Evolution 222:89-111.

Lunau K (2007) Stamens and mimic stamens as components of floral colour patterns. Botanische Jahrbücher für Systematik, Pflanzengeschichte und Pflanzengeographie 127:13-41.

Lunau K (2014) Visual ecology of flies with particular reference to colour vision and colour preferences. Journal of Comparative Physiology A 200:497-512.

Lunau K, Konzmann S, Bossems J, Harpke D (2016) A matter of contrast: Yellow flower colour constrains style length in *Crocus* species. PLoS ONE 11(4):e0154728.

Lunau K, Konzmann S, Winter L, Kamphausen V, Ren ZX (2017) Pollen and stamen mimicry: the alpine flora as a case study. Arthropod-Plant Interactions 11:427-447.

Lunau K, Maier EJ (1995) Innate colour preferences of flower visitors. Journal of Comparative Physiology A 177:1-19.

Lunau K, Papiorek S, Eltz T, Sazima M (2011) Avoidance of achromatic colours by bees provides a private niche for hummingbirds. Journal of Experimental Biology 214:1607-1612.

Lunau K, Verhoeven C (2017) Wie Bienen Blumen sehen - Falschfarbenaufnahmen von Blüten. Biologie in unserer Zeit 47:120-127.

Lunau K, Wacht S, Chittka L (1996) Colour choices of naive bumble bees and their implications for colour perception. Journal of Comparative Physiology A 178:477-489.

Lunau K, Wester P (2017) Mimicry and deception in pollination. In: Becard G (ed.) Advances in Botanical Research, Vol. 82, How Plants Communicate with their Biotic Environment; pp 259-279; Academic Press, Amsterdam.

Ma XK, Shi J, Bänziger H, Sun YN, Guo YY, Liu ZJ, Johnson SD, Luo YB (2015) The functional significance of complex floral colour pattern in a food-deceptive orchid. Functional Ecology 30:721-732.

Martínez-Harms J, Palacios AG, Márquez N, Estay P, Arroyo MTK, Mpodozis J (2010) Can red flowers be conspicuous to bees? *Bombus dahlbomii* and South American temperate forest flowers as a case in point. Journal of Experimental Biology 213:564-571.

Martínez-Harms J, Vorobyev M, Schorn J, Shmida A, Keasar T, Homberg U, Schmeling F, Menzel R (2012) Evidence of red sensitive photoreceptors in *Pygopleurus israelitus* (Glaphyridae:

Coleoptera) and its implications for beetle pollination in the southeast Mediterranean. Journal of Comparative Physiology A 198:451-463.

Medel R, Botto-Mahan C, Kalin-Arroyo M (2003) Pollinator mediated selection on the nectar guide phenotype in the Andean monkeyflower, *Mimulus luteus*. Ecology 84:1721-1732.

Milet-Pinheiro P, Ayasse M, Dötterl S (2015) Visual and olfactory floral cues of *Campanula* (Campanulaceae) and their significance for host recognition by an oligolectic bee pollinator. PLoS ONE 10(6):e0128577.

Myers N, Mittermeier RA, Mittermeier CG, da Fonseca GAB, Kent J (2000) Biodiversity hotspots for conservation priorities. Nature 403:853-858.

Neumeyer C (1981) Chromatic adaptation in the honey bee: Successive color contrast and color constancy. Journal of Comparative Physiology 144:543-553.

Oberrath R, Böhning-Gaese K (1999) Floral color change and the attraction of insect pollinators in lungwort (*Pulmonaria collina*). Oecologia 121:383-391.

Ohashi K, Makino TT, Arikawa K (2015) Floral colour change in the eyes of pollinators: testing possible constraints and correlated evolution. Functional Ecology 29:1144-1155.

Orbán LL, Plowright CMS (2014) Getting to the start line: how bumblebees and honeybees are visually guided towards their first floral contact. Insectes Sociaux 61:325-336.

Papiorek S, Junker RR, Alves-dos-Santos I, Melo GAR, Amaral-Neto LP , Sazima M, Wolowski M, Freitas L, Lunau K (2016) Bees, birds and yellow flowers: Pollinator-dependent convergent evolution of UV-patterns. Plant Biology 18:46-55.

Peitsch D, Fietz A, Hertel H, de Souza J, Ventura DF, Menzel R (1992) The spectral input systems of hymenopteran insects and their receptor-based colour vision. Journal of Comparative Physiology A, 170:23-40.

Peterson ML, Miller TJ, Kay KM (2015) An ultraviolet floral polymorphism associated with life history drives pollinator discrimination in *Mimulus guttatus*. American Journal of Botany 102:396-406.

Primack RB (1982) Ultraviolet patterns in flowers, or flowers as viewed by insects. Arnoldia 42: 139-146.

Ren ZX, Wang H, Bernhardt P, Li DZ (2014) Insect pollination and self-incompatibility in edible and/or medicinal crops in southwestern China, a global hotspot of biodiversity. American Journal of Botany 101:1700-1710.

Rodríguez-Gironés MA, Ruiz A (2016) toBeeView: A program for simulating the retinal picture of visual scenes on nonhuman eyes. Ecology and Evolution 6:7892-7900.

Rodríguez-Gironés MA, Santamaría L (2004) Why are so many bird flowers red? PLoS Biology 2(10):e350.

Rosen D, Barthlott W (1991) Ökologische Aspekte der Ultraviolett-Reflexion von Blumen in Mitteleuropa, besonders in der Eifel. Decheniana 144:72-112.

Schlangen K, Miosic S, Castro A, Freudmann K, Luczkiewicz M, Vitzthum F, Schwab W, Gamsjäger S, Musso M, Halbwirth H (2009) Formation of UV-honey guides in *Rudbeckia hirta*. Phytochemistry 70:889-898.

Shi J, Luo YB, Bernhardt P, Ran JC, Liu ZJ, Zhou Q (2009) Pollination by deceit in *Paphiopedilum barbigerum* (Orchidaceae): a staminode exploits the innate colour preferences of hoverflies (Syrphidae). Plant Biology 11:17-28.

Silberglied RE (1976) Visualization and recording of longwave ultraviolet reflection from natural objects. Functional Photography 11:20-29; 30-33.

Silberglied RE (1979) Communication in the ultraviolet. Annual Review of Ecology and Systematics 10:373-398.

Stavenga DG, van der Kooi CJ (2016) Coloration of the Chilean Bellflower, *Nolana paradoxa*, interpreted with a scattering and absorbing layer stack model. Planta 243:171-181.

Stockman A, Macleod D, Johnson N (1993) Spectral sensitivities of the human cones. Journal of the Optical Society of America 10:2491-2521.

Tao ZB, Ren ZX, Bernhardt P, Liang H, Li HD, Zhao YH, Wang H, Li DZ (2018) Does reproductive isolation reflect the segregation of color forms in the *Spiranthes sinensis* (Pers.) Ames complex (Orchidaceae) in the Chinese Himalayas? Ecology and Evolution; DOI:10.1002/ece3.4067

Tong ZY, Huang SQ (2016) Pre- and post-pollination interaction between six co-flowering *Pedicularis* species via heterospecific pollen transfer. New Phytologist 211:1452-1461.

Tong ZY, Huang SQ (2018) Safe sites of pollen placement: a conflict of interest between plants and bees? Oecologia 186:163-171.

Troscianko J, Stevens M (2015) Picture calibration and analysis toolbox – a free software suite for objectively measuring reflectance, colour and pattern. Methods in Ecology and Evolution 6:1320-1331.

Vasas V, Hanley D, Kevan PG, Chittka L (2017) Multispectral images of flowers reveal the adaptive significance of using long-wavelength-sensitive receptors for edge detection in bees. Journal of Comparative Physiology A 203:301-311.

Von Frisch K (1915) Der Farbensinn und Formensinn der Biene. Zoologische Jahrbücher Abteilung für allgemeine Zoologie und Physiologie 37:1-187.

Vorobyev M, Kunze J, Gumbert A, Giurfa M, Menzel R (1997) Flowers through the insect eyes. Israel Journal of Plant Science 45:93-102.

Vorobyev M, Marshall J, Osorio D, Hempel de Ibarra N, Menzel R (1999) Colourful objects through animal eyes. Colour Research and Application 26: 214-217.

Wang H, Li DZ (2005) Pollination biology of four *Pedicularis* species (Scrophulariaceae) in northwestern Yunnan, China. Annals of the Missouri Botanical Garden 92:127-138.

Weiss MR, Lamont BB (1997) Floral color change and insect pollination: A dynamic relationship. Israel Journal of Plant Science 45:185-200.

White TE, Dalrymple RL, Noble DWA, O'Hanlon JC, Zurek DB, Umbers KDL (2015) Reproducible research in the study of biological coloration. Animal Behaviour 106:51-57.

Wikipedia (2. März 2017). CIE 1931 color space. Accessed 9 March 2017 from Wikipedia: en.wikipedia.org/wiki/CIE_1931_color_space

Williams S, Dyer A (2007) A photographic simulation of insect vision. The Journal of Ophthalmic Photography 29:10-14.

Williams P, Tang Y, Yao J, Cameron S (2009) The bumblebees of Sichuan (Hymenoptera: Apidae, Bombini). Systematics and Biodiversity 7:101-189.

Wilmsen S, Gottlieb R, Junker RR, Lunau K (2017) Bumblebees require visual pollen stimuli to initiate and multimodal stimuli to complete a full behavioral sequence in close-range flower orientation. Ecology and Evolution 7:1384-1393.

Winter Y, Lopez J, von Helversen O (2003) Ultraviolet vision in a bat. Nature 425:612-614.

Wu ZY, Raven PH, Hong DY (2013) Flora of China Vol. 1: Introduction. Science Press, Beijing, China and Missouri Botanical Garden Press, St. Louis, USA.

Yan J, Wang G, Sui Y, Wang ML, Zhang L (2016) Pollinator responses to floral colour change, nectar, and scent promote reproductive fitness in *Quisqualis indica* (Combretaceae). Scientific Reports 6:24408.

Yan J, Wang ML, Zhang L (2018) Light induces petal color change in *Quisqualis indica* (Combretaceae). Plant Diversity 40:28-34.

Yang HB, Holmgren NH, Mill RR (1998) *Pedicularis*. In: Wu ZY, Raven PH, Hong DY (2013) Flora of China, Vol 18: Science Press, Beijing, China and Missouri Botanical Garden Press, St. Louis, USA.

Zhang C, Vereecken NJ, Wang LL, Tian B, Dafni A, Yang YP, Duan YW (2017) Are nectar guide colour changes a reliable signal to pollinators that enhances reproductive success? Plant Ecology & Diversity 10:89-96.

25

POLLEN CARRYOVER BETWEEN SEQUENTIAL FORAGING TRIPS BY A SOLITARY BEE: IMPLICATIONS FOR DISTANT OUTCROSSING

James H. Cane* and Byron Love

USDA-ARS Pollinating Insect Research Unit, Utah State University, Logan 84322-5310

Abstract—Animal pollination depends on foragers moving pollen between conspecific flowers and plants. Pollinating bees often exhibit patch fidelity and frequent grooming during a foraging trip, which results in intensely localized pollen carryover. If their host's seed dispersal is likewise limited, then spatial mosaics of small genetic neighbourhoods should result. However, spatial distributions of genetic markers of seeds often reveal infrequent long-distance gene flow mediated by pollinating bees, seemingly at odds with expectations of bees' local patch fidelity and suggesting an additional mechanism at work.

A greenhouse experiment that used nesting *Osmia californica* bees foraging at sunflowers tested the pollination fate of any bodily pollen that lingers on females even after they discharge a pollen load at their nest. Once a provisioning female had returned to her nest, we moved the nest with her inside from a cage with male-fertile sunflowers to a cage with male-sterile sunflowers. There females quickly resumed foraging. They nearly always pollinated and set seed at the first male-sterile flowerhead that they visited, the result of fertile bodily pollen carried over from the previous foraging trip. Every visit to a second male-sterile flowerhead set additional seed.

If, as seems likely, outbound nesting females sometimes switch between host patches, then this mechanism of pollen carryover could explain occasional distant pollen-mediated gene flow events, as summarized in a graphic model. Carryover should mitigate inbreeding depression in threatened plant populations, or conversely, sometimes contaminate seed crops, with relevance for spatially isolating foundation seed fields and coexistence of GMO and conventional crops.

Keywords: Asteraceae, grooming, Megachilidae, Osmia, pollen-mediated gene flow, pollination

INTRODUCTION

Among plant-animal mutualisms, both pollination and seed dispersal involve visitors that remove and transport propagules (pollen or seeds). Pollination then has an extra step, being the subsequent receipt of pollen at a receptive stigma. Bees are the predominant pollinators in most plant communities and croplands, conferring valuable outcrossing as they move between sequential plants during a foraging trip. However, females typically move among neighbouring conspecific flowers and plants during a foraging bout (e.g. Waddington 1979; Waser 1982; Thomson & Thomson 1989; Cresswell et al. 1995), an outcome of floral constancy and the reproductive advantages of minimizing travel time and energy spent. As a result, a foraging bee tends to thoroughly work a rewarding patch (or tree), moving only short distances between flowers (Waddington 1979; Thomson & Thomson 1989; Rasmussen & Brodsgaard 1992). Pollen flow should follow this foraging pattern, given the rapid decay in pollen carried over from previously visited flowers (e.g. Cresswell et al. 2002). The few available studies (all using bumble bees) do show that pollen transferred to a bee's body is quickly lost from the pollination pool after just a few subsequent flower visits (Thomson et al. 1986; Waser 1988; Thomson & Thomson 1989; Cresswell et al. 2002; Castellanos et al. 2003). To better appreciate the significance of a new pollen carryover mechanism, we first review the extent to which bees' foraging and transport traits constrain pollen-mediated gene flow to distances that are far short of bees' estimated flight ranges. We then present experimental evidence for a new mechanism of pollen carryover by bees.

Pollen carryover curves have been measured mostly for foraging bumble bees, for which carryover is often truncated or foreshortened by the bees' regular grooming and pollen packing while in a patch. Both actions further limit the spatial extent of gene flow by foragers. Periodic grooming largely terminates reproductive chances of most pollen (Thomson 1986; Holmquist et al. 2012), as a forager periodically sweeps loose surface pollen from her body to pack and transport it back to her nest. Residual dry pollen only lingers on bodily surfaces that cannot be groomed (e.g. Tepedino et al. 1999). Pollen in such "safe sites" nonetheless becomes buried by more recently deposited pollen, again limiting carryover (Thomson 1986). In addition, bumble bees, honey bees, and some other bees moisten groomed pollen with nectar while packing it on their hind legs, effectively terminating its ability to germinate (Parker et al. 2015). Most other non-parasitic bee taxa transport pollen dry, where it remains viable and sometimes well-placed for further pollination (e.g. abdominal venters of megachilids visiting Asteraceae flowers; Cane 2017). Collectively, these factors constrain pollen-mediated gene flow by bees. The genetic neighbourhoods resulting from

*Corresponding author: jim.cane2@gmail.com

pollen flow can extend several hundred meters for widely-spaced tropical rainforest trees pollinated by big-bodied, vagile bees (Dick et al. 2003; Ward et al. 2005; Jha & Dick 2010). More commonly, these genetic neighbourhoods are small, spanning a few meters for forbs (Rasmussen & Brodsgaard 1992; Fenster et al. 2003; Otero-Arnaiz et al. 2005) and row crops (St. Armand et al. 2000; Van Deynze et al. 2005).

Wherever pollen-mediated gene flow has been spatially well-sampled, though, population geneticists report right skewed leptokurtic distributions with a long tail, akin to rare but distant seed dispersal events (Clark et al. 2001). That tail represents distant albeit infrequent pollen flow events mediated by pollinators (Bradner et al. 1965; Rieger et al. 2002; Van Deynze et al. 2005). Distant gene flow can be problematic for agricultural seed production (Bradner et al. 1965), where varietal purity is paramount (Van Deynze et al. 2008; Greene et al. 2015). Conversely, rare plant conservation often benefits from distant outcrossing that can counter genetic drift and inbreeding depression (e.g. Paschke et al. 2002). These practical ramifications compel us to understand the mechanisms contributing to infrequent but distant pollen flow by foraging bees.

The potential significance of a new pollen carryover mechanism becomes apparent from a review of the spatial realities of bee foraging patterns and interfloral carryover during a foraging trip. Females range out to forage from a central nest, so their potential pollen flow distance is bounded by their flight range. Even for smaller solitary bees, maximum foraging distances ("M" in Fig. 1) are often a km or more (Zurbuchen et al. 2010b). However, bees' maximum flight ranges greatly overestimate the spatial scales of their realized contribution to pollen-mediated gene flow. For solitary bees, those that must forage distantly forego some progeny production (Zurbuchen et al. 2010a). Unique to honey bees, flight distances and directions are measured by decoding forager's waggle dances in the hive (Seeley 1995). Outbound foragers sometimes fly far (e.g. 5-9 kms) to reach distant rewarding patches, but only when nearer patches are wanting (Beekman & Ratnieks 2000). To minimize time and energy costs, they more typically forage much closer to home, ranging out 500m-3km (reviewed in Seeley 1995). Hence, most pollination by solitary and social bees is much closer to their nests than their maximum foraging ranges would imply (M vs. R in Fig. 1).

Even these reported radial foraging distances flown by bees greatly exceed the spatial scale of their conferred pollen-mediated gene flow. Although bees foraging in crop fields and orchards often range out hundreds of meters, measured pollen-mediated gene flow rarely exceeds a few meters (Handel 1983; St. Armand et al. 2000; Van Deynze et al. 2005). This highly localized pollen travel (G_1 in Fig. 1) results because bees apparently fly out and over the intervening landscape, then drop into a rewarding patch (or part of an agricultural field) to forage intensively among neighbouring, often conspecific flowers (L in Fig. 1). Studies supporting this contention have been limited to select social bees, as outbound flights of free-flying solitary species cannot be visually followed or experimentally manipulated. For honey bees, outbound foragers that were recruited to distant scented nectar feeders rarely investigated intervening scent stations placed along their route (Fig. 91 in von Frisch 1967). Extrapolating this discovery to scattered flower patches, recruited honey bees can be expected to fly past intervening flowers *en route* to the target floral patch. Unlike honey bees, bumble bees (and all solitary bees) forage independently without recruitment. Although bumble bees have been implicated in moving pollen up to 400 m (Schulke & Waser 2001), watching them fly between more distant floral patches generally fails. Only through use of harmonic radar has this limitation been remedied. Because only a big bee can tote the antenna mast, only the complete foraging trips of tagged bumble bees flying across extensive open landscapes have so far been tracked (Osborne et al. 1999; Woodgate et al. 2016). All but two similarly rigged large carpenter bees (*Xylocopa*) quickly lost their transceivers (Pasquet et al. 2008). Outbound experienced bumble bee foragers were seen to fly directly to distant (>200 m) small floral patches during "exploitation" flights (Woodgate et al. 2016); these foragers "were not searching for food *en route*" (Osborne et al. 1999), just as von Frisch reported for honey bees. These direct, non-stop outbound and return flights from nest to foraging patch largely explain why, despite these bees' extensive foraging ranges, their pollen-mediated gene flow is intensely localized, resulting in surprisingly small genetic neighbourhoods.

However, there remain the regular but rare distant pollen flow events that yield distant outcrossing. For the non-social

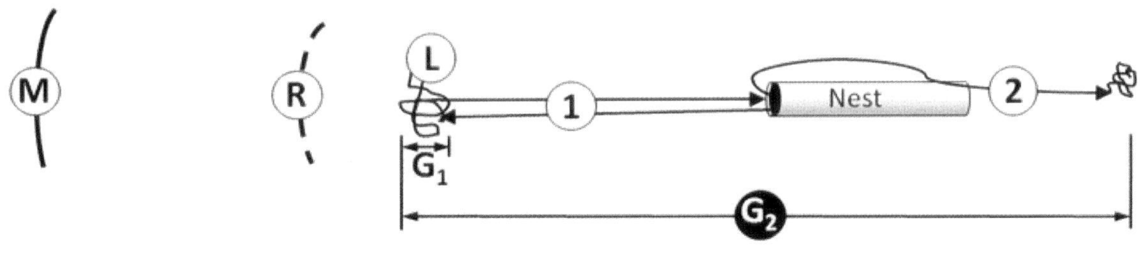

FIGURE 1. Graphic model depicting the spatial patterns of a bee's home range, two foraging trips and resulting pollen-mediated gene flow (not to scale). Labels are: M=maximum flight range, R=realized home foraging range, L= localized foraging bout in a flower patch, 1 & 2 = sequential foraging trips by nesting bee, G_1 = small genetic neighbourhood resulting from patch fidelity on one foraging trip, G_2 = pollen flow distance resulting from carryover into a sequential foraging trip.

bee *Megachile rotundata*, which is managed to pollinate alfalfa seed fields, St. Armand et al. (2000) found very short distances (<4 m) for most pollen-mediated gene flow. However, he and others (Bradner et al. 1965) also did find infrequent instances of distant travel by alfalfa pollen markers (out to 1000 m). Similarly, among wild patches of *Cucurbita foetidissima* pollinated by specialist *Peponapis* bees, most pollen-mediated gene flow was within patches, but 7% was between them (to 500 m; Kohn & Casper 1992). These instances may simply represent rare departures from these bees' usual behaviour during a foraging bout. Alternatively, long-distance pollen flow could represent a different mechanism, namely the carryover of bodily pollen from the previous foraging trip.

Consistent with this carryover mechanism, experiments with sequential foraging trips by honey bees in crop monocultures showed that departing foragers carry some viable pollen on their bodies (Hatjina et al. 1999) that continues to pollinate flowers (Free & Durrant 1966). In the latter study, apple pollen on bodies of outbound bees was acquired either on their previous foraging trip (carryover) or when foragers jostled against each other in the crowded hive. Pollen transfer during forager jostling was implied by experiments of DeGrandi-Hoffman et al. (1986), wherein outbound foragers picked up pollen from dead hive mates pinned at the hive entrance. In contrast, using more direct experiments, investigators who marked and released newly emerged, pollen-free workers into hives placed adjacent to cotton fields (Loper & DeGrandi-Hoffman 1994) found that, after 6 h, the marked hive bees had only acquired a few dozen pollen grains from returning foragers, none of it cotton. For social bees, it seems impossible to discern individual pollen carryover from pollen transfer during nestmate jostling. Most bees are not social but solitary, however, and so of course they have no nestmates to jostle. For a solitary bee, then, any bodily pollen on an outbound female must stem from her inability to completely groom and remove it while in her nest. This mechanism was shown for the solitary bee *Osmia cornifrons*, whose outbound females retained 10% of their bodily loads after off-loading at their nests (Matsumoto et al. 2009). However, its contribution to pollination was not tested.

If leftover pollen on bodies of outbound bees is to contribute to distant gene flow of their floral hosts, then on sequential foraging bouts, bees must sometimes visit a new floral patch ("2" in Fig. 1). Unfortunately, patch switching by free-flying non-social bees has proven logistically impossible to document at spatial scales of the field. However, freely foraging bumble bees tracked using harmonic radar did occasionally switch patches (Osborne et al. 1999; Woodgate et al. 2016). In each case, foragers switched patches on a subsequent outbound flight, not during a given trip. The combination of bodily pollen carryover from the previous foraging trip and switching flowering patches on some subsequent foraging trips could together account for the occasional long-distance gene flow events observed using genetic markers of bee-pollinated plants ("G_2" in Fig. 1). The objective of this study was to directly and experimentally test half of this hypothetical mechanism, namely that pollen being carried-over from a solitary bee's previous foraging trip can contribute to subsequent pollination and seed set.

MATERIALS AND METHODS

Study organisms

The experiment required a portable population of a solitary bee that could be induced to forage and nest in confinement using a suitable pollen host that could be brought to bloom concurrently with bee nesting. For this purpose, we chose a solitary cavity-nesting bee, *Osmia californica* Cresson, which is an Asteraceae specialist (Cripps & Rust 1989). These bees were progeny of a local population being propagated outdoors on their native perennial floral host, *Balsamorhiza sagittata* (Cane 2011). They readily nest in large cages and accept sunflowers as a pollen host (Cane 2017).

For a practical annual Asteraceae host bearing a male-sterile trait, we grew dwarfed, determinant commercial sunflower cultivars (*Helianthus annuus* L.) that are marketed for the cut-flower trade. The pollen-bearing variety was 'Sunspot'. The recipient male-sterile, pollen-less variety was 'Solita' (Johnny's Selected Seeds, Winslow, Maine). Each plant of both cultivars concurrently produced a single, similar-looking large yellow flowerhead. Experiments took place in an earth-floored glasshouse into which we sowed separate patches of the two sunflower cultivars, one within each of two neighbouring partitions (3.4 × 6.2 × 2.5 m) of a large hanging net cage. For a more familiar vocabulary, we shall refer to the sunflower's capitulum as a flowerhead, its individual flowers as florets, and its achenes as seeds.

Experiments

Cocoons of a wintered population of *O. californica* were incubated for a few days in early May so that adults were active during their local flight season. Each newly-emerged female was briefly chilled and given a unique thoracic paint mark. Females plus five males were then released into the cage with the pollen-bearing variety; 17 females commenced to forage and nest. Both cages were given a weathered, 49-hole drilled wooden nesting block with new paper straw inserts. Both nest blocks were placed in matching positions and orientations within their respective cages. Bees were monitored for the onset of nest cell provisioning, checked by viewing the nest's interior using an otoscope. Our intent was to create two visually similar nesting/foraging arenas between which we could individually shuttle actively provisioning bees while in their nesting straws, all without disturbing their foraging routines.

To shuttle bees between arenas, we awaited the return of a pollen-laden female to her nesting block. A female first walks into her nest head-first to disgorge her nectar load, then backs out, reverses direction on the nest block face, and backs into the straw to groom off her pollen load onto the growing provision mass. At that moment, we corked the front of her nesting straw, withdrew the occupied straw, and while holding it horizontally, we gingerly walked it to the nesting block in the adjoining cage with the pollen-less sunflowers. The straw was slipped into the same relative position in the provided block and uncorked to await the female's emergence. Thus, all females emerged having first groomed away their pollen loads at their nest, which we confirmed visually. With this design, seed set at the pollen-less variety would have to result from properly positioned loose pollen on a bee's body that was acquired from the pollen-bearing variety during the previous foraging trip.

Foraging females walked the flowerheads in arcs, following the narrow circle of newly-opened florets. We marked each female's landing and departure points with map pins on a pollen-less flowerhead while drawing her route and direction of travel on paper. After her departure, we enclosed the visited flowerhead in a paper bag for the rest of the experiment. We thereafter measured the length of arc that she walked so we could both estimate pollination efficacy per unit foraging effort (seeds per cm walked) as well as detect if there were declining seed returns as a visit progressed. It is impossible to know which individual florets females touched as they walked the ring of receptive florets. Any visit to a second flowerhead was similarly recorded. Each female was then removed, placed back in her nesting straw, and returned to the nest block in the cage with the pollen-bearing sunflowers. Each female was used only once.

Mature seeds (plump achenes) were harvested from individual flowerheads. The seeds along a bee's arcuate path around each flowerhead (marked by the pins) were extracted, collected and counted in sequential quarter arcs in the direction that the bee had walked. Any shorter remaining arcs were recorded separately. Achenes were rubbed free and sorted into two distinct classes: flat and flexible, or plump and hard (inset Fig. 4). X-radiographs revealed that only the plump hard achenes contained endosperm, a difference that we confirmed by sample dissection.

Statistics

The comparative likelihood of a female's pollen carryover resulting in pollination was tested using a simple 2×2 contingency table and logistic regression (Allison 2012), comparing the frequency of visits that yielded filled achenes in the bee's walking path versus any filled achenes of neighbouring florets untouched by bees. A hypothesis of diminishing seed returns was tested by paired t-tests on square roots of counts of seeds in two ways, comparing: 1) yields from the first ¼ arc walked with the 3rd arc walked on the first flowerhead visited, and 2) yields at the first flowerhead versus the second one visited. Seed yields per flowerhead were regressed on bees' walking distances (after square root transformation to satisfy normality; SigmaPlot, Systat Software Inc., San Jose, California).

RESULTS

Foraging and nesting behaviours appeared to be normal despite confinement to large cages in a glasshouse environment. After several days of orientation and feeding, females commenced nesting. They actively collected pollen loads, using rapid dorso-ventral drumming of their abdominal scopa (pollen-brush) against the day's ring of presenting florets on the flowerhead (Cane 2016). In the experiment, of the 17 females emerging from their translocated nests, 15 individuals flew directly to a male-sterile flowerhead to resume foraging, apparently undisturbed by their transfer. The other two flew to the cage walls and were not used. Most females lingered 30 secs or less before emerging, which combined with our straw removal and walking times between cages, falls within the norms for time female *O. lignaria* spent removing pollen (30 secs) and resting (15-60 secs) in glass nest tubes (Phillips & Klostermeyer 1978).

Transferred females visited 24 of the available flowerheads of the pollen-less sunflowers, walking a total of 3.4 m along the day's ring of presenting florets. Twelve females walked at least ¾ of an arc (path 14 ± 8 cm long, range 2 – 28 cm) while following the ring of receptive florets on a given flowerhead; seven females walked a complete circuit. Nine females visited a second flowerhead. All females took nectar; some were seen trying to actively collect pollen from pollen-less flowerheads by abdominal drumming (Cane 2017), but of course the florets bore no pollen. Each female with her nest was returned to the nesting cage, where they all resumed foraging and provisioning.

Bodily pollen carried over from the previous foraging trip frequently yielded filled achenes at male-sterile flowerheads (21 of 24 visits), whereas filled achenes never formed at untouched florets or from two of the three shortest walks (2-3 cm; Fig. 2). Visitation and carryover significantly improved the likelihood of seed set ($G = 20.5$, $P \leq 0.0001$). Overall, pollen carryover yielded 534 plump achenes (mean = 22 ± 36 achenes per flowerhead, median 7). Just three of the 24 visits accounted for 315 (59%) of the total set of plump achenes (Fig. 2). Converted to a bee's walking distance on a flowerhead, seed set averaged 1.7 ± 2 seeds per cm walked (median 0.8, range 0-6.4). However, distance walked per flowerhead (2-28 cm) did not predict seed yield (0-140 seeds; $r^2 = 0.05$, $P > 0.3$) although, of course, the shortest walks could not yield much seed (Fig. 2).

Our hypothesis of diminishing returns -- that progressively less seed would result as a female continued to forage at male sterile sunflowers -- gained only partial support from our small data set. At the first flowerhead visited, some seeds resulted from every final ¼ arc for the five bees that walked a complete circuit, so clearly bodily pollen remained. Comparing seed set along the 1st and 3rd ¼ arcs walked by ten

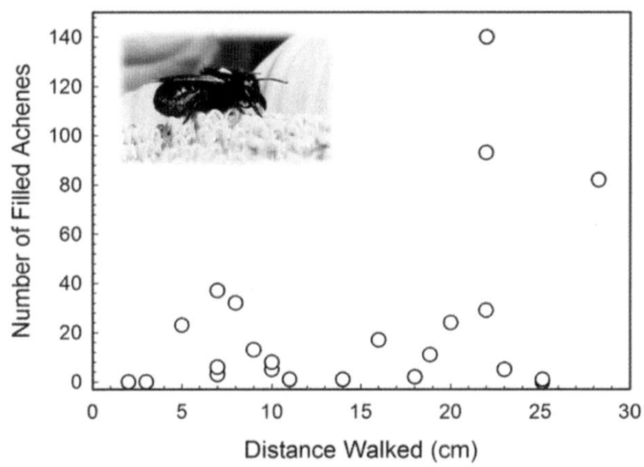

FIGURE 2. Scatterplot of plump achenes resulting for the distance walked by each transferred female *O. californica* visiting one of 24 male-sterile sunflower flowerheads (second visited flowerheads plotted independently). *Inset* Female *O. californica* transporting 'Sunspot' sunflower pollen in her ventral abdominal scopa (pollen-brush).

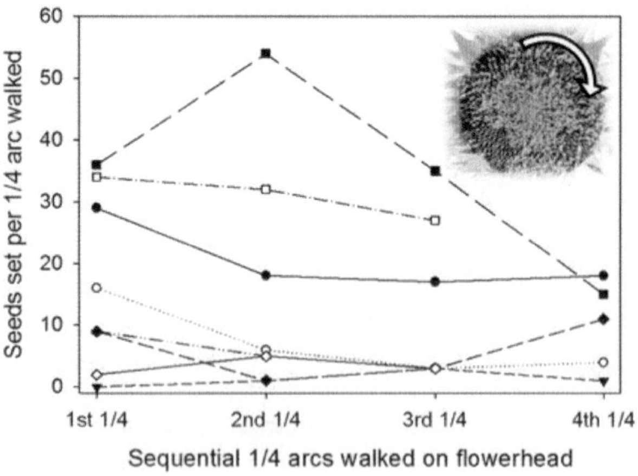

FIGURE 3. Changes in rates of achenes set by each female *O. californica* as she visited progressive ¼ arcs of receptive flowers on a single flowerhead. Only included are those eight females that walked at least ¾ of an arc and set five or more seeds. *Inset* Flowerhead with a broad ring of black, matured seeds resulting from a single *Osmia* visit. First ¼ arc of travel marked.

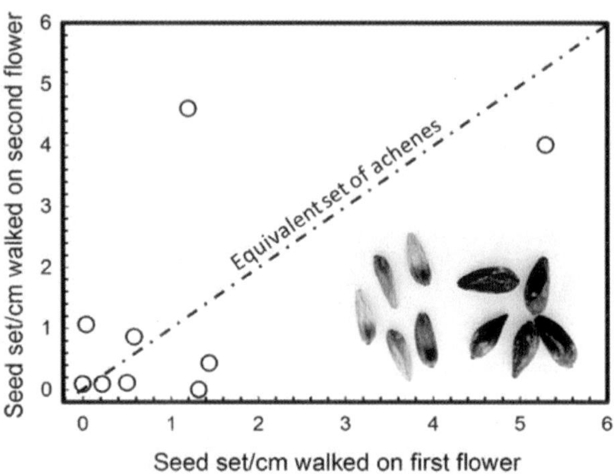

FIGURE 4. Scatterplot for the numbers of filled achenes set per cm walked by the nine female *O. californica* that visited two flowerheads sequentially, comparing the first visited flowerhead (x-axis) with the second (y-axis). Points below the dashed line indicate diminishing returns at the second flower. *Inset* Image comparing plump achenes filled with endosperm that were counted in these experiments (right) versus thin, flexible empty achenes.

females (Fig. 3), the first ¼-arcs walked yielded more seed overall (sum 135 vs. 97 seeds) and for the six most effective pollinators (sum 221 seeds). However, for 4 of the 9 visits to a second flowerhead (Fig. 4), seed set per cm walked on the second flowerhead exceeded that of the first. Thus, pollination continued for two flowerheads, but the order of flowerhead visitation did not predict resulting seed set (one-tailed $t = 1.9$, $P > 0.4$, $N = 9$).

DISCUSSION

The common observation that foraging solitary bees show local patch fidelity seems at odds with the infrequent but distant pollen-mediated gene flow revealed by genetic markers of some bee-pollinated flowering species. This paradox is readily reconciled if bodily pollen is carried over between sequential foraging trips and females occasionally exit their nest to forage in a different patch (Fig. 1). Nesting female *O. californica* apparently departed their nests with pollen from their preceding foraging trip(s), as we found that substantial seed resulted from their visits to male-sterile sunflowers, this despite the females having first groomed and deposited their pollen loads at their nests. Some seed resulted from most (89%) visitation bouts owing to this mechanism,

Just 3 of the 24 visits set 59% of the total seed in the experiment, a seemingly rare event. However, female solitary bees make many such foraging flights in a day. For instance, both *O. californica* and *O. montana* require 30 pollen trips to provision each nest cell (Torchio 1989). They provision one, sometimes two nest cells every day for several weeks, tempos matched by several other cavity-nesting *Osmia* (e.g. Maeta 1978; Phillips & Klostermeyer 1978; Frohlich & Tepedino 1986; Bosch 1994; Goodell 2003). For ground-nesting species, counts of provisioning trips per nest cell are difficult to know with confidence (as cell completion is hidden from the observer), but they also are thought to average several dozen flights per day (Neff 2008) during several weeks of nesting. Thus, a typical adult female solitary bee flies 200-400 foraging trips in her lifetime, affording 200-400 opportunities for pollen carryover between sequential foraging trips. Extrapolating from our study, 13% (3/24) of these trips should yield substantial pollination via pollen carried over from the preceding foraging trip, or an estimated 25-50 substantial carry-over events during each female's lifetime. If just a few of these trips involve outgoing bees changing patches from the previous foraging trip, then there would be ample opportunity for the occasional distant pollen-mediated gene flow events that have been revealed by genetic studies.

Some protocols and choices necessary to this experiment undoubtedly influenced carryover. First, we truncated foraging sequences after the second flowerhead visited, so as to reserve enough flowerheads for other bees' visits. Later, when processing mature achenes, we found little evidence of diminishing carryover pollination during these first two floral visits (Fig. 4), so pollination by carry-over pollen likely would have continued to three or more flowerheads. Hence, we surely underestimated the total possible carryover of pollen acquired during the previous foraging trip. Conversely, because the target sunflower variety lacked pollen (male-sterile), bees visiting these flowerheads did not groom (although some did drum as if there was pollen to collect). Grooming has been shown to sharply curtail -- but not eliminate -- further pollen carry-over by bumble bees (Thomson 1986; Thomson & Thomson 1989; Holmquist et al. 2012). The availability and exposure of pollen carried over from the previous foraging trip thus may be prolonged for bees visiting male-sterile (or monoecious or dioecious) flowers relative to more typical hermaphroditic flowers. However, as a guild of foraging bees daily depletes their host's standing crop of pollen, there should be a time each day when most flowers have been depleted but a small fraction are not yet visited. For instance, given a purely Poisson process, once the average flowerhead has been visited a third time, there would remain 5% unvisited flowerheads. At that time, foragers departing a pollen-bearing flowerhead encounter mostly pollen-less flowers, as we used here. Lastly, for females foraging at pollen-bearing flowerheads, pollen acquired earlier can be buried under newly acquired pollen, at least until the bee grooms. In our system, bees foraging at male-sterile sunflowers had no occasion to groom.

The attributes of the bees and flowers used in this experiment do seem ideal for pollen carryover to occur. The 3,000+ species of non-parasitic Megachilidae bees (like *Osmia*) all transport pollen dry in their abdominal pollen-brush (scopa). There it remains in play for pollination, especially for flowers like most of the huge plant family Asteraceae (>20,000 species) or others that instead produce panicles or umbels over which the foraging megachilid bee simply walks, often while drumming their scopa against the pollen-bearing surface (Cane 2017). In contrast with megachilids, all social bees and the other solitary bees transport pollen on their hind legs, some of them (bumble bees, stingless bees, honey bees) moistening the pollen with nectar to form a pellet. Every time these bees groom and pack pollen onto their hind legs, it is removed from circulation, which for bumble bees has been shown to hasten attenuation of pollen carryover during a foraging trip (Thomson 1986), and so probably from a prior foraging trip too. Many other flowering species, however, place and pick up pollen on bodily surfaces that bees have difficulty grooming (e.g. thoracic dorsum, proboscidial fossa, or between their leg bases). For such flowers, just where and how these bees groom and transport pollen are immaterial. Characterizing the magnitude of pollen carryover during sequential foraging flights by bees either receiving pollen placed in these "safe" sites (e.g. pollen deposited on a bee's thoracic dorsum by *Penstemon* flowers) (Tepedino et al. 1999) or transporting pollen in different scopal placements (e.g. hind leg) would help for interpreting the generality of the phenomenon found here with *Osmia* bees visiting sunflowers.

Pollen carryover to different floral patches might also be invoked for pollen residing overnight on the body of a female sleeping in her nest, since a forager's last and first foraging trips of sequential days probably target different patches of bloom. However, given the likely humid conditions within a nest (near saturation for a ground-nester; Cane, unpubl.), pollen of many flowering species held overnight would be expected to hydrate and expire by morning (Bassani et al. 1994). For this same reason, the minority of species that shed hydrated pollen (e.g. *Cucurbita*) have consistently short-lived pollen (Franchi et al. 2011). It is surprising, then, that Matsumoto et al. (2009) clearly demonstrated that apple pollen on bodies of cavity-nesting *O. cornifrons* remained viable for 12 days. The reproductive value of bodily pollen carried over into subsequent foraging days should be strongly influenced by intra-nest humidities and pollen species, and so will require further experiment.

Interestingly, gene flow promoted by this pollen carry-over mechanism is not diminished by a bees' localized foraging behaviour or patch fidelity during a foraging trip (Fig. 1). Moreover, it can multiply the effect of flight distance. For example, a bee that switches from a small patch 300 m east of its nest to one 200 m west of its nest will effectively move pollen ½ km, more than either foraging distance and far further than the short flights the female made between flowers while foraging in either floral patch (distance G_2 in Fig. 1). Actual gene flow frequencies should be functions of how often outbound foragers switch floral patches, the vector angle between patches, and the distances from the nest to each patch. In our experiment, only a minority of carry-over events set substantial seed, which was not related to either distance walked or flowerhead order. Collectively, these variables would result in sporadic distant gene flow events. This expectation accords with the data for distant gene flow in large-scale crop pollination studies, which do not fit an exponential decay function for the more distant gene flow events (e.g. Rieger et al. 2002).

This inter-trip mechanism of pollen carryover should contribute to agricultural yields, such as between rows of self-incompatible crops (e.g. almond; Bosch & Blas 1994) or into rows of male-sterile lines for hybrid seed production (e.g. sunflower; Parker & Frohlich 1983). It may also explain pollination of bee-pollinated rainforest trees. These trees are often widely scattered, self-incompatible mass bloomers (Ward et al. 2005) such as the Brazil nut (*Bertholletia excelsa*). This valuable Amazonian tree is largely self-incompatible; most of its seeds result from outcrossing (O'Malley et al. 1988). Its complex flowers are visited by large female non-social bees (Cavalcante et al. 2012) whose inter-tree movements are essential to its pollination and subsequent yield of harvested seeds.

The primary remaining question centres on the relative probability that bees switch between distant floral patches (or rows or trees) during a given foraging trip versus a subsequent foraging trip. The answer requires knowing the immediate travel histories of uniquely marked foragers seen flying into a patch, distinguishing between direct arrivals from another patch versus arrivals preceded by a nest visit, all over realistic distances (hundreds of meters). Observing and distinguishing these events have thus far eluded researchers but for the use of harmonic radar tracking of bumble bees. The second, more recent radar study with bumble bees (Woodgate et al. 2016) has revealed yet another mechanism – early, wide-ranging exploratory flights by young inexperienced foragers -- that also could result in some long-distance gene flow events between patches. Since all bees must learn to navigate within their home range to efficiently choose foraging patches (or a nesting site), it may be that such investigative young bees also confer infrequent, distant pollen flow. However, their possible contribution to pollination is not yet known. The mechanism of pollen-carryover between foraging trips by *O. californica* bees is shown here to regularly yield sunflower seed, sometimes abundantly. This mechanism reconciles the typical patch fidelity of bees during a foraging trip (Woodgate et al. 2016) and the rare but regular pollen-mediated gene flow events that occur over much greater distances, as revealed by genetic markers.

ACKNOWLEDGEMENTS

Devon Picklum contributed practical advice and later plant care in the greenhouse and in bee marking. Braden Jenkins helped follow foraging bees, and later, cleaned and sorted achenes. Jack Neff, Rebecca Irwin, T'ai Roulston and several anonymous reviewers all improved the manuscript.

Authors' Contributions

JC and BL conceived the ideas and designed methodology and collected the data; BL grew and maintained plantings and bees; JC analysed the data and wrote the manuscript. Both authors contributed critically to the drafts and gave final approval for publication.

References

Allison PD (2012) Logistic regression using SAS: Theory and application. SAS Institute, Cary, North Carolina USA.

Bassani M, Pacini E, Franchi GG (1994) Humidity stress responses in pollen of anemophilous and entomophilous species. Grana 33:146-150.

Beekman M, Ratnieks F (2000) Long-range foraging by the honey-bee, *Apis mellifera* L. Functional Ecology 14:490-496.

Bosch J (1994) The nesting behaviour of the mason bee *Osmia cornuta* (Latr.) with special reference to its pollinating potential (Hymenoptera, Megachilidae). Apidologie 25:84-93.

Bosch J, Blas M (1994) Foraging behaviour and pollination efficiency of *Osmia cornuta* and *Apis mellifera* on almond (Hymenoptera, Megachilidae and Apidae). Applied Entomology and Zoology 29:1-9.

Bradner NR, Frakes RV, Stephen WP (1965) Effects of bee species and isolation distance on possible varietal contamination in alfalfa. Agronomy Journal 57:247-248.

Cane JH (2011) Specialist *Osmia* bees forage indiscriminately among hybridizing *Balsamorhiza* floral hosts. Oecologia 167:107-116.

Cane JH (2017) Specialist bees collect Asteraceae pollen by distinctive abdominal drumming (*Osmia*) or tapping (*Melissodes, Svastra*). Arthropod-Plant Interactions 11:257-261.

Castellanos MC, Wilson P, Thomson JD (2003) Pollen transfer by hummingbirds and bumblebees, and the divergence of pollination modes in *Penstemon*. Evolution 57:2742-2752.

Cavalcante MC, Oliveira FF, Maues MM, Freitas BM (2012) Pollination requirements and the foraging behaviour of potential pollinators of cultivated Brazil Nut (*Bertholletia excelsa* Bonpl.) trees in central Amazon rainforest. Psyche 2012:1-9.

Clark JS, Lewis M, Horvath L (2001) Invasion by extremes: population spread with variation in dispersal and reproduction. The American Naturalist 157:537-554.

Cresswell JE, Bassom AP, Bell SA, Collins SJ, Kelly TB (1995) Predicted pollen dispersal by honey-bees and three species of bumble-bees foraging on oil-seed rape: a comparison of three models. Functional Ecology 9:829-841.

Cresswell JE, Osborne JL, Bell SA (2002) A model of pollinator-mediated gene flow between plant populations with numerical solutions for bumblebees pollinating oilseed rape. Oikos 98:375-384.

Cripps C, Rust RW (1989) Pollen preferences of seven *Osmia* species (Hymenoptera: Megachilidae). Environmental Entomology 18:133-138.

DeGrandi-Hoffman G, Hoopingarner R, Klomparens K (1986) Influence of honey bee (Hymenoptera: Apidae) in-hive pollen transfer on cross-pollination and fruit set in apple. Environmental Entomology 15:723-725.

Dick CW, Etchelecu G, Austerlitz F (2003) Pollen dispersal of tropical trees (*Dinizia excelsa*: Fabaceae) by native insects and African honeybees in pristine and fragmented Amazonian rainforest. Molecular Ecology 12:753-764.

Fenster CB, Vekemans X, Hardy OJ (2003) Quantifying gene flow from spatial genetic structure data in a metapopulation of *Chamaecrista fasciculata* (Leguminosae). Evolution 57:995-1007.

Franchi GG, Piotto B, Nepi M, Baskin CC, Baskin JM, Pacini E (2011) Pollen and seed desiccation tolerance in relation to degree of developmental arrest, dispersal, and survival. Journal of Experimental Botany 62:5267-5281.

Free JB, Durrant AJ (1966) The transport of pollen by honeybees from one foraging trip to the next. Journal of Horticultural Science 41:87-89.

Frohlich DR, Tepedino VJ (1986) Sex ratio, parental investment, and interparent variability in nesting success in a solitary bee. Evolution 40:142-151.

Goodell K (2003) Food availability affects *Osmia pumila* (Hymenoptera: Megachilidae) foraging, reproduction, and brood parasitism. Oecologia 134:518-527.

Greene SL, Kesoju SR, Martin RC, Kramer M (2015) Occurrence of transgenic feral alfalfa (*Medicago sativa* subsp. *sativa* L.) in alfalfa seed production areas in the United States. Plos One 10:e0143296.

Handel SN (1983) Contrasting gene flow patterns and genetic subdivision in adjacent populations of *Cucumis sativus* (Cucurbitaceae). Evolution 37:760-771.

Hatjina F, Free JB, Paxton RJ (1999) Hive-entrance pollen transfer devices to increase the cross-pollination potential of honey bees. II. Examination of three materials and pollen viability. Journal of Apicultural Research 38:3-9.

Holmquist K, Mitchell RJ, Karron JD (2012) Influence of pollinator grooming on pollen-mediated gene dispersal in *Mimulus ringens* (Phrymaceae). Plant Species Biology 27:77-85.

Jha S, Dick CW (2010) Native bees mediate long-distance pollen dispersal in a shade coffee landscape mosaic. Proceedings of the National Academy of Sciences 107:13760-13764.

Kohn JR, Casper BB (1992) Pollen-mediated gene flow in *Cucurbita foetidissima* (Cucurbitaceae). American Journal of Botany 79:57-62.

Loper GM, DeGrandi-Hoffman G (1994) Does in-hive pollen transfer by honey bees contribute to cross-pollination and seed set in hybrid cotton? Apidologie 25:94-102.

Maeta Y (1978) Comparative studies on the biology of the bees of the genus *Osmia* of Japan, with special reference to their managements for pollinations of crops (Hymenoptera: Megachilidae). Bulletin of the Tohoku National Agricultural Experiment Station 57:195-209.

Matsumoto S, Abe A, Maejima T (2009) Foraging behaviour of Osmia cornifrons in an apple orchard. Scientia Horticulturae 121:73-79.

Neff JL (2008) Components of nest provisioning behaviour in solitary bees (Hymenoptera : Apoidea). Apidologie 39:30-45.

O'Malley DM, Buckley DP, Prance GT, Bawa KS (1988) Genetics of Brazil nut (*Bertholletia excelsa* Humb. & Bonpl.: Lecythidaceae). 2. Mating system. Theoretical and Applied Genetics 76:929-932.

Osborne JL, Clark SJ, Morris RJ, Williams IH, Riley JR, Smith AD, Reynolds DR, Edwards AS (1999) A landscape-scale study of bumble bee foraging range and constancy, using harmonic radar. Journal of Applied Ecology 36:519-533.

Otero-Arnaiz A, Casas A, Hamrick JL (2005) Direct and indirect estimates of gene flow among wild and managed populations of *Polaskia chichipe*, an endemic columnar cactus in Central Mexico. Molecular Ecology 14:4313-4322.

Parker AJ, Tran JL, Ison JL, Bai JD, Weis AE, Thomson JD (2015) Pollen packing affects the function of pollen on corbiculate bees but not non-corbiculate bees. Arthropod-Plant Interactions 9:197-203.

Parker FD, Frohlich DR (1983) Hybrid sunflower pollination by a manageable composite specialist: the sunflower leafcutter bee

(Hymenoptera: Megachilidae). Environmental Entomology 12:576-581.

Paschke M, Abs C, Schmid B (2002) Effects of population size and pollen diversity on reproductive success and offspring size in the narrow endemic *Cochlearia bavarica* (Brassicaceae). American Journal of Botany 89:1250-1259.

Pasquet RS, Peltier A, Hufford MB, Oudin E, Saulnier J, Paul L, Knudsen JT, Herren HR, Gepts P (2008) Long-distance pollen flow assessment through evaluation of pollinator foraging range suggests transgene escape distances. Proceedings of the National Academy of Sciences 105:13456-13461.

Phillips JK, Klostermeyer EC (1978) Nesting behaviour of *Osmia lignaria propinqua* Cresson (Hymenoptera: Megachilidae). Journal of the Kansas Entomological Society 51:91-108.

Rasmussen IR, Brodsgaard B (1992) Gene flow inferred from seed dispersal and pollinator behaviour compared to DNA analysis of restriction site variation in a patchy population of *Lotus corniculatus* L. Oecologia 89:277-283.

Rieger MA, Lamond M, Preston C, Powles SB, Roush RT (2002) Pollen-mediated movement of herbicide resistance between commercial canola fields. Science 296:2386-2388.

Schulke B, Waser NM (2001) Long-distance pollinator flights and pollen dispersal between populations of *Delphinium nuttallianum*. Oecologia 127:239-245.

Seeley TD (1995) The wisdom of the hive. Harvard Univ. Press, Cambridge, MA.

St.Armand PC, Skinner DZ, Peaden RN (2000) Risk of alfalfa transgene dissemination and scale-dependent effects. Theoretical and Applied Genetics 101:107-114.

Tepedino VJ, Sipes SD, Griswold TL (1999) The reproductive biology and effective pollinators of the endangered beardtongue *Penstemon penlandii* (Scrophulariaceae). Plant Systematics and Evolution 219:39-54.

Thomson JD (1986) Pollen transport and deposition by bumble bees in *Erythronium:* influences of floral nectar and bee grooming. Journal of Ecology 74:329-341.

Thomson JD, Price MV, Waser NM, Stratton DA (1986) Comparative studies of pollen and fluorescent dye transport by bumble bees visiting *Erythronium grandiflorum*. Oecologia 69:561-566.

Thomson JD, Thomson BA (1989) Dispersal of *Erythronium grandiflorum* pollen by bumblebees: implications for gene flow and reproductive success. Evolution 43:657-661.

Torchio PF (1989) In-nest biologies and development of immature stages of three *Osmia* species (Hymenoptera: Megachilidae). Annals of the Entomological Society of America 82:599-615.

Van Deynze AE, Fitzpatrick SM, Hammon B, McCaslin MH, Putnam DH, Teuber LR, Undersander DJ (2008) Gene flow in alfalfa: biology, mitigation, and potential impact on production. Council for Agricultural Science and Technology Special Publ. no. 28, Ames, Iowa .

Van Deynze AE, Sundstrom FJ, Bradford KJ (2005) Pollen-mediated gene flow in California cotton depends on pollinator activity. Crop Science 45:1565-1570.

von Frisch K (1967) The Dance Language and Orientation of Bees. Harvard Univ. Press, Cambridge.

Waddington KD (1979) Flight patterns of three species of sweat bees (Halictidae) foraging at *Convolvulus arvensis*. Journal of the Kansas Entomological Society 52:751-758.

Ward M, Dick CW, Gribel R, Lowe AJ (2005) To self, or not to self... A review of outcrossing and pollen-mediated gene flow in neotropical trees. Heredity 95:246-254.

Waser NM (1982) A comparison of distances flown by different visitors to flowers of the same species. Oecologia 55:251-257.

Waser NM (1988) Comparative pollen and dye transfer by pollinators of *Delphinium nelsonii*. Functional Ecology 2:41-48.

Woodgate JL, Makinson JC, Lim KS, Reynolds AM, Chittka L (2016) Life-long radar tracking of bumblebees. Plos One 11:e0160333.

Zurbuchen A, Cheesman S, Klaiber J, Müller A, Hein S, Dorn S (2010a) Long foraging distances impose high costs on offspring production in solitary bees. Journal of Animal Ecology 79:674-681.

Zurbuchen A, Landert L, Klaiber J, Müller A, Hein S, Dorn S (2010b) Maximum foraging ranges in solitary bees: only few individuals have the capability to cover long foraging distances. Biological Conservation 143:669-676.

Permissions

All chapters in this book were first published in JPE, by Enviroquest Ltd; hereby published with permission under the Creative Commons Attribution License or equivalent. Every chapter published in this book has been scrutinized by our experts. Their significance has been extensively debated. The topics covered herein carry significant findings which will fuel the growth of the discipline. They may even be implemented as practical applications or may be referred to as a beginning point for another development.

The contributors of this book come from diverse backgrounds, making this book a truly international effort. This book will bring forth new frontiers with its revolutionizing research information and detailed analysis of the nascent developments around the world.

We would like to thank all the contributing authors for lending their expertise to make the book truly unique. They have played a crucial role in the development of this book. Without their invaluable contributions this book wouldn't have been possible. They have made vital efforts to compile up to date information on the varied aspects of this subject to make this book a valuable addition to the collection of many professionals and students.

This book was conceptualized with the vision of imparting up-to-date information and advanced data in this field. To ensure the same, a matchless editorial board was set up. Every individual on the board went through rigorous rounds of assessment to prove their worth. After which they invested a large part of their time researching and compiling the most relevant data for our readers.

The editorial board has been involved in producing this book since its inception. They have spent rigorous hours researching and exploring the diverse topics which have resulted in the successful publishing of this book. They have passed on their knowledge of decades through this book. To expedite this challenging task, the publisher supported the team at every step. A small team of assistant editors was also appointed to further simplify the editing procedure and attain best results for the readers.

Apart from the editorial board, the designing team has also invested a significant amount of their time in understanding the subject and creating the most relevant covers. They scrutinized every image to scout for the most suitable representation of the subject and create an appropriate cover for the book.

The publishing team has been an ardent support to the editorial, designing and production team. Their endless efforts to recruit the best for this project, has resulted in the accomplishment of this book. They are a veteran in the field of academics and their pool of knowledge is as vast as their experience in printing. Their expertise and guidance has proved useful at every step. Their uncompromising quality standards have made this book an exceptional effort. Their encouragement from time to time has been an inspiration for everyone.

The publisher and the editorial board hope that this book will prove to be a valuable piece of knowledge for researchers, students, practitioners and scholars across the globe.

List of Contributors

Ben A. Mensah and Andreas A. Kudom
Department of Entomology & Wildlife, University of Cape Coast, Cape Coast, Ghana

Brett M. Culbert and Jessica R. K. Forrest
Department of Biology, University of Ottawa, 30 Marie Curie, Ottawa, ON, K1N 6N5, Canada

Liliana Ramírez-Freire, Glafiro José Alanís Flores and Humberto Quiroz Martínez
Facultad de Ciencias Biológicas, Universidad Autónoma de Nuevo León, 66451 San Nicolás de los Garza, Nuevo León, México

Ricardo Ayala Barajas
Estación de Biología Chamela (Sede Colima) Instituto de Biología, UNAM

Carlos Gerardo Velazco Macías
Parques y Vida Silvestre. Av. Eugenio Garza Sada 6604, Col. Lagos del Bosque, 64890 Monterrey, Nuevo León, México

Aoife T. O'Rourke and Jane C. Stout
School of Natural Sciences, Botany Department, Trinity College Dublin, College Green, Dublin 2, Ireland
Trinity Centre for Biodiversity Research (TCBR), Trinity College Dublin, College Green, Dublin 2, Ireland

Úna Fitzpatrick
National Biodiversity Data Centre, Beechfield house, Carriganore WIT West Campus, County Waterford, Ireland

Peter Bernhardt
Department of Biology, Saint Louis University, St. Louis, MO 63103, USA

Retha Edens-Meier
Department of Educational Studies, Saint Louis University, St. Louis, MO 63103, USA

Eric Westhus
Center for Outcomes Research, Saint Louis University, St. Louis, MO 63103, USA

Nan Vance
Kooskia, ID 83539, USA

Milene Faria Vieira and Rúbia Santos Fonseca
Departamento de Biologia Vegetal, Universidade Federal de Viçosa, 36570-000 Viçosa, Minas Gerais, Brazil

B. J. Sampson, C. T. Pounders, C. T. Werle, T. R. Mallette, D. Larsen, L. Chatelain and K. C. Lee
USDA-ARS Thad Cochran Southern Horticultural Laboratory, Poplarville, MS 39470 USA

Sedonia D. Sipes, Aldwin Anterola and Daniel L. Nickrent
Department of Plant Biology Southern Illinois University Carbondale Illinois 62901-6509

Kara E. Huff Hartz and Hardik Amin
Department of Chemistry and Biochemistry Southern Illinois University Carbondale Illinois 62901-4409

Anne-Amélie C. Larue-Kontić and Robert R. Junker
University Salzburg, Department of Ecology and Evolution, Hellbrunnerstraße 34, 5020 Salzburg, Austria

Vernon G. Thomas
Department of Integrative Biology, College of Biological Science, University of Guelph, Guelph, Ontario N1G 2W1, Canada

Peter G. Kevan
School of Environmental Sciences, Ontario Agricultural College, University of Guelph, Guelph, N1G 2W1, Canada

Katharina Stein and Isabell Hensen
Martin-Luther-University Halle-Wittenberg, Institute for Biology/Geobotany and Botanical Garden, Am Kirchtor 1, 06108 Halle/Saale, Germany

Kristin Marie Lassen, Erik Dahl Kjær and Lene Rostgaard Nielsen
Department of Geosciences and Natural Resource Management, University of Copenhagen, Rolighedsvej 23, 1958 Frederiksberg C, Denmark

Moussa Ouédraogo
Centre National de Semences Forestières, Route de Kaya, 01 BP 2682 Ouagadougou, Burkina Faso

Yoko Luise Dupont
Institute of Bioscience, Aarhus University, Vejlsøvej 25, 8600 Silkeborg, Denmark

Arthur A. D. Broadbent
School of Environmental Sciences and University of East Anglia, Norwich Research Park, Norwich, UK

Andrew F. G. Bourke
School of Biological Sciences, University of East Anglia, Norwich Research Park, Norwich, UK

Andreas Burkart and Klaus Lunau
Institute of Sensory Ecology, Department Biology, Heinrich-Heine-University Düsseldorf, 40225 Düsseldorf, Germany

Clemens Schlindwein
Departamento de Botânica, Instituto de Ciêncais Biológicas, Universidade Federal de Minas Gerais Av. Antônio Carlos, 6627 - Pampulha - Caixa Postal 486, 31270-901, Belo Horizonte, MG, Brazil

Jakob Katzenberger and Dietmar Zacharias
Bremen University of Applied Sciences, Applied and Ecological Botany, Faculty 5, Neustadtswall 30, 28199 Bremen, Germany

J. Pérez-Balam, J. J. G Quezada-Euán and R. Alfaro-Bates
Departamento de Apicultura, Campus Ciencias Biológicas y Agropecuarias, Universidad Autónoma de Yucatán, Km. 15.5. carr. Mérida- Xmatkuil CP 97000, Mérida, Yucatán, México

S. Medina
Facultad de Matemáticas, Campus Ciencias Exactas, Universidad Autónoma de Yucatán, Mérida, Yucatán, México

L. McKendrick
School of Biological Sciences, Queen's University Belfast, MBC, 97 Lisburn Road, Belfast BT9 7BL, United Kingdom

A. Soro and R. J. Paxton
School of Biological Sciences, Queen's University Belfast, MBC, 97 Lisburn Road, Belfast BT9 7BL, United Kingdom
Institute fur Biology Martin- Luther-UniversitätHalle-Wittenberg, HoherWeg 8, D-06099 Halle (Saale), Germany

Jeff Ollerton and James R. Rouquette
Faculty of Arts, Science and Technology, University of Northampton, Avenue Campus, Northampton, NN2 6JD, UK

Tom D. Breeze
Centre for Agri-Environmental Research, School of Agriculture, Policy, and Development, University of Reading, Whiteknights, Reading, RG6 6AR, UK

Tadhg Carroll, Kate Rickard, Deborah Blake and Anita Diaz
Department of Life and Environmental Sciences, Faculty of Science and Technology, Bournemouth University, Poole, BH12 5BB

Elizabeth Franklin
Department of Life and Environmental Sciences, Faculty of Science and Technology, Bournemouth University, Poole, BH12 5BB
School of Environmental Sciences, University of Guelph, Guelph, Ontario, Canada, N1G 2W1

Keigo Takahashi
Graduate School of Science and Technology, Shinshu University, Matsumoto, Nagano 390-8621, Japan

Takao Itino
Department of Biology, Faculty of Science, Shinshu University, Matsumoto, Nagano 390-8621, Japan
Institute of Mountain Science, Shinshu University, Matsumoto, Nagano 390-8621, Japan

Emma Katz
Bates College, Program in Environmental Studies, Bates College, Lewiston, ME 04240, USA

Carla J. Essenberg
Bates College, Program in Environmental Studies, Bates College, Lewiston, ME 04240, USA
Department of Biology, Faculty of Science, Bates College, Lewiston, ME 04240, USA

Tan, Keng Hong
Mobula Research Sdn. Bhd., 20, Jalan Tan Jit Seng, 11200 Penang, Malaysia
Academy of Sciences Malaysia, MATRADE Tower, Jalan Sultan Haji Ahmad Shah, 50480 Kuala Lumpur, Malaysia

Tan, Lin Tze
Bachemer Strasse, 20, Bonn, 53179 Germany

Gidi Ne'eman
Department of Biology and Environment, Faculty of Science and Science Education, University of Haifa-Oranim, Oranim, Tivon 36006, Israel

Rina Ne'eman
Amirim 20115, Israel

Andrew C. McCall, Eric Thomson, Monica Edgerton and Skyler Jordan
Department of Biology, Denison University, Granville OH 43023 USA

Judith L. Bronstein
Department of Ecology and Evolutionary Biology, University of Arizona, Tucson AZ 85721 USA

Sarah Richman
Department of Ecology and Evolutionary Biology, University of Arizona, Tucson AZ 85721 USA
Present address: Department of Biology, University of Nevada, Reno; Reno NV 89557 USA

Christian Verhoeven and Klaus Lunau
Institute of Sensory Ecology, Department Biology, Heinrich-Heine-University Düsseldorf, Universitätsstr. 1, 40225 Düsseldorf, Germany

Zong-Xin Ren
Key Laboratory for Plant Diversity and Biogeography of East Asia, Kunming Institute of Botany, Chinese Academy of Sciences, Kunming, Yunnan 650201, People's Republic of China

James H. Cane and Byron Love
USDA-ARS Pollinating Insect Research Unit, Utah State University, Logan 84322-5310

Index